Lecture Notes in Mathematics   2108

Editors-in-Chief:
J.-M. Morel, Cachan
B. Teissier, Paris

Advisory Board:
Camillo De Lellis (Zürich)
Mario Di Bernardo (Bristol)
Alessio Figalli (Austin)
Davar Khoshnevisan (Salt Lake City)
Ioannis Kontoyiannis (Athens)
Gabor Lugosi (Barcelona)
Mark Podolskii (Aarhus)
Sylvia Serfaty (Paris and NY)
Catharina Stroppel (Bonn)
Anna Wienhard (Heidelberg)

For further volumes:
http://www.springer.com/series/304

Fondazione C.I.M.E., Firenze

C.I.M.E. stands for *Centro Internazionale Matematico Estivo*, that is, International Mathematical Summer Centre. Conceived in the early fifties, it was born in 1954 in Florence, Italy, and welcomed by the world mathematical community: it continues successfully, year for year, to this day.

Many mathematicians from all over the world have been involved in a way or another in C.I.M.E.'s activities over the years. The main purpose and mode of functioning of the Centre may be summarised as follows: every year, during the summer, sessions on different themes from pure and applied mathematics are offered by application to mathematicians from all countries. A Session is generally based on three or four main courses given by specialists of international renown, plus a certain number of seminars, and is held in an attractive rural location in Italy.

The aim of a C.I.M.E. session is to bring to the attention of younger researchers the origins, development, and perspectives of some very active branch of mathematical research. The topics of the courses are generally of international resonance. The full immersion atmosphere of the courses and the daily exchange among participants are thus an initiation to international collaboration in mathematical research.

C.I.M.E. Director	C.I.M.E. Secretary
Pietro ZECCA	Elvira MASCOLO
Dipartimento di Energetica "S. Stecco"	Dipartimento di Matematica "U. Dini"
Università di Firenze	Università di Firenze
Via S. Marta, 3	viale G.B. Morgagni 67/A
50139 Florence	50134 Florence
Italy	Italy
e-mail: zecca@unifi.it	e-mail: mascolo@math.unifi.it

For more information see CIME's homepage: http://www.cime.unifi.it

Aldo Conca • Sandra Di Rocco • Jan Draisma •
June Huh • Bernd Sturmfels • Filippo Viviani

Combinatorial Algebraic Geometry

Levico Terme, Italy 2013
Editors: Sandra Di Rocco, Bernd Sturmfels

In collaboration with

Aldo Conca
Dipartimento di Matematica
Universitá di Genova
Genova, Italy

Sandra Di Rocco
Department of Mathematics
KTH Royal Institute of Technology
Stockholm, Sweden

Jan Draisma
Department of Mathematics
and Computer Science
TU Eindhoven
Eindhoven, The Netherlands

June Huh
Department of Mathematics
University of Michigan at Ann Arbor
Ann Arbor, MI, USA

Bernd Sturmfels
Department of Mathematics
University of California, Berkeley
Berkeley, CA, USA

Filippo Viviani
Dipartimento di Matematica
Università Roma Tre
Roma, Italy

ISBN 978-3-319-04869-7   ISBN 978-3-319-04870-3 (eBook)
DOI 10.1007/978-3-319-04870-3
Springer Cham Heidelberg New York Dordrecht London

Lecture Notes in Mathematics ISSN print edition: 0075-8434
ISSN electronic edition: 1617-9692

Library of Congress Control Number: 2014939301

Mathematics Subject Classification (2010): 11H55, 13D02, 13P25, 14H10, 14M25, 16S37, 52B20, 62F10

© Springer International Publishing Switzerland 2014
This work is subject to copyright. All rights are reserved by the Publisher, whether the whole or part of the material is concerned, specifically the rights of translation, reprinting, reuse of illustrations, recitation, broadcasting, reproduction on microfilms or in any other physical way, and transmission or information storage and retrieval, electronic adaptation, computer software, or by similar or dissimilar methodology now known or hereafter developed. Exempted from this legal reservation are brief excerpts in connection with reviews or scholarly analysis or material supplied specifically for the purpose of being entered and executed on a computer system, for exclusive use by the purchaser of the work. Duplication of this publication or parts thereof is permitted only under the provisions of the Copyright Law of the Publisher's location, in its current version, and permission for use must always be obtained from Springer. Permissions for use may be obtained through RightsLink at the Copyright Clearance Center. Violations are liable to prosecution under the respective Copyright Law.
The use of general descriptive names, registered names, trademarks, service marks, etc. in this publication does not imply, even in the absence of a specific statement, that such names are exempt from the relevant protective laws and regulations and therefore free for general use.
While the advice and information in this book are believed to be true and accurate at the date of publication, neither the authors nor the editors nor the publisher can accept any legal responsibility for any errors or omissions that may be made. The publisher makes no warranty, express or implied, with respect to the material contained herein.

Printed on acid-free paper

Springer is part of Springer Science+Business Media (www.springer.com)

Preface

Combinatorics and Algebraic Geometry have enjoyed a fruitful interplay since the nineteenth century. Classical interactions include invariant theory, theta functions, and enumerative geometry. The aim of this volume is to introduce recent development in combinatorial algebraic geometry.

The five chapters of this book are based on the lectures delivered at the CIME-CIRM summer-school, Levico Terme, June 10–15, 2013.

We here regard algebraic geometry with a view towards applications, such as tensor calculus and algebraic statistics. A common theme is the study of algebraic varieties endowed with a rich combinatorial structure. Relevant techniques are polyhedral geometry, free resolutions, multilinear algebra, projective duality, and compactifications.

Aldo Conca offers an introduction to *Koszul Algebras and Their Syzygies*. Koszul algebras are fundamental in commutative algebra, and they have numerous applications in algebraic geometry. One application presented here is the study of Castelnuovo–Mumford regularity of projective varieties. Other results presented in this chapter concern syzygies of Koszul algebras, the Koszul property of Veronese algebras, and algebras in the theory of hyperspace arrangements.

Systems of polynomial equations in infinitely many variables arise naturally in applied algebraic geometry. Typically, these infinite-dimensional systems have a lot of symmetry, and, in favorable circumstances, one encounters *Noetherianity up to Symmetry*. Jan Draisma offers a glimpse on recent developments in this field. His chapter focuses on examples from algebraic statistics and on the combinatorics of well-quasi-ordered sets.

Maximum Likelihood Geometry studies the critical points of monomial functions over a variety inside the probability simplex. The number of complex critical points, known as its maximum likelihood degree, is a topological invariant. June Huh joined Bernd Sturmfels in writing a chapter, which introduces this theory and its statistical motivations. Many favorites from combinatorial algebraic geometry are featured: toric varieties, matroids, A-discriminants, Grassmannians, and determinantal varieties.

Sandra Di Rocco lectured on *Linear Toric Fibrations*, that is, toric varieties which are birational to projective toric bundles. On the combinatorial side, these correspond to Cayley polytopes, a class of highly structured lattice polytopes that has received much attention in the recent literature. This chapter presents geometrical phenomena, in algebraic geometry and neighboring fields, which are characterized by a Cayley structure.

Filippo Viviani takes the reader on *A Tour of Hermitian Symmetric Manifolds*. These are Hermitian manifolds which are homogeneous and such that every point has a symmetry preserving the Hermitian structure. Examples of such manifolds serve as moduli spaces in algebraic and analytic geometry. This chapter offers an introduction to several different perspectives from which Hermitian symmetric manifolds can be studied.

We thank the CIME foundation and the CIRM center for hosting the school and for their generous support. All of us had a wonderful time at Levico Terme. The beautiful scenery of Trentino made the mathematical interactions and the stimulating lectures even more enjoyable.

Stockholm, Sweden	Sandra Di Rocco
Berkeley, CA	Bernd Sturmfels
October 2013	

CIME activity is carried out with the collaboration and financial support of:

- INdAM (Istituto Nazionale di Alta Matematica)
- MIUR (Ministero dell'Istruzione, dell'Università e della Ricerca)
- Ente Cassa di Risparmio di Firenze

Contents

Koszul Algebras and Their Syzygies 1
Aldo Conca

Noetherianity up to Symmetry 33
Jan Draisma

Likelihood Geometry 63
June Huh and Bernd Sturmfels

Linear Toric Fibrations 119
Sandra Di Rocco

A Tour on Hermitian Symmetric Manifolds 149
Filippo Viviani

Koszul Algebras and Their Syzygies

Aldo Conca

Introduction

Koszul algebras, introduced by Priddy in [P], are positively graded K-algebras R whose residue field K has a linear free resolution as an R-module. Here linear means that the non-zero entries of the matrices describing the maps in the R-free resolution of K have degree 1. For example, if $S = K[x_1, \ldots, x_n]$ is the polynomial ring over a field K then K is resolved by the Koszul complex which is linear. In these lectures we deal with standard graded commutative K-algebras, that is, quotient rings of the polynomial ring S by homogeneous ideals. The program of the lectures is the following:

Lecture 1: Koszul algebras and Castelnuovo–Mumford regularity.
Lecture 2: Bounds for the degrees of the syzygies of Koszul algebras.
Lecture 3: Veronese algebras and algebras associated with collections of hyperspaces.

In the first lecture, based on the survey paper [CDR], we present various characterizations of Koszul algebras and strong versions of Koszulness. In the second lecture we describe recent results, obtained in cooperation with Avramov and Iyengar [ACI1, ACI2], on the bounds of the degrees of the syzygies of a Koszul algebra. Finally, the third lecture is devoted to the study of the Koszul property of Veronese algebras and of algebras associated with collections of hyperspaces and it is based on the papers [CDR, C].

A. Conca (✉)
Dipartimento di Matematica, Università degli Studi di Genova, Genova, Italy
e-mail: conca@dima.unige.it

1 Koszul Algebras and Castelnuovo–Mumford Regularity

1.1 Notation

Let K be a field and R be a commutative standard graded K-algebra, that is a K-algebra with a decomposition $R = \bigoplus_{i \in \mathbf{N}} R_i$ (as an Abelian group) such that $R_0 = K$, the vector space R_1 has finite dimension and $R_i R_j = R_{i+j}$ for every $i, j \in \mathbf{N}$. Let S be the symmetric algebra of R_1 over K. In other words, S is the polynomial ring $K[x_1, \ldots, x_n]$ where $n = \dim R_1$ and $x_1, \ldots, x_n$ is a K-basis of R_1. One has an induced surjection

$$S = K[x_1, \ldots, x_n] \to R \qquad (1)$$

of standard graded K-algebras. We call (1) the canonical presentation of R. Hence R is isomorphic to S/I where I is the kernel of the map (1). In particular, I is homogeneous and does not contain elements of degree 1. We say that I defines R. Denote by $\mathbf{m}_R$ the maximal homogeneous ideal of R. We may consider K as a graded R-module via the identification $K = R/\mathbf{m}_R$.

Unless otherwise stated, we will always assume that K-algebras are standard graded, modules and ideals are graded and finitely generated, and module homomorphisms have degree 0.

For an R-module $M = \bigoplus_{i \in \mathbf{Z}} M_i$ we denote by $\mathrm{HF}(M, i)$ the Hilbert function of M at i, that is

$$\mathrm{HF}(M, i) = \dim_K(M_i),$$

and by

$$H_M(z) = \sum_{i \in \mathbf{Z}} \dim_K(M_i) z^i \in \mathbf{Q}[[z]][z^{-1}]$$

the associated Hilbert series.

Given an integer $a \in \mathbf{Z}$ we will denote by $M(a)$ the graded R-module whose degree i component is M_{i+a}. In particular $R(-j)$ is a free R-module of rank 1 whose generator has degree j.

A minimal graded free resolution of M as an R-module is a complex of free R-modules

$$\mathbf{F} : \cdots \to F_{i+1} \xrightarrow{\phi_{i+1}} F_i \xrightarrow{\phi_i} F_{i-1} \to \cdots \to F_1 \xrightarrow{\phi_1} F_0 \to 0$$

such that:

1. $H_i(\mathbf{F}) = 0$ for $i > 0$,
2. $H_0(\mathbf{F}) \simeq M$,
3. $\phi_{i+1}(F_{i+1}) \subseteq \mathbf{m}_R F_i$ for every i.

Such a resolution exists and it is unique up to an isomorphism of complexes. We hence call it "the" minimal free (graded) resolution of M.

By definition, the i-th Betti number $\beta_i^R(M)$ of M as an R-module is the rank of F_i. Each F_i is a direct sum of shifted copies of R. The (i, j)-th graded Betti number $\beta_{ij}^R(M)$ of M is the number of copies of $R(-j)$ that appear in F_i. By construction one has

$$\beta_i^R(M) = \dim_K \operatorname{Tor}_i^R(M, K)$$

and

$$\beta_{ij}^R(M) = \dim_K \operatorname{Tor}_i^R(M, K)_j.$$

Here and throughout the notes an index on the right of a graded module denotes the homogeneous component of that degree.

The Poincaré series of M is defined as

$$P_M^R(z) = \sum_{i \in \mathbf{N}} \beta_i^R(M) z^i \in \mathbf{Q}[|z|],$$

and its bigraded version is

$$P_M^R(s, z) = \sum_{i \in \mathbf{N}, j \in \mathbf{Z}} \beta_{i,j}^R(M) s^j z^i \in \mathbf{Q}[s, s^{-1}][|z|].$$

We set

$$t_i^R(M) = \sup\{j : \beta_{ij}^R(M) \neq 0\}$$

where, by convention, $t_i^R(M) = -\infty$ if $F_i = 0$. By definition, $t_0^R(M)$ is the largest degree of a minimal generator of M. Two important invariants that measure the "growth" of the resolution of M as an R-module are the projective dimension

$$\operatorname{pd}_R(M) = \sup\{i : F_i \neq 0\} = \sup\{i : \beta_i^R(M) \neq 0\}$$

and the (relative) Castelnuovo–Mumford regularity

$$\operatorname{reg}_R(M) = \sup\{j - i : \beta_{ij}^R(M) \neq 0\} = \sup\{t_i^R(M) - i : i \in \mathbf{N}\}.$$

An R-module M has a linear resolution as an R-module if for some $d \in \mathbf{Z}$ one has $\beta_{ij}^R(M) = 0$ if $j \neq d + i$. Equivalently, M has a linear resolution as an R-module if it is generated by elements of degree $\operatorname{reg}_R(M)$.

We may as well consider M as a module over the polynomial ring S via the map (1). The absolute Castelnuovo–Mumford regularity is, by definition, the regularity $\operatorname{reg}_S(M)$ of M as an S-module. It has also a cohomological interpretation via local duality, see for example [EG, Sect. 1] or [BH, 4.3.1]. Denote by $H^i_{\mathfrak{m}_S}(M)$ the i-th local cohomology module with support on the maximal ideal of S. One has $H^i_{\mathfrak{m}_S}(M) = 0$ if $i < \operatorname{depth} M$ or $i > \dim M$ and

$$\operatorname{reg}_S(M) = \max\{j + i : H^i_{\mathfrak{m}_S}(M)_j \neq 0\}.$$

Both $\operatorname{pd}_R(M)$ and $\operatorname{reg}_R(M)$ can be infinite.

Example 1.1. Let $R = K[x]/(x^v)$ with $v > 1$. Then the minimal free resolution of K over R is:

$$\cdots \to R(-2v) \to R(-v-1) \to R(-v) \to R(-1) \to R \to 0$$

where the maps are multiplication by x or x^{v-1} depending on the parity. Hence $F_{2i} = R(-iv)$ and $F_{2i+1} = R(-iv - 1)$ so that $\operatorname{pd}_R(K) = \infty$ for every $v > 1$. Furthermore $\operatorname{reg}_R(K) = \infty$ if $v > 2$ and $\operatorname{reg}_R(K) = 0$ if $v = 2$.

Note that, in general, $\operatorname{reg}_R(M)$ is finite if $\operatorname{pd}_R(M)$ is finite. On the other hand, as we have seen in the example above, $\operatorname{reg}_R(M)$ can be finite even when $\operatorname{pd}_R(M)$ is infinite.

In the study of minimal free resolutions over R, the minimal free resolution $\mathbf{K}_R$ of the residue field K as an R-module plays a fundamental role. This is because

$$\operatorname{Tor}^R_*(M, K) = H_*(M \otimes_R \mathbf{K}_R)$$

and hence

$$\beta^R_{ij}(M) = \dim_K H_i(M \otimes_R \mathbf{K}_R)_j.$$

A very important role is played also by the Koszul complex $K(\mathfrak{m}_R)$ on a minimal system of generators of the maximal ideal $\mathfrak{m}_R$ of R. The Koszul complex is the typical example of a differential graded algebra, DG-algebra for short.

1.2 DG-Algebras

A graded algebra

$$C = \oplus_{i \geq 0} C_i$$

is graded-commutative if for every $a \in C_i$ and $b \in C_j$ one has:

$$ab = (-1)^{ij} ba$$

and furthermore

$$a^2 = 0$$

whenever i is odd.

A DG-algebra is a graded-commutative algebra $C = \oplus_{i \geq 0} C_i$ equipped with a linear differential

$$\partial : C \to C$$

of degree -1 (i.e. $\partial^2 = 0$ and $\partial(C_i) \subseteq C_{i-1}$) that satisfies the "twisted" Leibniz rule:

$$\partial(ab) = \partial(a)b + (-1)^i a \partial(b)$$

whenever $a \in C_i$.

The cycles $Z(C) = \ker \partial$, the boundaries $B(C) = \text{Image } \partial$ and the homology $H(C) = Z(C)/B(C)$ of a DG-algebra C inherit the algebra structure from C. Precisely, $Z(C)$ is a graded-commutative subalgebra of C, $B(C)$ is a (two-sided) graded ideal of $Z(C)$ and hence $H(C)$ is a (graded-commutative) algebra.

The component of $Z(C)$ of (homological) degree i is denoted by $Z_i(C)$. Similarly for the boundaries and the homology.

Given a DG-algebra C and a cycle $z \in Z_i(C)$ there is a canonical way to "kill" z in homology by adding a "variable" to C preserving the DG-algebra structure. If i is even then one considers $D = C[e]$ where e is an exterior variable (hence $e^2 = 0$) of degree $i + 1$ and extends the differential by setting $\partial(e) = z$. If i is odd then one considers $D = C[s]$ where s is a polynomial variable (or a divided power variable) of degree $i + 1$ and extends the differential by setting $\partial(s) = z$. By construction, the element $z \in Z(C) \subset Z(D)$ is now a boundary of the complex D and hence it is 0 in homology. Furthermore, by construction, $H_j(C) = H_j(D)$ for $j < i$. This process can clearly be iterated. One can, for instance, "kill" all the cycles in a given homological degree by adding variables.

1.3 Koszul Complex

The Koszul complex can be described in the following way. Let R be any ring and let $I = (a_1, \ldots, a_m)$ be an ideal of R. Consider R as a DG-algebra concentrated in degree 0 and the elements $a_1, \ldots, a_m$ as cycles of that complex. Then we add exterior variables $e_1, \ldots, e_m$ in degree 1 to R and obtain the DG-algebra

$$K(I, R) = R[e_1, \ldots, e_m] \text{ with } \partial(e_i) = a_i.$$

This is the Koszul complex associated with the ideal I and coefficients in the ring R. In other words, $K(I, R)$ is the exterior algebra $\bigwedge R^m$ equipped with the differential induced by $\partial(e_i) = a_i$ for $i = 1, \ldots, m$. If M is an R-module we then set

$$K(I, M) = K(I, R) \otimes_R M.$$

This is the Koszul complex associated with the ideal I with coefficients in M. Denote by $Z(I, M)$ the module of cycles of the complex $K(I, M)$ and similarly by $B(I, M)$ its boundaries, by $C(I, M)$ its cokernel and by $H(I, M)$ its homology. We will denote by $K_i(I, M)$ the component of homological degree i of $K(I, M)$. When the coefficients of the Koszul complex are taken in R we use a simplified notation

$$K(I) = K(I, R), \quad Z(I) = Z(I, R)$$

and so on.

By definition we have:

$$H_0(I) = R/I \text{ and } H_0(I, M) = M/IM.$$

Furthermore $H(I)$ is a (graded-commutative) algebra and $H(I, M)$ is a $H(I)$-module. In particular, $IH(I, M) = 0$.

It is well-known that the Koszul complex $K(I)$ is acyclic (and hence an R-free resolution of R/I) if (and only if in the local or standard graded setting) the chosen generators $a_1, \ldots, a_m$ of I form a regular sequence, see [BH, 1.6.14].

When $K(I)$ is not a free resolution one can nevertheless use the procedure of adding variables to kill homology degree by degree to obtain from $K(I)$ a free resolution of R/I as an R-module with a DG-algebra structure. In the local or standard graded setting this can be done in the following way.

1. Set $T_1 = K(I)$.
2. Choose a minimal system of generators of $H_1(T_1)$ and a set of cycles $z_{1,1}, \ldots, z_{1,u_1}$ representing them.
3. Add to T_1 a set of polynomial variables (or divided powers in positive characteristic) $Y_2 = \{s_{2,1}, \ldots, s_{2,u_1}\}$ in degree 2 and set

$$T_2 = T_1[s_{2,1}, \ldots, s_{2,u_1} : \partial(s_{2,i}) = z_{1,i}].$$

4. Choose a minimal system of generators of $H_2(T_2)$, a set of cycles $z_{2,1}, \ldots, z_{2,u_2}$ representing them.
5. Add to T_2 a set of exterior variables $Y_3 = \{e_{3,1}, \ldots, e_{3,u_2}\}$ in degree 3 and set

$$T_3 = T_2[e_{3,1}, \ldots, e_{3,u_2} : \partial(e_{3,i}) = z_{2,i}]$$

and so on. We obtain a DG-algebra $T = R[Y_1, Y_2, Y_3, \ldots]$ that is an R-free resolution of R/I which is (essentially) independent on the choices of the minimal system of generators of $H_i(T_i)$ and of the cycles representing them. It is called the Tate complex and we will denote it by

$$T(R, R/I).$$

We refer to [A] for a precise description of the construction and many beautiful results and questions concerning it. We just give two examples:

Example 1.2. Set $R = \mathbf{Q}[x]/(x^n)$, $a = \bar{x}$ and $I = (a)$. Then $T_1 = K(I) = R[e]$ with $\partial(e) = a$ and $z = a^{n-1}e$ generates the 1-cycles of $K(I)$. Hence the second round of Tate construction gives $T_2 = R[e, s]$ with $\partial(s) = a^{n-1}e$. It turns out that $T(R, R/I) = R[e, s]$ is a minimal resolution of R/I over R:

$$\cdots \to Rs^2e \to Rs^2 \to Rse \to Rs \to Re \to R \to 0$$

where the maps are given, alternatively, by multiplication with a and a^{n-1} up to non-zero scalars.

Example 1.3. Set $R = \mathbf{Q}[x, y]$ and $I = (x^2, xy)$. Then we have $T_1 = K(I) = R[e_1, e_2]$ with $\partial(e_1) = x^2$ and $\partial(e_2) = xy$. The cycle $ye_1 - xe_2$ generates $H_1(T_1)$. Hence the second round of Tate construction gives the DG-algebra $T_2 = R[e_1, e_2, s_1]$ with $\partial(s_1) = ye_1 - xe_2$. Now $H_2(T_2)$ is generated by $e_1e_2 - ys_1$. Hence the third round of Tate construction gives $T_3 = R[e_1, e_2, s_1, e_3]$ with $\partial(e_3) = e_1e_2 - ys_1$. Hence the beginning of the Tate complex is the following:

$$\cdots \to Re_3 \oplus Rs_1e_1 \oplus Rs_1e_2 \to Rs_1 \oplus Re_1e_2 \to Re_2 \oplus Re_1 \to R \to 0.$$

Note that the resolution in Example 1.2 is minimal while the one in Example 1.3 is not.

1.4 Auslander–Buchsbaum–Serre

We return to the graded setting and assume that R is a standard graded K-algebra. When is $\mathrm{pd}_R(M)$ finite for every M? The answer is given by the Auslander–Buchsbaum–Serre Theorem, a result that, in the words of Avramov [A1, p. 32], "really started everything". The graded variant of it is the following:

Theorem 1.4. *The following conditions are equivalent:*

(1) $\mathrm{pd}_R(M)$ *is finite for every R-module M,*
(2) $\mathrm{pd}_R(K)$ *is finite,*
(3) *the Koszul complex $K(\mathbf{m}_R)$ resolves K,*
(4) R *is a polynomial ring.*

When the conditions hold, then for every M one has $\mathrm{pd}_R(M) \leq \mathrm{pd}_R(K) = \dim R$.

Remark 1.5. The Koszul complex $K(\mathbf{m}_R)$ has three important features:

(1) it has finite length,

(2) it has a DG-algebra structure,
(3) the matrices describing its differentials have non-zero entries only of degree 1.

When R is not a polynomial ring the minimal free resolution $\mathbf{K}_R$ of K as an R-module does not satisfy condition (1) of Remark 1.5. Can $\mathbf{K}_R$ nevertheless satisfy conditions (2) and (3) of Remark 1.5?

For condition (2) the answer is yes: $\mathbf{K}_R$ has always a DG-algebra structure. Indeed a theorem, proved independently by Gulliksen and Schoeller, asserts that $\mathbf{K}_R$ is obtained by using the Tate construction. Furthermore results of Assmus, Tate, Gulliksen and Halperin clarify when $\mathbf{K}_R$ is finitely generated as an R-algebra. Again, we state the theorem in the graded setting and refer to [A, 6.3.5, 7.3.3, 7.3.4] for general statements and proofs.

Theorem 1.6. *Let R be a standard graded K-algebra. Let $T = T(R, K) = R[Y_1, Y_2, Y_3, \dots]$ be the Tate complex associated with $K = R/\mathbf{m}_R$. We have:*

(1) *T is the minimal graded resolution of K, i.e. $T = \mathbf{K}_R$.*
(2) *T is finitely generated as an R-algebra if and only if R is a complete intersection. In that case, T is generated by elements of degree at most 2, i.e. $Y_i = \emptyset$ for $i > 2$.*
(3) *If R is not a complete intersection then $Y_i \neq \emptyset$ for every i.*

Algebras R such that $\mathbf{K}_R$ satisfies condition (3) in Remark 1.5 are called Koszul:

Definition 1.7. The K-algebra R is Koszul if the matrices describing the differentials in the minimal free resolution $\mathbf{K}_R$ of K as an R-module have non-zero entries only of degree 1, that is, $\operatorname{reg}_R(K) = 0$ or, equivalently, $\beta_{ij}^R(K) = 0$ whenever $i \neq j$.

Koszul algebras were originally introduced by Priddy [P] in his study of homological properties of graded (non-commutative) algebras, see the volume [PP] of Polishchuk and Positselski for an overview and surprising aspects of the Koszul property. We collect below a list of important facts about Koszul commutative algebras. We always refer to the canonical presentation (1) of R as a quotient of the polynomial ring $S = \operatorname{Sym}(R_1)$ by the homogeneous ideal I. First we introduce a definition.

Definition 1.8. We say that R is quadratic if its defining ideal I is generated by quadrics (homogeneous polynomials of degree 2).

Definition 1.9. We say that R is G-quadratic if its defining ideal I has a Gröbner basis of quadrics with respect to some coordinate system of S_1 and some term order τ on S. In other words, R is G-quadratic if there exists a K-basis of S_1, say $x_1, \dots, x_n$ and a term order $\prec$ such that the initial ideal $\operatorname{in}_\prec(I)$ of I with respect to $\prec$ is generated by monomials of degree 2.

Remark 1.10. For a standard graded K-algebra R one has

$$\beta_{2j}^R(K) = \begin{cases} \beta_{1j}^S(R) & \text{if } j \neq 2 \\ \beta_{12}^S(R) + \binom{n}{2} & \text{if } j = 2 \end{cases}$$

and hence the resolution of K, as an R-module, is linear up to homological position 2 if and only if R is quadratic. In particular, if R is Koszul, then R is quadratic.

On the other hand there are algebras defined by quadrics that are not Koszul. For example if one takes

$$R = K[x, y, z, t]/(x^2, y^2, z^2, t^2, xy + zt)$$

then one has $\beta_{34}^R(K) = 5$ and hence R is not Koszul.

Remark 1.11. If I is generated by monomials of degree 2 with respect to some coordinate system of S_1, then a filtration argument, that we reproduce in Theorem 3.12, shows that R is Koszul. More precisely, for every subset Y of variables $R/(Y)$ has an R-linear resolution.

Remark 1.12. If I is generated by a regular sequence of quadrics, then R is Koszul. This follows from Theorem 1.6 because if R is a complete intersection of quadrics, then $\mathbf{K}_R$ is obtained from $K(\mathbf{m}_R)$ by adding polynomial variables in homological degree 2 and internal degree 2 to kill $H_1(K(\mathbf{m}_R))$. For example, if

$$R = \mathbf{Q}[x_1, x_2, x_3, x_4]/(x_1^2 + x_2^2, x_3 x_4)$$

then the Tate resolution of K over R is the DG-algebra

$$R[e_1, e_2, e_3, e_4, s_1, s_2]$$

with differential induced by $\partial(e_i) = x_i$ and $\partial(s_1) = x_1 e_1 + x_2 e_2$ and $\partial(s_2) = x_3 e_4$. Here the e_i's have internal degree 1 and the s_i's have internal degree 2.

Remark 1.13. If R is G-quadratic, then R is Koszul. This follows from Remark 1.11 and from the standard deformation argument showing that $\beta_{ij}^R(K) \leq \beta_{ij}^A(K)$ with $A = S/\operatorname{in}_\tau(I)$. For details see, for instance, [BC, Sect. 3].

Remark 1.14. On the other hand there are Koszul algebras that are not G-quadratic. One notes that an ideal defining a G-quadratic algebra must contain quadrics of "low" rank. For instance, if R is Artinian and G-quadratic then its defining ideal must contain the square of a linear form. But most Artinian complete intersections of quadrics do not contain the square of a linear form. For example, the ideal $I = (x^2 + yz, y^2 + xz, z^2 + xy) \subset S = \mathbf{C}[x, y, z]$ is a complete intersection not containing the square of a linear form and S/I is Artinian. Hence S/I is Koszul and not G-quadratic. See [ERT] for general results in this direction.

Remark 1.15. The Poincaré series $P_K^R(z)$ of K as an R-module can be irrational (even for rings with $R_3 = 0$), see [An]. However for a Koszul algebra R one has

$$P_K^R(z) H_R(-z) = 1 \qquad (2)$$

and hence $P_K^R(z)$ is rational. Indeed the equality (2) turns out to be equivalent to the Koszul property of R, see for instance [F].

Remark 1.16. A necessary (but not sufficient) numerical condition for R to be Koszul is that the formal power series $1/H_R(-z)$ has non-negative coefficients (indeed positive unless R is a polynomial ring). Another numerical condition is the following. Expand the formal power series $1/H_R(-z)$ as

$$\frac{\prod_{h \in 2\mathbf{N}+1}(1+z^h)^{e'_h}}{\prod_{h \in 2\mathbf{N}+2}(1-z^h)^{e'_h}}$$

with $e'_h \in \mathbf{Z}$. This can be done in a unique way, see [A, 7.1.1]. Furthermore set $e_h(R) = \#Y_h$ where Y_h is the set of variables that we add at the h-th iteration of the Tate construction of the minimal free resolution of K over R. The numbers $e_h(R)$ are called the deviations of R. If R is Koszul then $e'_h = e_h(R)$ for every h and hence $e'_h \geq 0$. More precisely, $e'_h > 0$ for every h unless R is a complete intersection.

For example, the Hilbert function of the ring in Remark 1.10 is

$$H(z) = 1 + 4z + 5z^2.$$

Expanding the series $1/H(-z)$ one sees that the coefficient of z^6 is negative. Furthermore the corresponding e'_3 is 0. So for two numerical reasons an algebra with Hilbert series $H(z)$ cannot be Koszul.

The following characterization of the Koszul property in terms of regularity is formally similar to the Auslander–Buchsbaum–Serre Theorem 1.4.

Theorem 1.17 (Avramov–Eisenbud–Peeva). *The following conditions are equivalent:*

(1) $\operatorname{reg}_R(M)$ *is finite for every R-module M,*
(2) $\operatorname{reg}_R(K)$ *is finite,*
(3) R *is Koszul.*

Avramov and Eisenbud proved in [AE] that every module M has finite regularity over a Koszul algebra R by showing that $\operatorname{reg}_R(M) \leq \operatorname{reg}_S(M)$. Avramov and Peeva showed in [AP] that if $\operatorname{reg}_R(K)$ is finite then it must be 0. Indeed they proved a more general result for graded algebras that are not necessarily standard.

We collect below further remarks, examples and questions relating the Koszul property and the existence of Gröbner bases of quadrics in various ways. Let us recall the following:

Definition 1.18. A K-algebra R is LG-quadratic (where the L stands for lifting) if there exist a G-quadratic algebra A and a regular sequence of linear forms $y_1, \ldots, y_c$ such that $R \simeq A/(y_1, \ldots, y_c)$.

We have the following implications:

$$\text{G-quadratic} \Rightarrow \text{LG-quadratic} \Rightarrow \text{Koszul} \Rightarrow \text{quadratic} \qquad (3)$$

As discussed in Remarks 1.10 and 1.16 the third implication in (3) is strict. The following remark, due to Caviglia, in connection with Remark 1.14 shows that also the first implication in (3) is strict.

Remark 1.19. Any complete intersection R of quadrics is LG-quadratic. Say

$$R = K[x_1, \ldots, x_n]/(q_1, \ldots, q_m)$$

is a complete intersection of quadrics. Then set

$$A = R[y_1, \ldots, y_m]/(y_1^2 + q_1, \ldots, y_m^2 + q_m)$$

and note that A is G-quadratic because the initial ideal of its defining ideal with respect to a lex term order satisfying $y_i > x_j$ for every i, j is $(y_1^2, \ldots, y_m^2)$. Furthermore $y_1, \ldots, y_m$ is a regular sequence in A because

$$A/(y_1, \ldots, y_m) \simeq R$$

and $\dim A - \dim R = m$.

In [Ca1, 1.2.6], [ACI1, 6.4] and [CDR, 12] it is asked whether a Koszul algebra is also LG-quadratic. The following example gives a negative answer to the question.

Example 1.20. Let

$$R = K[a, b, c, d]/(ac, ad, ab - bd, a^2 + bc, b^2).$$

The Hilbert series of R is

$$\frac{1 + 2z - 2z^2 - 2z^3 + 2z^4}{(1-z)^2}.$$

Also, R is Koszul as can be shown using a Koszul filtration argument, see Example 3.8. The h-polynomial does not change under lifting with regular sequences of linear forms. Hence to check that R is not LG-quadratic it is enough to check that there is no algebra with quadratic monomial relations and with h-polynomial equal to

$$h(z) = 1 + 2z - 2z^2 - 2z^3 + 2z^4.$$

In general, if J is an ideal in a polynomial ring A not containing linear forms and the h-polynomial of A/J is $1 + h_1 z + h_2 z^2 + \dots$ then J has codimension h_1 and exactly $\binom{h_1+1}{2} - h_2$ quadratic generators.

Now consider a quadratic monomial ideal J in a polynomial ring A with, say, n variables such that the h-polynomial of A/J is $h(z)$.

Such a J must have codimension 2 and 5 generators. So J is generated by 5 monomials chosen among the generators of $(a,b)(a,b,c,d,e,f,g)$ where $a, b, \dots, g$ are distinct variables. But an exhaustive CoCoA [CoCoA] computation shows that such a selection does not exist.

An interesting example of LG-quadratic algebra is the following:

Example 1.21. Let

$$R = K[a,b,c,d]/(a^2 - bc, d^2, cd, b^2, ac, ab).$$

The Hilbert series of R is

$$\frac{1 + 3z - 3z^3}{(1-z)}.$$

The h-polynomial $1 + 3z - 3z^3$ is not the h-polynomial of a quadratic monomial ideal in four variables. It is however the h-polynomial of a (unique up to permutations of the variables) quadratic monomial ideal in five variables, namely $U_1 = (a^2, b^2, ad, cd, be, ce)$. In six variables there is another quadratic monomial ideal with that h-polynomial. It is $U_2 = (a^2, ad, bd, be, ce, cf)$. Another one in seven variables, $U_3 = (ad, bd, ae, ce, bf, cg)$. And that is all. It turns out that R is LG-quadratic since it lifts to $K[a,b,c,d,e]/J$ with

$$J = (a^2 - bc + be, d^2, cd, b^2 + eb, ac, ab + ae)$$

and $\mathrm{in}_\prec(J)$ is U_1 (up to a permutation of the variables) where $\prec$ is the rev.lex. order associated with the total order of the variables $e > d > b > c > a$.

The ring of Example 1.20 is not LG-quadratic because of the obstruction in the h-polynomial, i.e., there are no quadratic monomial ideals with that h-polynomial. It would be interesting to identify a Koszul algebra with a non-obstructed h-polynomial that is not LG-quadratic.

2 Syzygies of Koszul Algebras

The second lecture is based on the results obtained jointly with Avramov and Iyengar and published in [ACI1, ACI2].

Given a standard graded K-algebra R with presentation

$$R = S/I$$

where $S = K[x_1, \ldots, x_n]$ and $I \subset S$ is a homogeneous ideal, we set

$$t_i^S(R) = \sup\{j : \beta_{ij}^S(R) \neq 0\}$$

i.e., $t_i^S(R)$ is the highest degree of a minimal i-th syzygy of R as an S-module. In particular, $t_0^S(R) = 0$ and $t_1^S(R)$ equals the highest degree of a minimal generator of I.

The starting point of the discussion is the following observation.

Remark 2.1. If I is generated by monomials of degree 2, then:

(1) $t_i^S(R) \leq 2i$ for every i,
(2) $\operatorname{reg}_S(R) \leq \operatorname{pd}_S(R)$,
(3) if $t_i^S(R) < 2i$ for some i then $t_{i+1}^S(R) < 2(i+1)$,
(4) $t_i^S(R) < 2i$ if $i > \dim S - \dim R$,
(5) $\beta_i^S(R) \leq \binom{\beta_1^S(R)}{i}$,
(6) $\operatorname{pd}_S(R) \leq \beta_1^S(R)$.

These inequalities are deduced from the (non-minimal) Taylor resolution of quadratic monomial ideals, see for instance [MS, 4.3.2].

Suppose now that, combining and iterating the following operations:

(a) changes of coordinates,
(b) the formation of initial ideals with respect to weights or term orders,
(c) liftings and specializations with regular sequences of degree 1,

the algebra R deforms to an algebra $R' = S'/J$ with S' a polynomial ring and J generated by quadratic monomials. Then R satisfies the inequalities (1), (2), (5) and (6) because the Betti numbers and the t_i's can only grow passing from R to R' and $\beta_1^S(R) = \beta_{12}^S(R)$ equals $\beta_1^{S'}(R') = \beta_{12}^{S'}(R')$. This observation suggests the following question:

Question 2.2. Are the bounds of Remark 2.1 valid for every Koszul algebra?

In [ACI1] we have proved that (1), (2), (3) and (4) hold for every Koszul algebra. As far as we know it is still open whether (5) and (6) hold as well for Koszul algebras. The inequality (1) for Koszul algebras [and its immediate consequence (2)] has a short proof that we present below, see Lemma 2.10.

In [ACI2] stronger limitations for the degrees of the syzygies of Koszul algebras are described (under mild assumptions on the characteristic of the base field). To explain the results in [ACI1] and [ACI2] we start from some general considerations concerning bounds on the t_i's.

For $S = K[x_1, \ldots, x_n]$ and $R = S/I$ (and no assumptions on R), a very basic question is whether one can bound $t_i^S(R)$ only in terms of $t_1^S(R)$ and the index i.

The answer is negative in this generality, but it is positive if one involves also the number of variables n. Indeed, in [BM] and [CaS] it is proved that

$$t_i^S(R) \leq (2t_1^S(R))^{2^{n-2}} - 1 + i. \tag{4}$$

Furthermore variations of the Mayr–Meyer ideals define algebras having a doubly exponential syzygy growth of the kind of the right-hand side of (4) (but with slightly different coefficients), see [BS, Ko]. So, without any further assumption, one cannot expect any better bound for the $t_i^S(R)$ in terms of $t_1^S(R)$ than the one derived from (4).

Things change drastically if either R is defined by monomials (i.e. I is a monomial ideal) or R is Koszul. Under these assumptions we have that:

$$t_i^S(R) \leq t_1^S(R)i \tag{5}$$

holds for every i. In particular

$$t_i^S(R) \leq 2i \text{ for every } i \tag{6}$$

holds for Koszul algebras since $t_1^S(R) = 2$.

When R is defined by monomials (5) is derived from the Taylor resolution, while when R is Koszul (6) is proved by Kempf [K], in [ACI1] and also in an unpublished manuscript of Backelin (Relations between rates of growth of homologies, unpublished manuscript, 1988).

One can ask whether the inequalities (5) and (6) are special cases of, or can be derived from, more general statements. We consider the following generalization of (5):

$$t_{i+j}^S(R) \leq t_i^S(R) + t_j^S(R) \text{ for every } i \text{ and } j. \tag{7}$$

No counterexample is known to us to the validity of (7) for algebras with monomial relations or for Koszul algebras.

Also, (7) for $j = 1$ holds for algebras defined by monomials, see [FG, 1.9] where the statement is proved when R is defined by monomials of degree 2 and [HS] for the general case. Furthermore in [EHU, 4.1] it is proved that (7) holds for algebras of Krull dimension at most 1 when $i + j = n$, see also [Mc] for related results.

Denote by

$$Z = \bigoplus_{i \geq 0} Z_i = \bigoplus_{i \geq 0} Z_i(\mathbf{m}_R),$$

$$B = \bigoplus_{i \geq 0} B_i = \bigoplus_{i \geq 0} B_i(\mathbf{m}_R),$$

and by

$$H = \bigoplus_{i \geq 0} H_i = \bigoplus_{i \geq 0} H_i(\mathbf{m}_R)$$

the modules of cycles, boundaries and homology of the Koszul complex $K(\mathbf{m}_R)$ associated with the maximal homogeneous ideal $\mathbf{m}_R$ of R. Similarly, $Z_i(\mathbf{m}_R, M)$ stands for the i-th cycles of $K(\mathbf{m}_R, M) = K(\mathbf{m}_R) \otimes M$ and so on. By construction,

$$t_i^S(R) = \sup\{j : (H_i)_j \neq 0\}. \tag{8}$$

For a Koszul algebra R we have shown in [ACI1] that the map

$$\wedge^i H_1 \to H_i \tag{9}$$

induced by the multiplicative structure on H is surjective in degree $2i$ and higher. Hence (6) (for Koszul algebras) is an immediate corollary of that assertion. The inequality (7) would follow from a similar statement regarding the multiplication map

$$H_i \otimes H_j \to H_{i+j}. \tag{10}$$

Indeed it would be enough to prove that the map (10) is surjective in degree $t_i^S(R) + t_j^S(R)$ and higher. Unfortunately we are not able to evaluate directly the cokernel of the map (10). Instead we can get some information by using the splitting map for Koszul cycles described originally in [BCR2] and rediscussed in [ACI2] from a more general perspective. Indeed in [ACI2, 2.2] it is proved that:

Theorem 2.3. *Let M be a graded R-module. For even i, j there is a natural map (of degree 0)*

$$\operatorname{Tor}_1^R(C_{i-1}, Z_j(\mathbf{m}_R, M)) \to H_{i+j}(\mathbf{m}_R, M)/H_i H_j(\mathbf{m}_R, M) \tag{11}$$

that is surjective provided R has characteristic 0 or large. Here C_{i-1} denotes the cokernel of the Koszul complex $K(\mathbf{m}_R)$ in homological position $i - 1$.

Taking $M = R$ one obtains a natural map:

$$\operatorname{Tor}_1^R(C_{i-1}, Z_j) = \operatorname{Tor}_1^R(B_{i-1}, B_{j-1}) \to H_{i+j}/H_i H_j$$

that is surjective in characteristic 0 or large. Note that $\operatorname{Tor}_1^R(B_{i-1}, B_{j-1})$ is a finite length module because the B_u's are free when localized at any non-maximal prime homogeneous ideal. In particular one has:

Proposition 2.4. *Set $T_{ij} = \operatorname{Tor}_1^R(B_{i-1}, B_{j-1})$. If R has characteristic 0 or large then*

$$t_{i+j}^S(R) \leq \max\{t_i^S(R) + t_j^S(R), \operatorname{reg}_S T_{ij}\} \tag{12}$$

where

$$\operatorname{reg}_S T_{ij} = \max\{v : (T_{ij})_v \neq 0\}.$$

In order to evaluate $\operatorname{reg}_S T_{ij}$ we have developed in [ACI2] a long and technically complicated inductive procedure. The results obtained take a simpler form in the Cohen–Macaulay case because, under such assumption, the $t_i^S(R)$'s form an increasing sequence. We have:

Theorem 2.5. *Let R be a Koszul and Cohen–Macaulay algebra of characteristic 0. Then*

$$t_{i+1}^S(R) \leq t_i^S(R) + t_1(R) = t_i^S(R) + 2 \tag{13}$$

and

$$t_{i+j}^S(R) \leq \max\{t_i^S(R) + t_j^S(R), t_{i-1}^S(R) + t_{j-1}^S(R) + 3\} \tag{14}$$

hold for every i and j.

Furthermore one also deduces:

Theorem 2.6. *If R is a Koszul algebra of characteristic 0 satisfying the Green–Lazarsfeld N_p condition for some $p > 1$ (i.e. $\beta_{ij}^S(R) = 0$ for every $i = 1, 2, \ldots, p$ and every $j > i+1$) then*

$$\operatorname{reg}_S(R) \leq 2 \left\lfloor \frac{\operatorname{pd}_S R}{p+1} \right\rfloor + 1 \tag{15}$$

where the "$+1$" can be omitted if $p+1$ divides $\operatorname{pd}_S R$.

For more general results the reader can consult [ACI2, Sect. 5].

Remark 2.7. The problem of bounding the regularity of Tor-modules has been studied in [EHU]. Let $S = K[x_1, \ldots, x_n]$, and let M and N be finitely generated graded S-modules. It is proved in [EHU] that if $\dim \operatorname{Tor}_1^S(M, N) \leq 1$, then for every i one has

$$\operatorname{reg}_S \operatorname{Tor}_i^S(M, N) \leq \operatorname{reg}_S(M) + \operatorname{reg}_S(N) + i. \tag{16}$$

Unfortunately, the formula (16) does not hold if we replace the polynomial ring S with a Koszul ring R. For example with

$$R = K[x, y]/(x^2 + y^2), \quad M = R/(\bar{x}) \quad \text{and } N = R/(\bar{y})$$

one has

$$\operatorname{reg}_R M = \operatorname{reg}_R N = 0 \quad \text{and} \quad \operatorname{Tor}_1^R(M, N) = K(-2)$$

so that has regularity

$$\operatorname{reg}_R \operatorname{Tor}_1^R(M, N) = 2.$$

Nevertheless variations of (16) (e.g. compute the Tor over R but regularity over S or add a correction term on the left depending on $\operatorname{reg}_S(R)$) might hold in general.

The regularity bound in Theorem 2.6 is much weaker than the logarithmic one obtained by Dao, Huneke and Schweig in [DHS, 4.8]. They showed that an algebra R with monomial quadratic relations and satisfying the property N_p for some $p > 1$ has a very low regularity compared with its embedding dimension. Their result asserts that for a given $p > 1$ there exist $f_p(x) \in \mathbf{R}[x]$ and $\alpha_p \in \mathbf{R}$ with $\alpha_p > 1$ (which are explicitly given in the paper) such that

$$\operatorname{reg}_S R \leq \log_{\alpha_p} f_p(n) \tag{17}$$

holds for every algebra R with quadratic monomial relations such that R has the property N_p and has embedding dimension n.

This type of bound cannot hold for Koszul algebras satisfying N_p no matter what $f_p(x) \in \mathbf{R}[x]$ and $\alpha_p \in (1, \infty)$ are. To show this, it is enough to describe a family of algebras $\{R_{p,m}\}$ with $p, m \in \mathbf{N}$ and $p > 1$ such that:

1. the algebra $R_{p,m}$ is Koszul and satisfies the N_p-property,
2. given p, the embedding dimension of $R_{p,m}$ is a polynomial function of m and the regularity of $R_{p,m}$ is linear in m.

For example, let $R_{p,m}$ be the p-th Veronese subalgebra of a polynomial ring in pm variables. Then $R_{p,m}$ is Koszul, it satisfies the N_p-property, it has regularity $(p-1)m$, and has embedding dimension equal to $\binom{pm+p-1}{p}$, see [BCR1].

Question 2.8. Consider the coordinate ring of the Grassmannian $G(2, n)$. It is defined by the Pfaffians of degree 2 (the 4×4 Pfaffians) in a $n \times n$ skew-symmetric generic matrix. We know that it is Koszul, it satisfies the N_2 condition (by work of Kurano [Ku]), it has regularity $n - 3$ and codimension $\binom{n-2}{2}$. Hence for this family the codimension is quadratic in the regularity. Does there exist a family like this (Koszul with the N_2 property) such that the codimension is linear in the regularity?

Remark 2.9. If we apply Theorem 2.3 in the case $R = S = K[x_1, \ldots, x_n]$ with K of characteristic 0 or large and M any graded module then we have a surjection

$$\operatorname{Tor}_1^S(C_{i-1}, Z_j(M)) \to H_{i+j}(\mathbf{m}_S, M)$$

because $H_i = 0$. Here we have set for simplicity $Z_j(M) = Z_j(\mathbf{m}_S, M)$. Since

$$\operatorname{Tor}_1^S(C_{i-1}, Z_i(M)) = H_i(\mathbf{m}_S, Z_j(M))$$

we obtain
$$\beta^S_{i,v}(Z_j(M)) \geq \beta^S_{i+j,v}(M) \qquad (18)$$
for all i, j, v and every M.

We conclude by presenting a short proof of the inequality $t^S_i(R) \leq 2i$ for Koszul algebras and a related question.

Lemma 2.10. *Let R be a Koszul algebra and $Z_i = Z_i(\mathbf{m}_R)$ the cycles of the Koszul complex $K(\mathbf{m}_R, R)$ associated with the maximal homogeneous ideal of R. Then $\mathrm{reg}_R(Z_i) \leq 2i$ for every i. In particular, $t^S_i(R) \leq 2i$ for every i.*

Proof. For $i > 0$ we have short exact sequences:
$$0 \to Z_i \to K_i \to B_{i-1} \to 0$$
$$0 \to B_{i-1} \to Z_{i-1} \to H_{i-1} \to 0.$$

Hence one has:
$$\mathrm{reg}_R(Z_i) = \mathrm{reg}_R(B_{i-1}) + 1$$
and
$$\mathrm{reg}_R(B_{i-1}) \leq \max\{\mathrm{reg}_R(Z_{i-1}), \mathrm{reg}_R(H_{i-1}) + 1\}.$$

Hence
$$\mathrm{reg}_R(Z_i) \leq \max\{\mathrm{reg}_R(Z_{i-1}) + 1, \mathrm{reg}_R(H_{i-1}) + 2\}.$$

Since $\mathbf{m}_R H_{i-1} = 0$ and R is Koszul we have
$$\mathrm{reg}_R(H_{i-1}) = t^R_0(H_{i-1}) \leq t^R_0(Z_{i-1}) \leq \mathrm{reg}_R(Z_{i-1}).$$

It follows that
$$\mathrm{reg}_R(Z_i) \leq \max\{\mathrm{reg}_R(Z_{i-1}) + 1, \mathrm{reg}_R(Z_{i-1}) + 2\} = \mathrm{reg}_R(Z_{i-1}) + 2.$$

Since $Z_0 = R$ one has $\mathrm{reg}_R(Z_0) = 0$ and it follows by induction that
$$\mathrm{reg}_R(Z_i) \leq 2i.$$

Since
$$t^S_i(R) = t^R_0(H_i) \leq t^R_0(Z_i) \leq \mathrm{reg}_R(Z_i) \leq 2i$$

we may conclude that

$$t_i^S(R) \leq 2i.$$

□

With the assumptions and notation of Lemma 2.10 one can ask:

Question 2.11. Does the inequality

$$\operatorname{reg}_R(Z_{i+j}) \leq \operatorname{reg}_R(Z_i) + \operatorname{reg}_R(Z_j)$$

hold for every i, j?

For similar questions and results the reader can consult [CM].

3 Veronese Rings and Algebras Associated with Families of Hyperspaces

In the third lecture we present two case studies: the Koszul properties of Veronese rings and of algebras associated with families of hyperspaces. The material we present is taken from [ABH, BF, B1, BaM, Ca, CC, CHTV, CTV, CRV, ERT].

3.1 Veronese Rings

We will use the following results whose proofs can be found, for example, in the survey paper [CDR].

Lemma 3.1. *Let R be a standard graded K-algebra. Let*

$$\mathbf{M} : \cdots \to M_i \to \cdots \to M_2 \to M_1 \to M_0 \to 0$$

be a complex of R-modules. Set $H_i = H_i(\mathbf{M})$. Then for every $i \geq 0$ one has

$$t_i^R(H_0) \leq \max\{\alpha_i, \beta_i\}$$

where $\alpha_i = \max\{t_j^R(M_{i-j}) : j = 0, \ldots, i\}$ and $\beta_i = \max\{t_j^R(H_{i-j-1}) : j = 0, \ldots, i-2\}$. Moreover one has

$$\operatorname{reg}_R(H_0) \leq \max\{\alpha, \beta\}$$

where $\alpha = \sup\{\operatorname{reg}_R(M_j) - j : j \geq 0\}$ and $\beta = \sup\{\operatorname{reg}_R(H_j) - (j+1) : j \geq 1\}$.

Theorem 3.2. *Let A be a standard graded K-algebra, $J \subset A$ a homogeneous ideal and $B = A/J$. Then:*

(1) *If $\operatorname{reg}_A(B) \leq 1$ and A is Koszul, then B is Koszul.*
(2) *If $\operatorname{reg}_A(B)$ is finite and B is Koszul, then A is Koszul.*

We apply now Theorem 3.2 to prove that the Veronese subrings of a Koszul algebra are Koszul.

Let R be a standard graded K-algebra. Let $c \in \mathbf{N}$ and

$$R^{(c)} = \oplus_{j \in \mathbf{Z}} R_{jc}$$

be the c-th Veronese subalgebra of R. Similarly, given a graded R-module M one defines

$$M^{(c)} = \oplus_{j \in \mathbf{Z}} M_{jc}.$$

The formation of the c-th Veronese submodule can be seen as an exact functor from the category of graded R-modules and maps of degree 0 to the category of graded $R^{(c)}$-modules and maps of degree 0. For $u = 0, \ldots, c-1$ consider the Veronese submodules

$$V_u = \oplus_{j \in \mathbf{Z}} R_{jc+u}$$

of R. Note that V_u is an $R^{(c)}$-module generated in degree 0 and that for $a \in \mathbf{Z}$ one has

$$R(-a)^{(c)} = V_u(-\lceil a/c \rceil)$$

where $u = 0$ if $a \equiv 0 \bmod(c)$ and $u = c - r$ if $a \equiv r \bmod(c)$ and $0 < r < c$.

Theorem 3.3. *Let R be a Koszul algebra. Then for every $c \in \mathbf{N}$ one has:*

(1) *$R^{(c)}$ is Koszul.*
(2) *$\operatorname{reg}_{R^{(c)}}(V_u) = 0$ for every $u = 0, \ldots, c-1$, i.e. the Veronese submodules of R have a linear resolution over $R^{(c)}$.*

Proof. Denote by A the ring $R^{(c)}$. We first prove assertion (2). Indeed we prove by induction on i that $t_i^A(V_u) \leq i$ for every i. There is nothing to prove in the case $i = 0$. For $i > 0$, observe that, since R is Koszul, one has $\operatorname{reg}_R \mathbf{m} = 1$ and by induction one has that $\operatorname{reg}_R \mathbf{m}^u = u$. Now let $M = \mathbf{m}_R^u(u)$ so that $\operatorname{reg}_R(M) = 0$ and $M^{(c)} = V_u$. Consider the minimal free resolution $\mathbf{F}$ of M over R and apply the functor $-^{(c)}$. We get a complex $\mathbf{G} = \mathbf{F}^{(c)}$ of A-modules such that $H_0(\mathbf{G}) = V_u$, $H_j(\mathbf{G}) = 0$ for $j > 0$ and $G_j = F_j^{(c)}$ is a direct sum of copies of $R(-j)^{(c)}$. Applying Lemma 3.1 and the inductive assumption we get $t_i^A(V_u) \leq i$ as required.

To prove that A is Koszul we consider the minimal free resolution $\mathbf{F}$ of K over R and apply $-^{(c)}$. We get a complex $\mathbf{G} = \mathbf{F}^{(c)}$ of A-modules such that $H_0(\mathbf{G}) = K$,

$H_j(\mathbf{G}) = 0$ for $j > 0$ and $G_j = F_j^{(c)}$ is a direct sum of copies of $V_u(-\lceil j/c \rceil)$. Hence $\operatorname{reg}_A(\mathbf{G}_j) = \lceil j/c \rceil$ and applying Lemma 3.1 we obtain

$$\operatorname{reg}_A(K) \leq \sup\{\lceil j/c \rceil - j : j \geq 0\} = 0.$$

□

We also have:

Theorem 3.4. *Let R be a standard graded K-algebra. Then:*

(1) *The Veronese subalgebra $R^{(c)}$ is Koszul for $c \gg 0$.*
(2) *If $R = S/I$ with $S = K[x_1, \ldots, x_n]$, then $R^{(c)}$ is Koszul for every c such that*

$$c \geq \max\{t_i^S(R)/(1+i) : i \geq 0\}.$$

Proof. Let $\mathbf{F}$ be the minimal free resolution of R as an S-module. Set $B = S^{(c)}$ and note that B is Koszul because of Theorem 3.3. Then $\mathbf{G} = \mathbf{F}^{(c)}$ is a complex of B-modules such that $H_0(\mathbf{G}) = R^{(c)}$, $H_j(\mathbf{G}) = 0$ for $j > 0$. Furthermore $G_i = F_i^{(c)}$ is a direct sum of shifted copies of the Veronese submodules V_u. Using Theorem 3.3 we get the bound $\operatorname{reg}_B(G_i) \leq \lceil t_i^A(R)/c \rceil$. Applying Lemma 3.1 we get

$$\operatorname{reg}_B(R^{(c)}) \leq \max\{\lceil t_i^S(R)/c \rceil - i : i \geq 0\}.$$

Hence for $c \geq \max\{t_i^S(R)/(1+i) : i \geq 0\}$ one has $\operatorname{reg}_B(R^{(c)}) \leq 1$ and we conclude from Theorem 3.2 that $R^{(c)}$ is Koszul. □

Remark 3.5. (1) In [ERT, 2] it is proved that if $c \geq (\operatorname{reg}_S(R) + 1)/2$, then $R^{(c)}$ is even G-quadratic. See [Sh] for other interesting results in this direction.

(2) Backelin proved in [B1] that $R^{(c)}$ is Koszul if $c \geq \operatorname{Rate}(R)$. Here $\operatorname{Rate}(R)$ is defined as

$$\sup_{i>0}\{(t_{i+1}^R(K) - 1)/i\}$$

and it is finite. It measures the deviation from the Koszul property in the sense that $\operatorname{Rate}(R) \geq 1$ with equality if and only if R is Koszul.

3.2 Strongly Koszul Algebras

A powerful tool for proving that an algebra is Koszul is a typical "divide and conquer" strategy that can be formulated in the following way, see [CTV] and [CRV]:

Definition 3.6. A Koszul filtration of a K-algebra R is a set $\mathscr{F}$ of ideals of R such that:

(1) Every ideal $I \in \mathscr{F}$ is generated by elements of degree 1.
(2) The zero ideal 0 and the maximal ideal $\mathbf{m}_R$ are in $\mathscr{F}$.
(3) For every $I \in \mathscr{F}$, $I \neq 0$, there exists $J \in \mathscr{F}$ such that $J \subset I$, I/J is cyclic and $\mathrm{Ann}(I/J) = J : I \in \mathscr{F}$.

One easily proves:

Lemma 3.7. *Let $\mathscr{F}$ be a Koszul filtration of a standard graded K-algebra R. Then one has:*

(1) $\mathrm{reg}_R(R/I) = 0$ and R/I is Koszul for every $I \in \mathscr{F}$.
(2) R is Koszul.

Example 3.8. Let

$$R = K[a,b,c,d]/(ac, ad, ab - bd, a^2 + bc, b^2).$$

We have seen in Example 1.20 that R is not LG-quadratic. We show now that R is Koszul by constructing a Koszul filtration. Indeed, there is a Koszul filtration based on the given system of coordinates, i.e. a Koszul filtration whose ideals are generated by residue classes of variables. Here it is:

$$\mathscr{F} = \{(a,b,c,d), (a,c,d), (c,d), (a,c), (c), (a), 0\}.$$

To check that it is a Koszul filtration we observe that in R one has:

$$(a,c,d) : (a,b,c,d) = (a,b,c,d)$$
$$(c,d) : (a,c,d) = (a,b,c,d)$$
$$(c) : (c,d) = (a,c)$$
$$(c) : (a,c) = (a,c,d)$$
$$0 : (a) = (c,d)$$
$$0 : (c) = (a)$$

The following notion is very natural for algebras with a canonical coordinate system (e.g. in the toric case).

Definition 3.9. An algebra R is strongly Koszul if there exists a basis X of R_1 such that for every $Y \subset X$ and for every $x \in X \setminus Y$ there exists $Z \subseteq X$ such that $(Y) : x = (Z)$.

This definition of strongly Koszul is taken from [CDR] and is slightly different from the one given in [HHR]. In [HHR] it is assumed that the basis X of R_1 is totally ordered and in the definition one adds the requirement that x is larger than every element in Y.

Remark 3.10. If R is strongly Koszul with respect to a basis X of R_1 then the set $\{(Y) : Y \subseteq X\}$ is obviously a Koszul filtration.

We have:

Theorem 3.11. *Let $R = S/I$ with $S = K[x_1, \ldots, x_n]$ and $I \subset S$ be an ideal generated by monomials of degrees $\leq d$. Then $R^{(c)}$ is strongly Koszul for every $c \geq d - 1$.*

The proof of Theorem 3.11 is based on the fact that the Veronese ring $R^{(c)}$ is a direct summand of R and that computing the colon ideal of monomial ideals in a polynomial ring is a combinatorial operation. Let us single out an interesting special case:

Theorem 3.12. *Let $S = K[x_1, \ldots, x_n]$ and let $I \subset S$ be an ideal generated by monomials of degree 2. Then S/I is strongly Koszul.*

The results presented for Veronese rings and Veronese modules have their analogous in the multigraded setting, see [CHTV]. We discuss below the bigraded case.

Let

$$S = K[x_1, \ldots, x_n, y_1, \ldots, y_m]$$

with $\mathbf{Z}^2$-grading induced by the assignment $\deg(x_i) = (1, 0)$ and $\deg(y_i) = (0, 1)$. For every $c = (c_1, c_2)$ we look at the diagonal subalgebra

$$S_\Delta = \oplus_{a \in \Delta} S_a$$

where

$$\Delta = \{ic : i \in \mathbf{Z}\}.$$

The algebra S_Δ is nothing but the Segre product of the c_1-th Veronese ring of $K[x_1, \ldots, x_n]$ and the c_2-th Veronese ring of $K[y_1, \ldots, y_m]$. For a $\mathbf{Z}^2$-graded standard K-algebra $R = S/I$ with $I \subset S$ a bigraded ideal we may consider the associated diagonal algebra

$$R_\Delta = \oplus_{a \in \Delta} R_a$$

and similarly for modules. One has:

Theorem 3.13. (1) *For every $(a, b) \in \mathbf{Z}^2$ the S_Δ-submodule $S(-a, -b)_\Delta$ of S has a linear resolution.*

(2) *For every $\mathbf{Z}^2$-standard graded algebra R one has that R_Δ is Koszul for "large" Δ (i.e. $c_1 \gg 0$ and $c_2 \gg 0$). One can give explicit bounds in terms of the bigraded Betti numbers of R as an S-module.*

Let I be a homogeneous ideal of $S = K[x_1, \ldots, x_n]$ generated by elements $f_1, \ldots, f_r$ of degree d. The Rees ring

$$\mathrm{Rees}(I) = \bigoplus_{i \in \mathbf{N}} I^i = S[f_1 t, \ldots, f_r t] \subset S[t]$$

is a bigraded K-algebra. Its component of degree (i, j) is

$$\mathrm{Rees}(I)_{(i,j)} = (I^j)_{jd+i}$$

It is easy to check that $\mathrm{Rees}(I)$ is a standard bigraded algebra. It can be seen as a quotient ring of $S[y] = S[y_1, \ldots, y_r]$ bigraded by $\deg(x_i) = (1, 0)$ and $\deg(y_i) = (0, 1)$. Then we may apply Theorem 3.13 and we get that

Corollary 3.14. *There exist integers c_0 and e_0 (depending on I) such that for every $c \geq c_0$ and $e \geq e_0$ the K-subalgebra of S generated by the vector space $(I^e)_{ed+c}$ is Koszul.*

If one has information or bounds on the bigraded resolution of $\mathrm{Rees}(I)$ an $S[y]$-module then Corollary 3.14 can be formulated more precisely. One of the few families of ideals I for which the resolution of $\mathrm{Rees}(I)$ is known are the complete intersections. If $f_1, \ldots, f_r$ form a regular sequence then

$$\mathrm{Rees}(I) = S[y]/I_2 \begin{pmatrix} y_1 & y_2 & \cdots & y_r \\ f_1 & f_2 & \cdots & f_r \end{pmatrix}$$

and $\mathrm{Rees}(I)$ is resolved by the Eagon–Northcott complex. Then applying the principle described above to this specific case one has:

Theorem 3.15. *Let $f_1, \ldots, f_r$ be a regular sequence of elements of degree d in $S = K[x_1, \ldots, x_n]$ and $I = (f_1, \ldots, f_r)$. For $c, e \in \mathbf{N}$ set $A = K[(I^e)_{ed+c}]$. Then:*

(1) *If $c \geq d/2$ then A is quadratic.*
(2) *if $c \geq d(r-1)/r$ then A is Koszul.*

See [CHTV] for details of the proofs of Theorem 3.15.

Example 3.16. With $r = n$, $d = 2$ and $f_i = x_i^2$ for every $i = 1, \ldots, n$ we have that the toric algebra

$$K[x^a : a \in \mathbf{N}^n, |a| = 2 + c \text{ and } \max(a) \geq 2]$$

is quadratic for every c and Koszul for $c \geq 2(n-1)/n$.

Given integers n, d, s we set

$$\mathbf{PV}(n, d, s) = K[x^a : a \in \mathbf{N}^n, |a| = d \text{ and } \#\{i : a_i > 0\} \leq s].$$

This is called the pinched Veronese generated by the monomials in n variables, of total degree d and supported on at most s variables.

Question 3.17. For which values of n, d, s is the algebra $\mathbf{PV}(n, d, s)$ quadratic or Koszul? Not all of them are quadratic, for instance $\mathbf{PV}(4, 5, 2)$ is defined, according to CoCoA [CoCoA], by 168 quadrics and 12 cubics.

The algebra of Example 3.16 for $c = n - 2$ coincides with the pinched Veronese $\mathbf{PV}(n, n, n - 1)$. Hence $\mathbf{PV}(n, n, n - 1)$ is quadratic for every n and Koszul for $n > 3$. For $n = 3$ we have that

$$\mathbf{PV}(3, 3, 2) = K[x^3, x^2y, x^2z, xy^2, xz^2, y^3, y^2z, yz^2, z^3]$$

is quadratic. The argument above does not answer the question whether $\mathbf{PV}(3, 3, 2)$ is a Koszul algebra. This turns out to be a difficult question on its own. In [Ca] and [CC] it is proved that:

Theorem 3.18. *The pinched Veronese* $\mathbf{PV}(3, 3, 2)$ *is Koszul. The same holds for the generic projection of the Veronese surface of* $\mathbf{P}^9$ *to* $\mathbf{P}^8$.

It is not clear whether $\mathbf{PV}(3, 3, 2)$ is G-quadratic. The Koszul property of a toric ring is equivalent to the Cohen–Macaulay property of intervals of the underlying poset, see [PRS, 2.2]. Recently Tancer has shown that the intervals of the poset associated with $\mathbf{PV}(n, n, n - 1)$ are shellable for $n > 3$, see [Ta]. It is not clear whether the same is true for $n = 3$.

3.3 Koszul Algebras Associated with Hyperspace Configurations

Another interesting family of Koszul algebras with relations to combinatorics arises in the following way. Let $V = V_1, \ldots, V_m$ be a collection of subspaces of the space of linear forms in the polynomial ring $K[x_1, \ldots, x_n]$. Denote by $A(V)$ the K-subalgebra of $K[x_1, \ldots, x_n]$ generated by the elements of the product $V_1 \cdots V_m$. We have:

Theorem 3.19. *The algebra* $A(V)$ *is Koszul.*

We outline the proof of Theorem 3.19. Denote by R the polynomial ring $K[x_1, \ldots, x_n]$ and set $d_i = \dim V_i$. Consider auxiliary variables $y_1, \ldots, y_m$ and the Segre product:

$$S = K[y_i x_j : i = 1, \ldots, m, j = 1, \ldots, n]$$

of $K[y_1,\ldots,y_m]$ with R. Set

$$B(V) = K[y_1 V_1,\ldots,y_m V_m].$$

and

$$T = K[t_{ij} : i = 1,\ldots,m, j = 1,\ldots,n].$$

Note that $B(V)$ is a K-subalgebra of S. We give degree $e_i \in \mathbf{Z}^m$ to $y_i x_j$ and to t_{ij} so that S, T and $B(V)$ are $\mathbf{Z}^m$-graded. Let

$$\Delta = \{(a,a,\ldots,a) \in \mathbf{Z}^m : a \in \mathbf{Z}\}.$$

By construction, the diagonal algebra

$$B(V)_\Delta = \bigoplus_{b \in \mathbf{Z}^m} B(V)_b$$

coincides with $K[V_1 \cdots V_m y_1 \cdots y_m]$ and hence

$$B(V)_\Delta = A(V).$$

For $i = 1,\ldots,m$ let $\{f_{ij} : j = 1,\ldots,d_i\}$ be a basis of V_i and complete it to a basis of R_1 with elements $\{f_{ij} : j = d_i + 1,\ldots,n\}$ (no matter how). Set $T(V) = K[t_{ij} : 1 \leq i \leq m, 1 \leq j \leq d_i]$. We have presentations:

$$\phi : T \to S \qquad \text{with } t_{ij} \to y_i f_{ij} \text{ for all } i, j$$

$$\phi' : T(V) \to B(V) \text{ with } t_{ij} \to y_i f_{ij} \text{ for all } i \text{ and } 1 \leq j \leq d_i$$

We have:

Lemma 3.20. *Suppose that* $\ker \phi'$ *has a Gröbner basis (with respect to some term order $>$) of elements of degrees bounded above by* $(1,1,\ldots,1) \in \mathbf{Z}^m$. *Then* $A(V)$ *is Koszul.*

Proof. Set $I = \ker \phi'$. Applying Δ to the presentation $B(V) = T(V)/I$ we obtain $A(V) = T(V)_\Delta / I_\Delta$. Now $T(V)_\Delta$ is a multiple Segre product and hence it is strongly Koszul (the argument is similar to the one for the Veronese case) and by assumption $\text{in}_>(I_\Delta)$ is generated by a subset of the semigroup generators of $T(V)_\Delta$. But then $T(V)/\text{in}_>(I_\Delta)$ is Koszul because of the strongly Koszul property of $T(V)_\Delta$. Hence $T(V)_\Delta/I_\Delta$ is Koszul by Gröbner deformation. □

Since, by construction, $\ker \phi' = \ker \phi \cap T(V)$, a Gröbner basis of $\ker \phi'$ can be obtained from a lexicographic Gröbner basis of $\ker \phi$ by elimination. Therefore, combining this point of view with Lemma 3.20 we have that Theorem 3.19 is a corollary of:

Lemma 3.21. *The ideal* $\ker \phi$ *has a universal Gröbner basis whose elements have degrees bounded above by* $(1, 1, \ldots, 1) \in \mathbf{Z}^m$.

Observe that ϕ is a presentation of the Segre product S but with respect to a non-necessarily monomial basis. Hence $\ker \phi$ is obtained from the ideal $I_2(t)$ of the 2-minors of the $m \times n$ matrix

$$t = (t_{ij})$$

by a change of coordinates preserving the $\mathbf{Z}^m$-graded structure. Since the Hilbert function does not change under taking initial ideals, it is enough to prove the following (very strong) assertion:

Lemma 3.22. *Every ideal of T that has the $\mathbf{Z}^m$-graded Hilbert function of $I_2(t)$ is generated in degrees bounded above by* $(1, 1, \ldots, 1) \in \mathbf{Z}^m$.

Lemma 3.22 has been proved by Cartwright and Sturmfels [CS] using multi-graded generic initial ideals and a result proved in [C]. This approach has been generalized in [CDG] to identify universal Gröbner bases of ideals of maximal minors of matrices of linear forms hence generalizing the classical result of Bernstein, Sturmfels and Zelevinsky [BZ, SZ], see also [K]. In detail, the group $\mathrm{GL}_n(K)^m$ acts as the group of $\mathbf{Z}^m$-graded K-algebra automorphisms on T by linear substitution (row by row). An ideal $I \subset T$ is Borel-fixed if it is invariant under the action of the Borel subgroup $B_n(K)^m$ of $\mathrm{GL}_n(K)^m$. Here $B_n(K)$ is the group of upper triangular matrices. In [CDG] it has been proved that

Lemma 3.23. *If $J \subset T$ is Borel-fixed and radical then every ideal I with the $\mathbf{Z}^m$-graded Hilbert function of J is generated in degrees bounded above by* $(1, 1, \ldots, 1) \in \mathbf{Z}^m$.

Summing up, to conclude the proof of Theorem 3.19 it is enough to prove that:

Lemma 3.24. *The ideal $I_2(t)$ has the $\mathbf{Z}^m$-graded Hilbert function of the radical and Borel-fixed ideal J generated by the monomials $t_{i_1 j_1} \cdots t_{i_k j_k}$ satisfying the following conditions:*

$$1 \leq i_1 < \cdots < i_k \leq m,$$
$$1 \leq j_1, \ldots, j_k \leq n,$$
$$j_1 + \cdots + j_k \geq n + k.$$

This is done in [C] by proving that J is indeed the multigraded generic initial ideal $\mathrm{gin}(I_2(t))$ of $I_2(t)$. The inclusion $J \subseteq \mathrm{gin}(I_2(t))$ is a consequence of the following Lemma 3.25. The other inclusion is proved by checking that J is pure with codimension and degree equal those of $I_2(t)$.

Lemma 3.25. Let $V_1, \ldots, V_m$ be subspaces of the vector space of the linear forms R_1. If $\sum_{i=1}^{m} \dim V_i \geq n + m$ then $\dim \prod_{i=1}^{m} V_i < \prod_{i=1}^{m} \dim V_i$, i.e. there is a non-trivial linear relation among the generators of the product $\prod_{i=1}^{m} V_i$ obtained by multiplying K-bases of the V_i's.

One can also prove directly that $T/I_2(t)$ and T/J have the same $\mathbf{Z}^m$-graded Hilbert function in the following way. For every $a = (a_1, \ldots, a_m) \in \mathbf{N}^m$ we show that the vector space dimensions of $T/I_2(t)$ and of T/J in multidegree a are equal. By induction on m we may assume that a has full support, i.e. $a_i > 0$ for all i. Since $T/I_2(t)$ is the Segre product of $K[y_1, \ldots, y_m]$ and $K[x_1, \ldots, x_n]$, a K-basis of $T/I_2(t)$ in degree a is given by the monomials of the form $\prod y_i^{a_i} p$ where p is a monomial in the x's of degree $\sum a_i$. It follows that the dimension of $T/I_2(t)$ in degree a equals:

$$\binom{n-1+a_1+a_2+\cdots+a_m}{n-1}.$$

Given a monomial p in the t_{ij}'s of multidegree a we set $M_i(p) = \max\{j : t_{ij}|p\}$. It is easy to see that:

$$p \in J \iff \sum_{i=1}^{m} M_i(p) \geq n + m.$$

For a fixed $c = (c_1, \ldots, c_m) \in \{1, \ldots, n\}^m$ the cardinality of the set of the monomials p in the t_{ij}'s with degree a and $M_i(p) = c_i$ is given by

$$\prod_{i=1}^{m} \binom{c_i - 1 + a_i - 1}{c_i - 1}.$$

Therefore the dimension of T/J in multidegree a is given by:

$$\sum_{c} \prod_{i=1}^{m} \binom{c_i - 1 + a_i - 1}{c_i - 1}$$

where the sum is extended to all the $c = (c_1, \ldots, c_m) \in \{1, \ldots, n\}^m$ with $c_1 + \cdots + c_m < n + m$. Replacing $n - 1$ with n and $c_i - 1$ with c_i, we have to prove the following identity:

$$\binom{n + a_1 + a_2 + \cdots + a_m}{n} = \sum_{c} \prod_{i=1}^{m} \binom{c_i + a_i - 1}{c_i} \qquad (19)$$

where the sum is extended to all the $c = (c_1, \ldots, c_m) \in \mathbf{N}^m$ with $c_1 + \cdots + c_m \leq n$. The equality (19) is a specialization ($v = m + 1$, $b_i = a_i$ for $i = 1, \ldots, m$ and $b_{m+1} = 1$) of the following identity:

$$\binom{n + b_1 + b_2 + \cdots + b_v - 1}{n} = \sum_c \prod_{i=1}^{v} \binom{c_i + b_i - 1}{c_i} \qquad (20)$$

where the sum is extended to all the $c = (c_1, \ldots, c_v) \in \mathbf{N}^v$ with $c_1 + \cdots + c_v = n$.

Now the identity (20) is easy: both the left and right side of it count the number of monomials of total degree n in a set of variables which is a disjoint union of subsets of cardinality $b_1, b_2, \ldots, b_v$.

Acknowledgements We thank Giulio Caviglia, Alessio D'Alì, Emanuela De Negri and Dang Hop Nguyen for their valuable comments and suggestions upon reading preliminary versions of the present notes and Christian Krattenthaler for suggesting the proof of formula (19).

References

[An]   D. Anick, A counterexample to a conjecture of Serre. Ann. Math. **115**, 1–33 (1982)

[ABH]   A. Aramova, Ş. Bărcănescu, J. Herzog, On the rate of relative Veronese submodules. Rev. Roum. Math. Pures Appl. **40**, 243–251 (1995)

[A]   L.L. Avramov, Infinite free resolutions, in *Six Lectures on Commutative Algebra (Bellaterra, 1996)*. Progress in Mathematics, vol. 166 (Birkhäuser, Basel, 1998), pp. 1–118

[A1]   L.L. Avramov, Local algebra and rational homotopy, in *Homotopie algebrique et algebre locale (Luminy, 1982)*, ed. by J.-M. Lemaire, J.-C. Thomas. Asterisque, vols. 113 and 114 (Soc. Math. France, Paris, 1984), pp. 15–43

[AP]   L.L. Avramov, I. Peeva, Finite regularity and Koszul algebras. Am. J. Math. **123**, 275–281 (2001)

[ACI1]   L.L. Avramov, A. Conca, S. Iyengar, Free resolutions over commutative Koszul algebras. Math. Res. Lett. **17**, 197–210 (2010)

[ACI2]   L.L. Avramov, A. Conca, S. Iyengar, Subadditivity of syzygies of Koszul algebras (2013). Preprint [arXiv:1308.6811]

[AE]   L.L. Avramov, D. Eisenbud, Regularity of modules over a Koszul algebra. J. Algebra **153**, 85–90 (1992)

[B1]   J. Backelin, On the rates of growth of the homologies of Veronese subrings, in *Algebra, Algebraic Topology, and Their Interactions (Stockholm, 1983)*. Lecture Notes in Mathematics, vol. 1183 (Springer, Berlin, 1986), pp. 79–100

[BF]   J. Backelin, R. Fröberg, Koszul algebras, Veronese subrings, and rings with linear resolutions. Rev. Roum. Math. Pures Appl. **30**, 85–97 (1985)

[BM]   D. Bayer, D. Mumford, What can be computed in algebraic geometry?, in *Computational Algebraic Geometry and Commutative Algebra (Cortona, 1991)*. Sympos. Math., vol. XXXIV (Cambridge University Press, Cambridge, 1993), pp. 1–48

[BS]   D. Bayer, M. Stillman On the complexity of computing syzygies. Computational aspects of commutative algebra. J. Symb. Comput. **6**(2–3), 135–147 (1988)

[BaM]   S. Bărcănescu, N. Manolache, Betti numbers of Segre-Veronese singularities. Rev. Roum. Math. Pures Appl. **26**, 549–565 (1981)

[BZ]   D. Bernstein, A. Zelevinsky, Combinatorics of maximal minors. J. Algebr. Comb. **2**(2), 111–121 (1993)
[BC]   W. Bruns, A. Conca, Gröbner bases and determinantal ideals, in *Commutative Algebra, Singularities and Computer Algebra (Sinaia, 2002)*. NATO Science Series II: Mathematics, Physics and Chemistry, vol. 115 (Kluwer Academic, Dordrecht, 2003), pp. 9–66
[BCR1]   W. Bruns, A. Conca, T. Römer, Koszul homology and syzygies of Veronese subalgebras. Math. Ann. **351**, 761–779 (2011)
[BCR2]   W. Bruns, A. Conca, T. Römer, Koszul cycles, in *Combinatorial Aspects of Commutative Algebra and Algebraic Geometry*, ed. by G. Floystad et al. Abel Symposia, vol. 6 (2011)
[BH]   W. Bruns, J. Herzog, in *Cohen-Macaulay Rings*, Revised edn. Cambridge Studies in Advanced Mathematics, vol. 39 (Cambridge University Press, Cambridge, 1998)
[CS]   D. Cartwright, B. Sturmfels, The Hilbert scheme of the diagonal in a product of projective spaces. Int. Math. Res. Not. IMRN **9**, 1741–1771 (2010)
[Ca1]   G. Caviglia, Koszul algebras, Castelnuovo-Mumford regularity and generic initial ideal. Ph.D. Thesis, University of Kansas, 2004
[Ca]   G. Caviglia, The pinched Veronese is Koszul. J. Algebr. Comb. **30**, 539–548 (2009)
[CC]   G. Caviglia, A. Conca, Koszul property of projections of the Veronese cubic surface. Adv. Math. **234**, 404–413 (2013)
[CaS]   G. Caviglia, E. Sbarra, Characteristic-free bounds for the Castelnuovo-Mumford regularity. Compos. Math. **141**(6), 1365–1373 (2005)
[CoCoA] CoCoA Team, CoCoA: A System for Doing Computations in Commutative Algebra. Available at http://cocoa.dima.unige.it
[C]   A. Conca, Linear spaces, transversal polymatroids and ASL domains. J. Algebr. Comb. **25**(1), 25–41 (2007)
[CDR]   A. Conca, E. De Negri, M.E. Rossi, Koszul algebras and regularity, in *Commutative Algebra* (Springer, New York, 2013), pp. 285–315
[CDG]   A. Conca, E. De Negri, E. Gorla, Universal Gröbner bases for maximal minors Int. Math. Res. Not. (to appear) [arXiv:1302.4461]
[CHTV]   A. Conca, J. Herzog, N.V. Trung, G. Valla, Diagonal subalgebras of bigraded algebras and embeddings of blow-ups of projective spaces. Am. J. Math. **119**, 859–901 (1997)
[CM]   A. Conca, S. Murai, Regularity bounds for Koszul cycles. Proc. Am. Math. Soc. [arXiv:1203.1783] (to appear)
[CRV]   A. Conca, M.E. Rossi, G. Valla, Gröbner flags and Gorenstein algebras. Comp. Math. **129**, 95–121 (2001)
[CTV]   A. Conca, N.V. Trung, G. Valla, Koszul propery for points in projectives spaces. Math. Scand. **89**, 201–216 (2001)
[DHS]   H. Dao, C. Huneke, J. Schweig, Bounds on the regularity and projective dimension of ideals associated to graphs. J. Algebr. Comb. **38**(1), 37–55 (2013) [arXiv:1110.2570]
[EG]   D. Eisenbud, S. Goto, Linear free resolutions and minimal multiplicity. J. Algebra **88**, 89–133 (1984)
[ERT]   D. Eisenbud, A. Reeves, B. Totaro, Initial ideals, Veronese subrings, and rates of algebras. Adv. Math. **109**, 168–187 (1994)
[EHU]   D. Eisenbud, C. Huneke, B. Ulrich, The regularity of Tor and graded Betti numbers. Am. J. Math. **128**, 573–605 (2006)
[FG]   O. Fernandez, P. Gimenez, Regularity 3 in edge ideals associated to bipartite graphs. J. Algebr. Comb. (2012) [arXiv:1207.5553] (to appear)
[F]   R. Fröberg, Koszul algebras, in *Advances in Commutative Ring Theory*. Proceedings of the Fez Conference, 1997. Lectures Notes in Pure and Applied Mathematics, vol. 205 (Marcel Dekker Eds., New York, 1999)
[HHR]   J. Herzog, T. Hibi, G. Restuccia, Strongly Koszul algebras. Math. Scand. **86**, 161–178 (2000)

[HS]   J. Herzog, H. Srinivasan, A note on the subadditivity problem for maximal shifts in free resolutions [arXiv:1303.6214]
[K]   M.Y. Kalinin, *Universal and comprehensive Gröbner bases of the classical determinantal ideal.* Zap. Nauchn. Sem. S.-Peterburg. Otdel. Mat. Inst. Steklov. (POMI) 373 (2009). Teoriya Predstavlenii, Dinamicheskie Sistemy, Kombinatornye Metody. **XVII**, 134–143, 348; Translation in J. Math. Sci. (N.Y.) **168**(3), 385–389 (2010).
[Ko]   J. Koh, Ideals generated by quadrics exhibiting double exponential degrees. J. Algebra **200**(1), 225–245 (1998)
[Ku]   K. Kurano, Relations on Pfaffians. I. Plethysm formulas. J. Math. Kyoto Univ. **31**(3), 713–731 (1991)
[Mc]   J. McCullough, A polynomial bound on the regularity of an ideal in terms of half of the syzygies. Math. Res. Lett. **19**(3), 555–565 (2012) [arXiv:1112.0058]
[MS]   E. Miller, B. Sturmfels, in *Combinatorial Commutative Algebra.* Graduate Texts in Mathematics, vol. 227 (Springer, Berlin, 2004)
[PRS] I. Peeva, V. Reiner, B. Sturmfels, How to shell a monoid. Math. Ann. **310**(2), 379–393 (1998)
[P]   S.B. Priddy, Koszul resolutions. Trans. Am. Math. Soc. **152**, 39–60 (1970)
[PP]   A. Polishchuk, L. Positselski, in *Quadratic Algebras.* University Lecture Series, vol. 37 (American Mathematical Society, Providence, 2005)
[Sh]   T. Shibuta, Gröbner bases of contraction ideals. J. Algebr. Comb. **36**(1), 1–19 (2012) [arXiv:1010.5768]
[SZ]   B. Sturmfels, A. Zelevinsky, Maximal minors and their leading terms. Adv. Math. **98**(1), 65–112 (1993)
[Ta]   M. Tancer, Shellability of the higher pinched Veronese posets. J. Algebr. Comb. (to appear) [arXiv:1305.3159]

Noetherianity up to Symmetry

Jan Draisma

1 Kruskal's Tree Theorem

All finiteness proofs in these lecture notes are based on a beautiful combinatorial theorem due to Kruskal. In fact, the special case of that theorem known as Higman's Lemma suffices for all of those proofs. But, hoping that Kruskal's Tree Theorem will soon find further applications in infinite-dimensional algebraic geometry, I have decided to prove the theorem in its full strength. Original sources for Kruskal's Tree Theorem and Higman's Lemma are [Kru60] and [Hig52], respectively. We follow closely the beautiful proof in [NW63]. Throughout we use the notation $\mathbb{N} := \{1, 2, \ldots\}$, $\mathbb{Z}_{\geq 0} := \{0, 1, \ldots\}$, and $[n] := \{1, \ldots, n\}$ for $n \in \mathbb{Z}_{\geq 0}$. In particular, we have $[0] = \emptyset$.

The main concept is that of a well-partial-order on a set S. This is a partial order $\leq$ with the property that for any infinite sequence $s_1, s_2, \ldots$ of elements of S there exists a pair of indices $i < j$ with $s_i \leq s_j$. Arguing by contradiction one then proves that there exists an index i such that $s_j \geq s_i$ holds for infinitely many indices $j > i$. Take the first such index i_1, and retain only the term s_{i_1} together with the infinitely many terms s_j with $j > i_1$ and $s_j \geq s_{i_1}$. Among these pick an index $i_2 > i_1$ in a similar fashion, etc. This leads to the conclusion that in a well-partially-ordered set any infinite sequence has an infinite ascending subsequence $s_{i_1} \leq s_{i_2} \leq \ldots$ with $i_1 < i_2 < \ldots$.

Examples of well-partial-orders are partial orders on finite sets, and well-orders (which are linear well-partial-orders). If two sets S, T are both equipped with well-partial-orders, then the componentwise partial order on the Cartesian product $S \times T$

J. Draisma (✉)
Department of Mathematics and Computer Science, Eindhoven University of Technology, Eindhoven, The Netherlands

Centrum Wiskunde en Informatica, Amsterdam, The Netherlands
e-mail: j.draisma@tue.nl

defined by $(s, t) \leq (s', t')$ if and only if $s \leq s'$ and $t \leq t'$ is again a well-partial-order. Indeed, in an infinite sequence $(s_1, t_1), (s_2, t_2), \ldots$ there is an infinite subsequence of indices where the s_i increase weakly and in that subsequence there exist a pair of indices $i \leq j$ where in addition to $s_i \leq s_j$ also the inequality $t_i \leq t_j$ holds.

Repeatedly applying this Cartesian-product construction with all factors equal to the non-negative integers $\mathbb{Z}_{\geq 0}$ one obtains the statement that the componentwise order on $\mathbb{Z}_{\geq 0}^n$ is a well-partial-order. This fact, known as Dickson's Lemma, can be used to prove Hilbert's Basis Theorem. In a similar fashion we shall use Kruskal's Tree Theorem to prove Noetherianity of certain rings up to symmetry. Before stating and proving Kruskal's Tree Theorem, we first discuss the following special case.

Lemma 1.1. *For any well-partial-order on a set S the partial order on the set of finite multi-subsets of S defined by $A \leq B$ if and only if there exists an injective map $f : A \to B$ with $a \leq f(a)$ for all $a \in A$ is a well-partial-order.*

Proof. Suppose that it is not. Then there exists an infinite sequence $A_1, A_2, \ldots$ of finite multi-subsets of S such that $A_i \not\leq A_j$ for all pairs of indices $i < j$. Such a sequence is called a bad sequence, and we may assume that it is minimal in the following sense. First, the cardinality $|A_1|$ of A_1 is minimal among all bad sequences. Second, $|A_2|$ is minimal among all bad sequences starting with A_1, etc.

As the empty multi-set is smaller than all other multi-sets, none of the multi-sets A_i is empty, so we may choose an element a_i from each A_i and define $B_i := A_i \setminus \{a_i\}$. There exists an infinite subsequence $i_1 < i_2 < \ldots$ where $a_{i_1} \leq a_{i_2} \leq \ldots$. Now the desired contradiction will follow by considering the sequence

$$A_1, A_2, \ldots, A_{i_1 - 1}, B_{i_1}, B_{i_2}, \ldots.$$

Indeed, no A_i with $i \leq i_1 - 1$ is less than or equal to A_j with $i < j \leq i_1 - 1$. But neither is any A_i with $i \leq i_1 - 1$ less than or equal to any B_j with $j \in \{i_1, i_2, \ldots\}$, or else we would have $A_i \leq B_j \leq A_j$, with the inclusion map witnessing the second inequality. Finally, no relation $B_i \leq B_j$ holds with $i, j \in \{i_1, i_2, \ldots\}$ and $i < j$. Indeed, a map $B_i \to B_j$ witnessing that inequality could be extended to a map $A_i \to A_j$ witnessing $A_i \leq A_j$ by mapping a_i to a_j. We conclude that the new sequence is bad, but this contradicts the minimality of $|A_{i_1}|$ among all bad sequences starting with $A_1, \ldots, A_{i_1 - 1}$. □

The general case of Kruskal's Tree Theorem concerns the set of (isomorphism classes of) finite, rooted trees whose vertices are labelled with elements of a fixed partially ordered set S. We call such objects S-labelled trees. A partial order on S-labelled trees is defined recursively as follows; see also Fig. 1. Suppose that T is an S-labelled tree with root r and suppose that T branches at r into trees $B_1, \ldots, B_p$ whose roots are the children of r; $p = 0$ is allowed here, and renders void one of the conditions that follow. We say that T is less than or equal to a second S-labelled tree T' if the latter has a vertex v (not necessarily its root) where T' branches into trees $B'_1, \ldots, B'_q$ (rooted at children of v), such that the S-label of v is at least

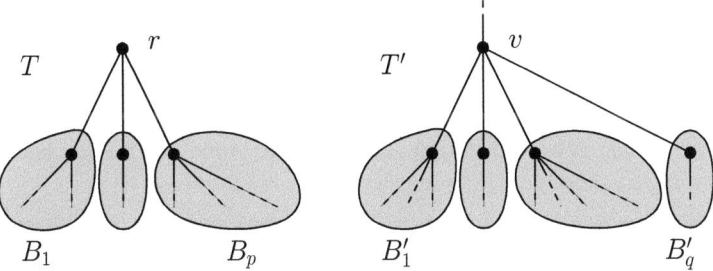

Fig. 1 If the S-label of r is at most that of v, and if $B_i \leq B'_{\pi(i)}$ for some injective $\pi : [p] \to [q]$, then $T \leq T'$

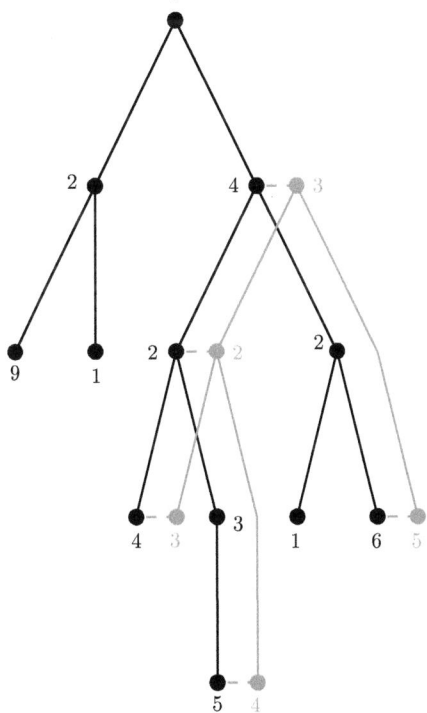

Fig. 2 The $\mathbb{N}$-labelled tree in *gray* is smaller than the $\mathbb{N}$-labelled tree in *black* in Kruskal's order

that of r and such that there exists an injective map π from $[p]$ into $[q]$ with $B_i \leq B'_{\pi(i)}$ for all i. Unfolding this recursive definition one finds that the one-dimensional CW-complex given by the tree T can then be homeomorphically embedded into that of T' in such a way that each vertex u of T gets mapped into a vertex of T' whose S-label is at least that of u; see Fig. 2.

Theorem 1.2 (Kruskal's Tree Theorem). *For any non-empty well-partially-ordered set S, the set of S-labelled trees is well-partially-ordered by the partial order just defined.*

Note that the lemma is, indeed, a special case of this theorem, obtained by restricting to star trees with a root (labelled with some fixed, irrelevant element of S) connected directly to all its leaves.

Proof. The proof is very similar to that of the lemma. Assume, for a contradiction, the existence of a bad sequence $T_1, T_2, \ldots$, which we may take minimal in the sense that the cardinality of the vertex set of T_i is minimal among all bad sequences starting with $T_1, \ldots, T_{i-1}$. At its root, T_i branches into a finite multi-set R_i of smaller S-labelled trees. Let R be the set-union of all R_i as i runs through $\mathbb{N}$, so that R is a set of S-labelled trees. If R contained a bad sequence, then it would contain a bad sequence $B_{i_1}, B_{i_2}, \ldots$ with $B_{i_j} \in R_{i_j}$ and $i_1 < i_2 < \ldots$. Then, as in the proof of the lemma, one shows that the sequence $T_1, \ldots, T_{i_1-1}, B_{i_1}, B_{i_2}, \ldots$ would be a bad sequence of trees contradicting the minimality of the original sequence. Hence R is well-partially-ordered.

Consider a subsequence $T_{k_1}, T_{k_2}, \ldots$ with $k_1 < k_2 < \ldots$ for which the root labels of the T_{k_i} weakly increase in S. Applying the lemma to the well-partially-ordered set R, we find that there exist $j < l$ such that $R_{k_j} \leq R_{k_l}$. But then $T_{k_j} \leq T_{k_l}$ (by mapping the root of T_{k_j} to the root of T_{k_l} and R_{k_j} suitably into R_{k_l}), a contradiction. □

We now formulate Higman's Lemma, which is a useful direct consequence of Kruskal's Tree Theorem. Given a partially ordered set S, define a partial order on the set $S^* = \bigcup_p S^p$ of finite sequences over S as follows. A sequence $(s_1, \ldots, s_p)$ is less than or equal to a sequence $(s'_1, \ldots, s'_q)$ if there exists a strictly increasing map $\pi : [p] \to [q]$ satisfying $s_i \leq s'_{\pi(i)}$ for all i.

Theorem 1.3 (Higman's Lemma). *For any well-partially-ordered set S, the partial order on S^* just defined is a well-partial-order.*

Proof. Encode a sequence $(s_1, \ldots, s_p)$ in S^* as an S-labelled tree with root 1 labelled s_1, with a single child 2 labelled s_2, etc. Under this encoding the partial order on S^* agrees with that on S-labelled trees, so that Kruskal's theorem implies Higman's Lemma. □

2 Equivariant Gröbner Bases

Just like, through a leading term argument, Dickson's Lemma implies Hilbert's Basis Theorem, Higman's Lemma implies the central finiteness result that all our later proofs build upon. What follows is certainly not the most general setting, but it will suffice our purposes. For much more on this theme see [Coh67, Coh87, AH07, AH08, DS06, HS12, HMdC13, LSL09].

Let X be a (typically infinite) set of variables, and let Mon denote the free commutative monoid of monomials in those variables. Let $\leq$ be a monomial order on Mon, i.e., a well-order that satisfies the additional condition that $u \leq v \Rightarrow uw \leq vw$ for all $u, v, w \in$ Mon. Let Π be a (typically non-commutative) monoid acting from the left on Mon by means of monoid endomorphisms and assume that the action preserves strict inequalities, i.e., $u < v$ implies $\pi u < \pi v$ for all $\pi \in \Pi$. In particular, π acts by means of an injective map on Mon. Moreover, we have $\pi u \geq u$ since otherwise the sequence $u > \pi u > \pi^2 u > \ldots$ would contradict that $\leq$ is a well-order.

Let K be a field and denote by $K[X] = K$Mon the ring of polynomials in the variables in X, or, equivalently, the monoid algebra over K of Mon. The action of Π on Mon extends to an action on $K[X]$ by means of ring endomorphisms preserving 1. For a non-zero $f \in K[X]$ denote by $\mathrm{lm}(f) \in$ Mon the largest monomial with a non-zero coefficient in f. As the action of Π preserves the (strict) monomial order, we have $\mathrm{lm}(\pi f) = \pi \mathrm{lm}(f)$ in addition to the usual properties of lm. In other words, the map lm from $K[X] \setminus \{0\}$ to Mon is Π-equivariant, and this motivates the terminology in the following definition.

Definition 2.1. Let I be a Π-stable ideal in $K[X]$. Then a Π-Gröbner basis, or equivariant Gröbner basis if Π is clear from the context, of I is a subset B of I with the property that for any $f \in I$ there exists a $g \in B$ and a $\pi \in \Pi$ with $\mathrm{lm}(\pi g) | \mathrm{lm}(f)$.

The set B is an equivariant Gröbner basis of I if and only if the union $\Pi B = \{\pi g \mid \pi \in \Pi, g \in B\}$ of the Π-orbits of elements of B is an ordinary Gröbner basis of I (except that it will typically not be finite). Then, in particular, ΠB generates I as an ideal; and we also say that B generates I as a Π-stable ideal.

We do not require that an equivariant Gröbner basis be finite, but finite ones will of course be the most useful ones to us. To formulate a criterion guaranteeing the existence of finite equivariant Gröbner bases we define the Π-divisibility relation on Mon by $u |_\Pi v$ if and only if there exists a $\pi \in \Pi$ such that πu divides v. This relation is reflexive (take $\pi = 1$), transitive [if $v = u' \cdot \pi u$ and $w = v' \cdot \sigma v$, then $w = (v'\sigma u') \cdot (\sigma \pi) u$], and antisymmetric (if $\pi u | v$ and $\sigma v | u$ then $u \leq \pi u \leq v$ and $v \leq \sigma v \leq u$ so that $u = v$).

Proposition 2.2 ([HS12]). *Every Π-stable ideal $I \subseteq K[X]$ has a finite Π-Gröbner basis if and only if $|_\Pi$ is a well-partial-order.*

Proof. For the "only if" part observe that if $u_1, u_2, \ldots$ were a bad sequence of monomials, then the Π-stable ideal generated by them, i.e., the smallest Π-stable ideal containing them, would not have a finite equivariant Gröbner basis. For the "if" part let I be a Π-stable ideal in $K[X]$. Let M denote the set of $|_\Pi$-minimal elements of $\{\mathrm{lm}(f) \mid f \in I \setminus \{0\}\}$. As $|_\Pi$ is a well-partial-order, M is finite, say $M = \{u_1, \ldots, u_p\}$. Choose $f_1, \ldots, f_p \in I \setminus \{0\}$ with $\mathrm{lm}(f_i) = u_i$. Then $\{f_1, \ldots, f_p\}$ is a Π-Gröbner basis of I. □

The main example that we shall use has $X := \{x_{ij} \mid i \in [k], j \in \mathbb{N}\}$ and $\Pi :=$ Inc($\mathbb{N}$), the monoid of maps $\mathbb{N} \to \mathbb{N}$ that are strictly increasing in the standard order on $\mathbb{N}$. This monoid acts on X by $\pi x_{ij} = x_{i\pi(j)}$; the action extends multiplicatively to an action on Mon and linearly to an action by ring endomorphisms on the polynomial ring $R := K[X] = K[(x_{ij})_{ij}]$. There exist monomial orders $\leq$ for which $u < v$ implies $\pi u < \pi v$; for instance, the lexicographic order with $x_{ij} < x_{i'j'}$ if $i < i'$ or $i = i'$ and $j < j'$.

Theorem 2.3 ([Coh87, HS12]). *Fix a natural number k. Then any* Inc($\mathbb{N}$)*-stable ideal I in the ring $K[x_{ij} \mid i \in [k], j \in \mathbb{N}]$ has a finite* Inc($\mathbb{N}$)*-Gröbner basis with respect to any monomial order preserved by* Inc($\mathbb{N}$)*. In particular, any* Inc($\mathbb{N}$)*-stable ideal I in that ring is generated, as an ideal, by finitely many* Inc($\mathbb{N}$)*-orbits of polynomials.*

Proof. By Proposition 2.2 it suffices to prove that $|_{\mathrm{Inc}(\mathbb{N})}$ is a well-partial-order. To this end, we shall apply Higman's Lemma to $S = \mathbb{Z}_{\geq 0}^k$ with the componentwise partial order, which is a well-partial-order by Dickson's Lemma. Encode a monomial u in the variables x_{ij} as a word $(s_1, \ldots, s_p)$ in S^* as follows: p is the largest value of the column index j for which some variable x_{ij} appears in u, and $(s_j)_i$ is the exponent of x_{ij} in u. Now given any sequence $u_1, u_2, \ldots$ of monomials, by Higman's Lemma there exist indices $m < l$ such that the sequences s, s' encoding u_m, u_l satisfy $s \leq s'$. This means that there exists a strictly increasing map $\pi : [p] \to [p']$, with p, p' the lengths of s, s', such that $s_j \leq s'_{\pi(j)}$ for all $j \in [p]$. Extend π in any manner to a strictly increasing map $\mathbb{N} \to \mathbb{N}$. Then the exponent of any variable x_{ij} in πu_m equals 0 if $j \notin \pi([p])$ and $(s_{\pi^{-1}j})_i \leq (s'_j)_i$ otherwise. This proves that $\pi u_m | u_l$, as desired. □

The second statement in Theorem 2.3 has several consequences. One is that any ascending chain $I_1 \subseteq I_2 \subseteq \ldots$ of Inc($\mathbb{N}$)-stable ideals in R stabilises at some finite index n: $I_n = I_{n+1} = \ldots$; we express this fact by saying that R is Inc($\mathbb{N}$)-Noetherian. This implies that R is Sym(∞)-Noetherian, where the group Sym(∞) := $\bigcup_{j \in \mathbb{N}}$ Sym($[j]$) is obtained by embedding Sym($[j]$) into Sym($[j+1]$) as the stabiliser of $j+1$ and where $\pi \in$ Sym(∞) acts on x_{ij} by $\pi x_{ij} = x_{i\pi(j)}$. Indeed, the Sym(∞)-orbit of any polynomial f contains the Inc($\mathbb{N}$)-orbit of f, and hence any Sym(∞)-stable ideal is also Inc($\mathbb{N}$)-stable. Note that one can also replace the countable group Sym(∞) by the uncountable group of all permutations of $\mathbb{N}$, because the two have exactly the same orbits on R.

Example 2.4. In contrast to these beautiful positive results, consider the set $X = \{y_{ij} \mid i, j \in \mathbb{N}\}$ with $\Pi = $ Inc($\mathbb{N}$)-action given by $\sigma y_{ij} = y_{\sigma(i)\sigma(j)}$. We claim that $|_\Pi$ is not a well-partial-order. Indeed, consider the sequence of monomials

$$y_{12}y_{21}, y_{12}y_{23}y_{31}, \ldots$$

encoding directed cycles on two, three, etc. vertices. Any $\pi \in$ Inc($\mathbb{N}$) maps such a monomial to a monomial representing another directed cycle of the same length.

Since no larger cycle contains a smaller cycle as a subgraph, this is a bad sequence of monomials. The same argument shows that $K[X]$ is not $\mathrm{Sym}(\infty)$-Noetherian. Similar counterexamples exist for the action of $\mathrm{Sym}(\infty) \times \mathrm{Sym}(\infty)$ on X given by $(\pi, \sigma) y_{ij} = y_{\pi(i)\sigma(j)}$.

We now return to the general setting of equivariant Gröbner bases, without the assumption that $|_\Pi$ is a well-partial-order. These bases can sometimes be computed by a Π-equivariant version of Buchberger's algorithm. The halting criterion in this equivariant Buchberger algorithm is the following equivariant version of Buchberger's criterion involving S-polynomials.

Proposition 2.5 (Equivariant Buchberger Criterion). *Let B be a subset of a Π-stable ideal I in $K[X]$. Then B is a Π-Gröbner basis of I if and only if for all $f, g \in B$ and all $\sigma, \tau \in \Pi$ the ordinary S-polynomial $S(\sigma f, \tau g)$ gives remainder 0 upon division by ΠB.*

This criterion follows immediately from the ordinary Buchberger criterion applied to ΠB—indeed, while most textbooks assume a finite number of variables, division-with-remainder and Buchberger's criterion apply to infinitely many variables as well; the crucial ingredient is the fact that the monomial order is a well-order. Unfortunately, since Π is typically infinite, checking whether B is an equivariant Gröbner basis using the equivariant Buchberger criterion may be an infinite task, even when B is finite. But in many cases of interest this task can be reduced to a finite task as follows.

Assume, first, that for any two polynomials $f, g \in K[X]$ the Cartesian product $\Pi f \times \Pi g$ of the Π-orbits of f and g is the union of finitely many diagonal orbits $\Pi(\sigma_i f, \tau_i g) = \{(\pi \sigma_i f, \pi \tau_i g) \mid \pi \in \Pi\}$, $i = 1, \ldots, r$, where $r \in \mathbb{N}$ and $\sigma_1, \tau_1, \ldots, \sigma_r, \tau_r \in \mathrm{Inc}(\mathbb{N})$ are allowed to depend on f, g. Then we would like to check only whether the S-polynomials $S(\sigma_i f, \tau_i g)$ reduce to zero upon division by ΠB, and conclude that all $S(\sigma f, \tau g)$ reduce to zero. For this we would like that $S(\pi \sigma_i f, \pi \tau_i g) = \pi S(\sigma_i f, \tau_i g)$, because letting π act on the reduction of $S(\sigma_i f, \tau_i g)$ to zero yields a reduction of $S(\pi \sigma_i f, \pi \tau_i g)$ to zero. This desired Π-equivariance of S-polynomials does not follow from the assumptions so far, but it does follow if we make the further assumption that each $\pi \in \Pi$ preserves least common multiples, i.e., that $\mathrm{lcm}(\pi u, \pi v) = \pi \, \mathrm{lcm}(u, v)$ for all $u, v \in \mathrm{Mon}$. This is, in particular, the case if π maps variables to variables.

Theorem 2.6. *Assume that Cartesian products $\Pi f \times \Pi g$ of Π-orbits on $K[X]$ are unions of finitely many diagonal Π-orbits, and assume that Π preserves least common multiples of monomials. Let S be a finite subset of $K[X]$ and consider the following algorithm:*

(1) *Set $B := S$ and $P := \binom{S}{2} \cup \{(f, f) \mid f \in S\}$, where $\binom{S}{2}$ is the set of pairs of distinct elements from S;*
(2) *If $P = \emptyset$, then stop, otherwise pick $(f, g) \in P$ and remove it from P.*
(3) *Choose $r \in \mathbb{N}$, $\sigma_1, \tau_1, \ldots, \sigma_r, \tau_r \in \Pi$ such that $\Pi f \times \Pi g = \bigcup_{i=1}^{r} \Pi(\sigma_i f, \tau_i g)$.*

(4) *For each $i = 1, \ldots, r$ do the following: reduce $S(\sigma_i f, \tau_i g)$ modulo ΠB, and if the remainder h is non-zero, then add h to B and consequently add $B \times \{h\}$ to P.*
(5) *Return to step 2.*

If and when this algorithm terminates, then B is a Π-Gröbner basis for the Π-stable ideal generated by S. Moreover, if $|_\Pi$ is a well-partial-order, then this algorithm does terminate.

As argued above, all but the last sentence of this theorem follows from the ordinary Buchberger criterion. The last sentence follows as for the ordinary Buchberger algorithm: if the algorithm does not terminate, then an infinite number of non-zero remainders $h_1, h_2, \ldots$ are added. If $|_\Pi$ is a well-partial-order, then there exist $i < j$ with $\mathrm{lm}(h_i)|_\Pi \mathrm{lm}(h_j)$, which means that h_j was not reduced with respect to h_i, a contradiction.

One point to stress is that in initialising the pair set P also pairs of two identical polynomials (f, f), $f \in S$ need to be added; and that similarly, when adding a remainder h to B, also the pair (h, h) needs to be added to P. Indeed, already the Π-stable ideal generated by a single polynomial can be interesting, as the following example shows.

Example 2.7. Let $X = \{x_i \mid i \in \mathbb{N}\} \cup \{y_{ij} \mid i, j \in \mathbb{N}, i > j\}$ and let $\Pi = \mathrm{Inc}(\mathbb{N})$ act on X by $\pi x_i = x_{\pi(i)}$ and $\pi y_{ij} = y_{\pi(i)\pi(j)}$. Set $S := \{y_{21} - x_2 x_1\}$ and let I denote the Π-stable ideal generated by S. We would like to compute an equivariant Gröbner basis of the elimination ideal $I \cap K[y_{ij} \mid i > j]$. To this end, we choose the lexicographic monomial order with $x_1 < x_2 < \ldots$ and $y_{ij} < y_{kl}$ if either $i < k$ or $i = k$ and $j < l$, and with $y_{kl} < x_i$ for all i, k, l. Note that Π preserves the strict monomial order and least common multiples.

To apply the equivariant Buchberger algorithm we must further check that Cartesian products of Π-orbits on $K[X]$ are finite unions of diagonal Π-orbits. For this, let f, g be elements of $K[X]$ and let p, q be such that all variables in f, g have indices contained in $[p], [q]$, respectively. Then $\sigma f, \tau g$ depend only on the restrictions of σ and τ to $[p], [q]$ respectively. Enumerate all (finitely many) pairs $(\sigma_i : [p] \to \mathbb{N}, \tau_i : [q] \to \mathbb{N})$, $i = 1, \ldots, r$ of increasing maps for which the union of $\mathrm{im}(\sigma_i)$ and $\mathrm{im}(\tau_i)$ equals some interval $[t] = \{1, \ldots, t\}$, necessarily with $t \leq p + q$. Extend these σ_i and τ_i arbitrarily to elements of Π. We claim that for any pair $\sigma, \tau \in \Pi$ we have $(\sigma f, \tau g) = (\pi \sigma_i f, \pi \tau_i g)$ for some $i \in [r]$ and some $\pi \in \Pi$. Indeed, there exists a unique i for which there exists an (again, unique) increasing map $\pi : [t] = \mathrm{im}(\sigma_i) \cup \mathrm{im}(\tau_i) \to \sigma([p]) \cup \tau([q])$ such that the restrictions of σ, τ to $[p], [q]$ equal the restrictions of $\pi \circ \sigma_i, \pi \circ \tau_i$ to $[p], [q]$, respectively. Extend π in any manner to an element of Π and we find $(\sigma f, \tau g) = (\pi \sigma_i f, \pi \tau_i g)$, as desired.

This means that we can apply the equivariant Buchberger algorithm, but without the guarantee that it terminates, since $|_\Pi$ is not a well-partial-order (adapt Example 2.4 to see this). It turns out the algorithm does terminate, though, and yields the following equivariant Gröbner basis (after self-reduction):

$$B = \{x_1 x_2 - y_{21},$$
$$x_3 y_{21} - x_2 y_{31}, \ x_3 y_{21} - x_1 y_{32}, \ x_2 y_{31} - x_1 y_{32},$$
$$x_1^2 y_{32} - y_{31} y_{21},$$
$$y_{43} y_{21} - y_{41} y_{32}, \ y_{42} y_{31} - y_{41} y_{32}.\}$$

Since the monomial order is an elimination order, we conclude that $I \cap K[y_{ij} \mid i > j]$ has a Π-Gröbner basis given by the last two binomials. In particular, that ideal is generated, as an $\mathrm{Inc}(\mathbb{N})$-stable ideal, by these binomials. The result that we have just proved by computer has first appeared as a theorem in [dLST95]. This example gives the ideal of the so-called second hypersimplex, or, with a slight modification, of the Gaussian one-factor model. The k-factor model for $k = 2$ and higher will be the subject of Sect. 6.

3 Equivariant Noetherianity

In this section we establish a number of constructions of equivariantly Noetherian rings and spaces. For some of the material see [Dra10].

Given a ring R (always commutative, with 1) and a monoid Π with a left action on R by means of (always unital) endomorphisms we say that R is Π-Noetherian, or equivariantly Noetherian if Π is clear from the context, if every chain $I_1 \subseteq I_2 \subseteq \ldots$ of Π-stable ideals in R eventually stabilises, that is, if there exists an $n \in \mathbb{N}$ for which $I_n = I_{n+1} = \ldots$. This is equivalent to the condition that any Π-stable ideal I is generated, as an ideal, by finitely many Π-orbits $\Pi f_1, \ldots, \Pi f_s$. We then say that $f_1, \ldots, f_s$ generate I as a Π-stable ideal.

We have seen a major example in Sect. 2, namely, for any fixed natural number k and any field K the ring $K[x_{ij} \mid i \in [k], j \in \mathbb{N}]$ with its action of $\mathrm{Inc}(\mathbb{N})$ on the second index is $\mathrm{Inc}(\mathbb{N})$-Noetherian.

There are several constructions of new equivariantly Noetherian rings from existing ones. The first and most obvious is that if R is Π-Noetherian and $I \subseteq R$ is a Π-stable ideal, then R/I is Π-Noetherian: any chain of Π-stable ideals in R/I lifts to a chain of Π-stable ideals in R containing I, and the first chain stabilises exactly when the second chain does.

A second construction takes a Π-Noetherian ring R to the polynomial ring $R[x]$ in a variable x, where Π acts only on the coefficients from R. The standard proof of Hilbert's Basis Theorem, say from [Lan65], generalises word by word from trivial Π to general Π.

It is not true, in general, that a subring of an equivariantly Noetherian ring is equivariantly Noetherian. Indeed, this is already not true for ordinary Noetherianity, where Π is the trivial monoid. However, the following construction proves that certain well-behaved subrings of equivariantly Noetherian rings are again equivariantly Noetherian. Suppose that S is a subring of R with the property that R

splits as a direct sum $S \oplus M$ of S-modules. If J is an ideal in S and I is the ideal in R generated by J, then we claim that $S \cap I = J$—indeed, any element f of $S \cap I$ can be written as $f = \sum_i f_i g_i$ with the f_i elements of J and the g_i elements of R. Applying the S-linear projection $\pi : R \to S$ along M to both sides yields $f = \sum_i f_i \pi(g_i) \in J$, as claimed. If, moreover, the monoid Π acts on R and stabilises S, then I is Π-stable if J is. We conclude that if R is Π-Noetherian, then any chain $I_1 \subseteq I_2 \subseteq \ldots$ of Π-stable ideals in S generates such a chain $J_1 \subseteq J_2 \subseteq \ldots$ in R, and $J_n = J_{n+1} = \ldots$ implies that $I_n = J_n \cap S = J_{n+1} \cap S = I_{n+1} = \ldots$; so S is Π-Noetherian.

A particularly important example of this situation is the following proposition, due to Kuttler. Suppose that a group H acts on R by means of ring automorphisms, and that the action of H commutes with that of Π, i.e., for every $\pi \in \Pi$ and $h \in H$ and $f \in R$ we have $\pi h f = h \pi f$. Then the ring $R^H := \{f \in R \mid Hf = \{f\}\}$ of H-invariant elements of R is stable under the action of Π.

Proposition 3.1. *If on the one hand R is Π-Noetherian and on the other hand R splits as a direct sum of irreducible $\mathbb{Z}H$-modules, then R^H is also Π-Noetherian.*

Proof. By the discussion preceding the proposition, we need only prove that R splits as a direct sum $R^H \oplus M$ of R^H-modules. For this, split R as a direct sum $\oplus_i M_i$ of irreducible $\mathbb{Z}H$-modules M_i. Then R^H is the direct sum of the M_i with trivial H-action, and we set M equal to the direct sum of the M_i with non-trivial H-action. We want to show that $fM_i \subseteq M$ for every $f \in R^H$ and every $M_i \subseteq M$. To this end, let $\rho : R \to R^H$ be the projection along M and consider the map $M_i \to R^H$ sending m to $\rho(fm)$. By invariance of f this map is H-equivariant, and by irreducibility of M_i its kernel is either $\{0\}$ or all of M_i. But in the first case, the non-trivial H-module M_i would be embedded into the trivial H-module R^H, which is impossible. Hence that kernel is all of M_i, and $fM_i \subseteq M$. □

In our applications, R will typically be an algebra over some field K, and H will act K-linearly. Then it suffices that R splits as a direct sum of irreducible KH-modules (as one can infer from the proposition by replacing H by the group $K^* \times H$).

We give several applications of this proposition. First, we have seen in Example 2.4 that the ring of polynomials in the entries y_{ij} of an $\mathbb{N} \times \mathbb{N}$-matrix is not Inc($\mathbb{N}$)-Noetherian. The following corollaries show that interesting quotients of such rings are Inc($\mathbb{N}$)-Noetherian.

Corollary 3.2. *Let k be a natural number. Consider the homomorphism $\psi : K[y_{\mathbf{m}} \mid \mathbf{m} \in \mathbb{N}^k] \to K[x_{ij} \mid i \in [k], j \in \mathbb{N}]$ sending $y_{\mathbf{m}}$ to $\prod_{i=1}^k x_{i,m_i}$. The kernel of ψ is generated by finitely many Inc($\mathbb{N}$)-orbits of polynomials, and the quotient $K[(y_{\mathbf{m}})_{\mathbf{m}}]/\ker \psi$ is Inc($\mathbb{N}$)-Noetherian.*

In more geometric language, that quotient is the coordinate ring of k-dimensional infinite-by-infinite-by-...-by-infinite tensors of rank one.

Proof. The first statement follows from the standard fact that the ideal of the variety of rank-one tensors is generated by the quadrics $y_{m_0 m_1} y_{m'_0 m'_1} - y_{m_0 m'_1} y_{m'_0 m_1}$, where m_0, m'_0 are multi-indices of length equal to some $\ell \leq k$ and m_1, m'_1 are multi-indices of length $k - \ell$. The entries of the multi-indices m_0, m'_0, m_1, m'_1 taken together form a set of cardinality at most $2k$, and this implies that each quadric of the form above is obtained by applying some element of $\mathrm{Inc}(\mathbb{N})$ to one of the finitely many such quadrics with all indices in the interval $[2k]$.

The second statement follows from the proposition (or rather the discussion preceding it): the quotient is isomorphic, as a ring with $\mathrm{Inc}(\mathbb{N})$-action, to the subring $S = \mathrm{im}\,\psi$ of $K[x_{ij} \mid i \in [k], j \in \mathbb{N}]$. The ring S consists of all monomials in the x_{ij} that involve equally many variables, counted with their exponents, from all of the k rows of the $k \times \mathbb{N}$-matrix $(x_{ij})_{ij}$. If one writes M for the vector space complement of S spanned by all other monomials, then M is an S-module, and the fact that $K[(x_{ij})_{ij}]$ is $\mathrm{Inc}(\mathbb{N})$-Noetherian implies that S is. $\square$

Alternatively, if K is infinite, then one can characterise S as the set of H-invariants, where H is the subgroup of $(K^*)^k$ consisting of k-tuples with product 1 and where h acts on x_{ij} by $hx_{ij} = h_i x_{ij}$. Each monomial outside S spans an irreducible, non-trivial H-module, and the proposition implies that S is $\mathrm{Inc}(\mathbb{N})$-Noetherian.

A substantial generalisation of Corollary 3.2, which applies to a wide class of monomial maps into $K[(x_{ij})_{ij} \mid i \in [k], j \in \mathbb{N}]$, is proved in [DEKL13]. For stabilisation of appropriate lattice ideals, see [HMdC13].

The next corollary shows that determinantal quotients of the coordinate ring of infinite-by-infinite matrices are $\mathrm{Inc}(\mathbb{N})$-Noetherian, provided that the field has characteristic zero.

Corollary 3.3. *For any natural number k and any field K of characteristic zero, the quotient of the ring $K[y_{ij} \mid i, j \in \mathbb{N}]$ by the ideal I_k generated by all $(k+1) \times (k+1)$-minors of the matrix $(y_{ij})_{ij}$ is $\mathrm{Inc}(\mathbb{N})$-Noetherian.*

Note that the set of these determinants is the union of finitely many $\mathrm{Inc}(\mathbb{N})$-orbits of equations, so that the corollary implies that any $\mathrm{Inc}(\mathbb{N})$-stable ideal containing I_k is generated by finitely many $\mathrm{Inc}(\mathbb{N})$-orbits.

Proof. Let the group $H = \mathrm{GL}_k$ act on the ring $K[x_{il} \mid i \in \mathbb{N}, l \in [k]]$ by $hx_{il} := (xh)_{il}$, where xh is the product of the $\mathbb{N} \times k$-matrix x with variable entries and the $k \times k$-matrix h. Similarly, let H act on the ring $K[z_{lj} \mid l \in [k], j \in \mathbb{N}]$ by $hz_{lj} := (h^{-1}z)_{lj}$. Note that both actions commute with the action of $\mathrm{Inc}(\mathbb{N})$ on the indices i, j, respectively. Let R be the polynomial ring $K[x_{il}, z_{lj} \mid i, j \in \mathbb{N}, l \in [k]]$, equipped with the natural $\mathrm{Inc}(\mathbb{N})$-action and H-action. Classical invariant theory tells us that rings, like R, on which H acts as an algebraic group split into a direct sum of irreducible KH-modules; here we use that $\mathrm{char}\, K = 0$. So we may apply Proposition 3.1.

The First Fundamental Theorem [GW09, Theorem 5.2.1] for H states that the algebra R^H of H-invariant elements of the ring R is generated by all pairings $p_{ij} :=$

$\sum_l x_{il} z_{lj} = (xz)_{ij}$. The Second Fundamental Theorem [GW09, Theorem 12.2.12] states that the kernel of the homomorphism $K[(y_{ij})_{ij}] \mapsto R$ determined by $y_{ij} \mapsto p_{ij}$ is precisely I_k. Thus the quotient by I_k is isomorphic, as a K-algebra with $\mathrm{Inc}(\mathbb{N})$-action, to R^H. This proves the corollary. □

Similar results are obtained by taking other rings with group actions where the invariants and the polynomial relations among them are known. Here is an example, which first appeared in [Dra10].

Corollary 3.4. *For any natural number k and any field K of characteristic zero, the kernel of the homomorphism $\psi : K[y_\mathbf{m} \mid \mathbf{m} \in \mathbb{N}^k, m_1 < \ldots < m_k] \to K[x_{ij} \mid i \in [k], j \in \mathbb{N}]$ sending $y_\mathbf{m}$ to the determinant of $x[\mathbf{m}]$, the $k \times k$-submatrix of x obtained by taking the columns indexed by $\mathbf{m}$, is generated by finitely many $\mathrm{Inc}(\mathbb{N})$-orbits; and the quotient of $K[(y_\mathbf{m})_\mathbf{m}]$ by $\ker \psi$ is $\mathrm{Inc}(\mathbb{N})$-Noetherian.*

Proof. Let $H = \mathrm{SL}_k$, the group of $k \times k$-matrices of determinant 1, act on $K[(x_{ij})_{ij}]$ by $hx_{ij} = (h^{-1}x)_{ij}$. The First Fundamental Theorem for SL_n says that the $k \times k$-minors of x generate the invariant ring of H, and the Second Fundamental Theorem says that the Plücker relations among those determinants, which can be covered by finitely many $\mathrm{Inc}(\mathbb{N})$-orbits, generate the ideal of all relations. Now proceed as in the previous case. □

We remark that Alexei Krasilnikov showed that the Noetherianity of this corollary does not hold when $\mathrm{char}\, K = 2$ and $k = 2$. However, a weaker form of Noetherianity, which we introduce now, does hold.

Let X be a topological space equipped with a right action of a monoid Π by means of continuous maps $X \to X$. Then we call X equivariantly Noetherian, or Π-Noetherian, if every chain $X = X_0 \supseteq X_1 \supseteq X_2 \supseteq \ldots$ of closed, Π-stable subsets stabilises. If R is a K-algebra with a left action of Π by means of K-algebra endomorphisms, then for any K-algebra A the set $X := R(A) := \mathrm{Hom}_K(R, A)$ of A-valued points of R is a topological space with respect to the Zariski topology in which closed sets are defined by the vanishing of elements of R. Moreover, the monoid Π acts from the right on $R(A)$ by $(p\pi)(r) = p(\pi r)$. If R is Π-Noetherian, then $R(A)$ is Π-Noetherian in the topological sense. Conversely, if $R(A)$ is Π-Noetherian in the topological sense for every K-algebra A, then R is Π-Noetherian—indeed, just take A equal to R, so that the map that takes closed sets to vanishing ideals is a bijection. However, topological Noetherianity of, say, $R(K)$ does not necessarily imply Noetherianity of R. An example of this phenomenon is given by Krasilnikov's example: the ring $K[\det x[i, j] \mid i, j \in \mathbb{N}, i < j]$, where x is a $[2] \times \mathbb{N}$-matrix of variables, is not $\mathrm{Inc}(\mathbb{N})$-Noetherian if $\mathrm{char}\, K = 2$, but its set of K-valued points is—indeed, this set of points is the image of $K^{[2] \times \mathbb{N}}$ under the $\mathrm{Inc}(\mathbb{N})$-equivariant map sending a matrix to the vector of its 2×2-determinants. Since $K^{[2] \times \mathbb{N}}$ is $\mathrm{Inc}(\mathbb{N})$-Noetherian, so is its image.

More generally, Π-equivariant images of Π-Noetherian topological spaces are Π-Noetherian, and so are Π-stable subsets with the induced topology. Another construction that we shall make much use of is the following.

Proposition 3.5. *Let G be a group with a right action by homeomorphisms on a topological space X. Let Π be a submonoid of G and let Z be a Π-stable subset of X. Assume that Z is Π-Noetherian with the induced topology. Then $Y := ZG = \bigcup_{g \in G} Zg \subseteq X$ is G-Noetherian with the induced topology.*

Proof. Let $Y = Y_1 \supseteq Y_2 \supseteq Y_3 \supseteq \ldots$ be a chain of G-stable closed subsets of Y. Then each $Z_i := Y_i \cap Z$ is Π-stable and closed, hence by Π-Noetherianity of Z there exists an n with $Z_n = Z_{n+1} = \ldots$. By definition of Y, for each $y \in Y_i$ there exist a $g \in G$ and a $z \in Z$ with $y = zg$, and by G-stability of Y_i we have $z = yg^{-1} \in Z_i$. This means that Y_i can be recovered from Z_i as $Y_i = Z_i G$, and hence the chain $Y_1 \supseteq Y_2 \supseteq Y_3 \supseteq \ldots$ stabilises at Y_n, as well. □

4 Chains of Varieties

In the remainder of these notes we study various chains of interesting embedded finite-dimensional varieties, for which we want to prove that from some member of the chain on, all equations for later members come from those of earlier members by applying symmetry. To use the infinite-dimensional techniques from the previous chapters, we first pass to a projective limit, prove that the limit is defined by finitely many orbits of equations, and from this fact we derive the desired result concerning the finite-dimensional varieties. In this short section we set up the required framework for this, again without trying to be as general as possible. Most of this material is from [Dra10].

Thus let K be a field and let $R_1, R_2, \ldots$ be commutative K-algebras with 1. The algebra R_i plays the role of coordinate ring of the ambient space of the i-th variety in our chain. Assume that the R_i are linked by (unital) ring homomorphisms $\iota_i : R_i \to R_{i+1}$ and $\pi_i : R_{i+1} \to R_i$ satisfying $\pi_i \circ \iota_i = 1_{R_i}$. Then we can form the K-algebra $R_\infty := \bigcup_{i \in \mathbb{N}} R_i$ with respect to the inclusions ι_i; the use of the π_i will become clear later.

Suppose, next, that are given ideals $I_i \subseteq R_i$ such that π_i maps I_{i+1} into I_i and ι_i maps I_i into I_{i+1}. The ideal I_i plays the role of defining ideal of the i-th variety in our chain. Writing $S_i := R_i/I_i$ we find that the ι_i, π_i induce inclusions $S_i \to S_{i+1}$ and surjections $S_{i+1} \to S_i$, respectively, and we set $I_\infty := \bigcup_i I_i$ and $S_\infty := \bigcup_i S_i$, which also equals R_∞/I_∞.

Assume, next, that a group G_i acts on R_i from the left by means of K-algebra automorphisms, and that we are given embeddings $G_i \to G_{i+1}$ that render both ι_i and π_i equivariant with respect to G_i. Suppose furthermore that each I_i is G_i-stable, which expresses that the i-th variety has the same symmetries as imposed on the ambient space. We form the group $G_\infty := \bigcup_i G_i$, which acts on $R_\infty, I_\infty, S_\infty$ by means of automorphisms.

For any K-algebra A, we write $R_i(A), S_i(A), R_\infty(A), S_\infty(A)$ for the sets of A-valued points of these algebras, i.e., for the set of homomorphisms $R_i \to A$, etc. As customary in algebraic geometry, for a p in these point sets, we write $f(p)$

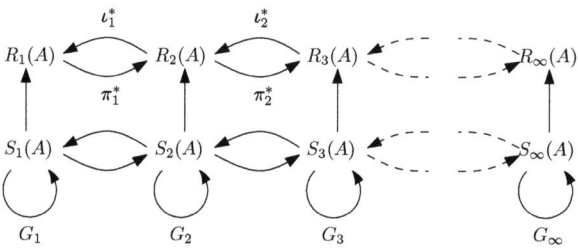

Fig. 3 Chains of varieties

rather than $p(f)$ for the evaluation of p on an element f in the corresponding algebra. These sets are topological spaces with respect to the Zariski topology, in which closed sets are of the form $\{p \in R_i(A) \mid J(p) = \{0\}\}$ for some ideal J in R_i, and similarly for the other algebras. On these topological spaces G_i or G_∞ acts by means of homeomorphisms. Our set-up so far is summarised in the diagram of Fig. 3, where ι^*, π^* are the pull-backs of ι and π, respectively.

The relation $\iota^* \circ \pi^* = 1$ implies that ι^* is surjective (and not just dominant) and that π^* is injective; indeed, the latter is a closed embedding. Still, π^* is needed only a bit later.

The topological space $R_\infty(A)$ is canonically the same as the projective limit, in the category of topological spaces, of the spaces $R_i(A)$ with their Zariski topologies: First, at the level of sets, an A-valued point p of R_∞ gives rise, by composition with the embeddings $R_i \to R_\infty$, to homomorphisms $p_i := R_i \to A$ for all $i \in \mathbb{N}$. The resulting sequence $(p_1, p_2, \ldots)$ has the property that the pull-back of ι_i^* maps p_{i+1} to p_i, i.e., it is a point of the inverse limit $\varprojlim_i R_i(A)$ of sets. Conversely, a point of this inverse limit gives homomorphisms $p_i : R_i \to A$ such that $p_{i+1} \circ \iota_i = p_i$, and together these define a homomorphism $R_\infty \to A$. Second, the projective limit topology on $R_\infty(A)$ is the weakest topology that renders all maps $R_\infty(A) \to R_i(A)$ continuous. This means, in particular, that sets given by the vanishing of a single element of $R_i \subseteq R_\infty$ must be closed, and so must intersections of these, which are sets given by the vanishing of an ideal in R_∞. This shows that the projective limit topology on $R_\infty(A)$ has at least as many closed sets as the Zariski topology, and the converse is also clear since the maps $R_\infty(A) \to R_i(A)$ are continuous in the Zariski topology.

The basic result that we shall use throughout the rest of the notes is the following, where the use of the π_i becomes apparent.

Proposition 4.1. *Let $i_0 \in \mathbb{N}$ and assume that the set $S_\infty(A)$ is characterised inside $R_\infty(A)$ by the vanishing of all $gf \in R_\infty$ with $g \in G_\infty$ and $f \in I_{i_0}$. Then for $i \geq i_0$ the set $S_i(A)$ is characterised by the vanishing of all functions of the form $\pi_i \cdots \pi_{l-1} gf$ with $l \geq i$, $g \in G_l$, and $f \in I_{i_0}$.*

Proof. That these functions vanish on $S_i(A)$ follows from the inclusion $I_{i_0} \subseteq I_l$, the fact that G_l stabilises I_l, and the fact that π_j maps I_{j+1} into I_j. Conversely, suppose

Noetherianity up to Symmetry

that $p_i \in R_i(A)$ is a zero of all functions in the proposition, and let $p = (p_1, p_2, \ldots)$ be the point of $R_\infty(A)$ obtained by setting $p_{j+1} := \pi_j^* p_j$ for $j \geq i$ and $p_j := \iota_j^* p_{j+1}$ for $j < i$. Then p is a point in $S_\infty(A)$ by the assumed characterisation of the latter set: any $g \in G_\infty$ lies in G_l for some l which we may take larger than i, and for $f \in I_{i_0}$ we have

$$(gf)(p) = (gf)(p_l) = (gf)(\pi_{l-1}^* \cdots \pi_i^* p_i) = (\pi_{l-1} \cdots \pi_i gf)(p_i) = 0,$$

as desired. In particular, this means that p_i lies in $S_i(A)$. □

It would be more elegant to characterise $S_i(A)$ for $i \geq i_0$ as the common vanishing set of $G_i I_{i_0}$, i.e., of all functions of the form gf with $f \in I_{i_0}$ and $g \in G_i$. For this we introduce an additional condition. For $l \geq i$ write $\iota_{il} : R_i \to R_l$ for the composition $\iota_{l-1} \cdots \iota_i$ and $\pi_{li} : R_l \to R_i$ for the composition $\pi_i \cdots \pi_{l-1}$. The condition that we want is:

For all indices l, i_0, i_1 with $l \geq i_0, i_1$ and for all $g \in G_l$ there exist an index $j \leq i_0, i_1$ and group elements $g_0 \in G_{i_0}, g_1 \in G_{i_1}$ such that

$$\pi_{li_1} g \iota_{i_0 l} = g_1 \iota_{ji_1} \pi_{i_0 j} g_0 \qquad (*)$$

holds as an equality of homomorphisms $R_{i_0} \to R_{i_1}$.

This guarantees that the functions $\pi_{li} gf = \pi_{li} g \iota_{i_0 l} f$ from the proposition can be written as $g_1 \iota_{ji} \pi_{i_0 j} g_0 f$ for some $j \leq i, i_0$ and $g_0 \in G_{i_0}$ and $g_1 \in G_i$. Since I_{i_0} is G_{i_0}-stable and $\pi_{i_0 j}$ maps I_{i_0} into $I_j \subseteq I_{i_0}$ the latter expression is an element of $G_i I_{i_0}$.

The discussion so far concerned an arbitrary, fixed K-algebra A. In several applications we shall just take A equal to K, and the conclusion is that the point sets $S_i(A)$ for $i \geq i_0$ are defined set-theoretically by equations coming from I_{i_0} using symmetry. However, if one assumes that $G_\infty I_{i_0}$ generates I_∞, then the assumption in the proposition holds for all K-algebras A, hence so does the conclusion. From this one can conclude that for $i \geq i_0$ the functions featuring in the proposition generate the ideal I_i. Under the additional assumption $(*)$ one finds that $G_i I_{i_0}$ generates the ideal I_i. We conclude this section with a well-known example which paves the way for the treatment of the k-factor model in the next section.

Example 4.2. Fix a natural number k. For $n \in \mathbb{N}$ let R_n be the polynomial ring over K in the $\binom{n+1}{2}$ variables $y_{ij} = y_{ji}$ with $i, j \leq n$. Let ι_n be the natural inclusion $R_n \to R_{n+1}$ and let π_n be the projection $R_{n+1} \to R_n$ mapping all variables to zero that have one or both indices equal to $n+1$. Then we have $\pi_n \iota_n = 1$ as required. Let $I_n \subseteq R_n$ be the ideal of polynomials vanishing on all symmetric $n \times n$-matrices over K of rank at most k. Then ι_n maps I_n into I_{n+1} since the upper-left $n \times n$-block of an $(n+1) \times (n+1)$-matrix of rank at most k has itself rank at most k, and π_n maps I_{n+1} into I_n since appending a zero last row and column to any matrix yields a matrix of the same rank. Let $G_n := \mathrm{Sym}(n)$ act on R_n by $g y_{ij} = y_{g(i), g(j)}$, and embed G_n into G_{n+1} as the stabiliser of $n+1$. Then G_n stabilises I_n and the maps π_n and ι_n are G_n-equivariant. So we are in the situation discussed in this section.

Even the additional assumption (∗) holds. Indeed, consider the effect of appending $l - i_1$ zero rows and columns to a symmetric $i_1 \times i_1$-matrix, then simultaneously permuting rows and columns, and finally forgetting the last $l - i_0$ rows and columns of which, say, m come from the zero rows and columns introduced in the first step. Set $j := i_1 - (l - i_0 - m)$, which is the number of rows and columns of the original matrix surviving this operation. Then the same effect is obtained by first permuting (with g_1) rows and columns such that the $i_1 - j$ rows and columns to be forgotten are in the last $i_1 - j$ positions, then forgetting those rows and columns, then appending $l - i_1 - m = i_0 - j$ zero rows and columns, and finally suitably permuting rows and columns with a $g_0 \in \mathrm{Sym}(i_0)$.

It is known, of course, that if K is infinite, then I_n is generated by all $(k+1) \times (k+1)$-minors of the matrix $(y_{ij})_{ij}$. This implies that I_∞ is generated by $G_\infty I_{2k+2}$; here $2k+2$ is the smallest size where representatives of all $\mathrm{Sym}(\mathbb{N})$-orbits of minors of an infinite symmetric matrix can be seen. But conversely, if through some other method (computational or otherwise) one can prove that I_∞ is indeed generated by $G_\infty I_{2k+2}$, then by the discussion above this implies that I_n is generated by $G_n I_{2k+2}$ for all $n \geq 2k+2$. Using the equivariant Gröbner basis techniques from Sect. 2 one can prove such a statement automatically for small k, say $k = 1$ and $k = 2$, much like we have done in Example 2.7.

5 The Independent Set Theorem

To appreciate the results of this section—though not to understand the proofs—one needs some familiarity with Markov bases and their relation to toric ideals. We formulate and prove the independent set theorem from [HS12], first conjectured in [SHS07], directly at the level of ideals.

Fix a natural number m, let Γ be a subset of $2^{[m]}$ (thought of as a hypergraph with vertex set $[m]$; see Fig. 4 for an example), and fix an infinite field K; what follows will, in fact, not depend on K. To any m-tuple $r = (r_1, \ldots, r_m) \in \mathbb{N}^{[m]}$ of natural numbers we associate the polynomial ring

$$R_r := K[y_{i_1,\ldots,i_m} \mid (i_1, \ldots, i_m) \in [r_1] \times \cdots \times [r_t]]$$

in $\prod_{t \in [m]} r_t$-many variables, and the polynomial ring

$$Q_r := K[x_{F,(i_t)_{t \in F}} \mid F \in \Gamma \text{ and } \forall t \in F : i_t \in [r_t]].$$

Furthermore, we define $I_r \subseteq R_r$ as the kernel of the homomorphism $R_r \to Q_r$ mapping $y_{(i_t)_{t \in [m]}}$ to $\prod_{F \in \Gamma} x_{F,(i_t)_{t \in F}}$. On R_r, Q_r acts the group $G_r := \prod_t \mathrm{Sym}([r_t])$ by permutations of the variables, and the homomorphism defining I_r is G_r-equivariant. We want to let some of the r_t, namely, those with t in a given

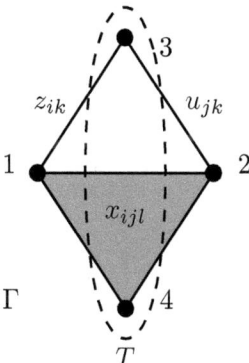

Fig. 4 A hypergraph Γ with parameters x_{ijl}, z_{ik}, u_{jk} and independent set T

subset $T \subseteq [m]$, tend to infinity, and conclude that the ideals I_r stabilise up to G_r-symmetry. To put this statement in the context of Sect. 4, we give the r_t with $t \notin T$ some fixed values, and take the $r_t, t \in T$ all equal, say to $n \in \mathbb{N}$. The corresponding rings and ideals are called R_n, Q_n, I_n. We have inclusions $\iota_n : R_n \to R_{n+1}$ obtained by inclusion of variables and projections $\pi_n : R_{n+1} \to R_n$ obtained by mapping all variables to zero that have at least one T-labelled index equal to $n + 1$. This maps I_{n+1} into I_n because there is a compatible homomorphism $Q_{n+1} \to Q_n$ setting the relevant variables equal to zero. As in Sect. 4 we write R_∞, I_∞ for the union of all R_n, I_n.

The group $\mathrm{Sym}(n)^T$ acts on R_n, Q_n, I_n, but in fact the independent set theorem only needs one copy of $\mathrm{Sym}(n)$ acting diagonally. The additional assumption (∗) holds by the same reasoning as in Example 4.2.

Theorem 5.1 (Independent Set Theorem [HS12].). *Suppose that $T \subseteq [m]$ is an independent set in Γ, i.e., that T intersects any $F \in \Gamma$ in at most one element. Then there exists an n_0 such that I_n is generated by $\mathrm{Sym}(n) I_{n_0}$ for all $n \geq n_0$.*

The condition that T is an independent set cannot simply be dropped. For instance, if $m = 3$ and $\Gamma = \{\{1,2\},\{2,3\},\{3,1\}\}$ (the model of *no three-way interaction*), then if $r_1 = n$ tends to infinity for fixed r_2, r_3, then the ideal stabilises (see [AT03] for the case of $r_2 = r_3 = 3$); but if $r_1 = r_2 = n$ both tend to infinity and r_1 is fixed, say, to 2, then the ideal does not stabilise [DS98].

Example 5.2. As an example take $m = 4$ and $\Gamma = \{124, 13, 23\}$, where 124 is short-hand for $\{1, 2, 4\}$, etc. We write

$$y_{ijkl}, x_{ijl}, z_{ik}, u_{jk} \text{ instead of } y_{i_1,i_2,i_3,i_4}, x_{124,(i_1,i_2,i_4)}, x_{13,(i_1,i_3)}, x_{23,(i_2,i_3)},$$

respectively; see Fig. 4. The ideal I_n is the kernel of the homomorphism sending y_{ijkl} to $x_{ijl} z_{ik} u_{jk}$. Take $T = \{3, 4\}$, an independent set in Γ. The ideal I_∞ contains obvious quadratic binomials such as

$$y_{ijkl}y_{ijk'l'} - y_{ijkl'}y_{ijk'l}$$

with $i \in [r_1], j \in [r_2], k, l, k', l' \in \mathbb{N}$. Indeed, the first monomial maps to $x_{ijl}z_{ik}u_{jk} \cdot x_{ijl'}z_{ik'}u_{jk'}$, and the second monomial maps to $x_{ijl'}z_{ik}u_{jk} \cdot x_{ijl}z_{ik'}u_{jk'}$, which is the same thing.

These obvious binomials generalise verbatim to the general case, where they read

$$y_{jkl}y_{jk'l'} - y_{jkl'}y_{jk'l} \tag{1}$$

where now j runs over the finite set $\prod_{t \in [m] \setminus T}[r_t]$, k, k' run over $\mathbb{N}^S$ and l, l' run over $\mathbb{N}^{T \setminus S}$ for some S that runs over all subsets of T. Indeed, for any variable $x = x_{F, (i_t)_{t \in F}}$ at most one $t \in F$ lies in T. If such a t exists for x and lies in S, then whether x appears in the image of a variable y_{jkl} does not depend on the value of l. But disregarding that third index, the two monomials above are the same. A similar reasoning for the case where $t \in T \setminus S$ and for the case where $F \cap T = \emptyset$ shows that x has the same exponent in the image of both monomials in the binomial above.

By Proposition 4.1 relating chains to infinite-dimensional varieties, we are done if we can prove that I_∞ is generated by $\mathrm{Sym}(\infty)I_n$ for some finite n. Let $J_n, J_\infty \subseteq I_n, I_\infty$, respectively, be the ideals generated by all quadratic binomials as in (1). We claim that J_∞ is generated by $\mathrm{Sym}(\infty)J_{2|T|}$. Indeed, for any binomial as above, the set of all indices appearing in k, l, k', l' has cardinality $n \leq 2|T|$, and there exists a bijection in $\mathrm{Sym}(\infty)$ mapping its support bijectively onto $[n]$, witnessing that the binomial lies in $\mathrm{Sym}(\infty)J_{2|T|}$. The remainder of the proof consists of showing that the quotient R_∞/J_∞ is, in fact, $\mathrm{Sym}(\infty)$-Noetherian. To this end, we introduce a new polynomial ring

$$P := K\left[y'_{jtq} \mid j \in \prod_{t \in [m] \setminus T}[r_t], t \in T, \text{ and } q \in \mathbb{N}\right],$$

where the y'_{jtq} are new variables, and consider the subring R'_∞ of P generated by all monomials $m_{ji} := \prod_{t \in T} y'_{jti_t}$ with j as before and $i \in \mathbb{N}^T$. The monomials m_{ji} satisfy the binomial relations (1) (for all splittings of i into two subsequences k and l), and it is known that these binomials generate the ideal of relations among the m_{ji}—indeed, the ring R'_∞ is the coordinate ring of the Cartesian product of $\prod_{t \in [m] \setminus T} r_t$-many copies of the variety of pure $|T|$-dimensional tensors. Thus we have an isomorphism $R'_\infty \cong R_\infty/J_\infty$, and we want to show that R'_∞ is $\mathrm{Sym}(\infty)$-Noetherian. The enveloping polynomial ring $P \supseteq R'_\infty$ is $\mathrm{Sym}(\infty)$-Noetherian by Theorem 2.3 (only the index q of the variables y'_{jtq} is unbounded), but passing to a subring one may, in general, lose Noetherianity. However, let H be the torus in $(K^*)^T$ consisting of T-tuples of non-zero scalars whose product is 1. Then the m_{ji} are H-invariant, and these monomials generate the ring of H-invariant polynomials (if, as we may assume, K is infinite). Hence by Proposition 3.1 we may conclude that R'_∞ is $\mathrm{Sym}(\infty)$-Noetherian, and this concludes the proof of the independent set theorem.

6 The Gaussian k-Factor Model

This section discusses finiteness results for a model from algebraic statistics known as the Gaussian k-factor model. General stabilisation results for this model were first conjectured in [DSS07], and for 1 factor established prior to that in [dLST95]. For 2 factors, a positive-definite variant was established in [DX10], and an ideal-theoretic variant in [BD11]. The ideal-theoretic version for more factors is open, but the set-theoretic version was established in [Dra10].

The Gaussian k-factor model consists of all covariance matrices for a large number n of jointly Gaussian random variables consistent with the hypothesis that those variables can be written as linear combinations of a small number k of hidden factors plus independent, individual noise. Algebraically, let R_n be the K-algebra of polynomials in variables $y_{ij} = y_{ji}$ with $i, j \in [n]$, and let P_{kn} be the K-algebra of polynomials in the variables x_{il}, $i \in [n], l \in [k]$ and further variables $z_1, \ldots, z_n$. Let I_{kn} be the kernel of the homomorphism $\phi_{kn} : R_n \to P_{kn}$ that maps y_{ij} to the (i, j)-entry of the matrix

$$x \cdot x^T + \mathrm{diag}(z_1, \ldots, z_n),$$

where we interpret x as an $[n] \times [k]$-matrix of variables. Set $S_{kn} := R_n/I_{kn}$; the set $S_{kn}(K) \subseteq K^{\binom{n}{2}}$ of K-valued points of S_{kn} is (the Zariski closure of) the Gaussian k-factor model. Observe that ϕ_{kn} is $\mathrm{Sym}([n])$-equivariant, so that I_{kn} is $\mathrm{Sym}([n])$-stable. We are in the setting of Sect. 4, with the map $\pi_n : R_{n+1} \to R_n$ mapping $y_{i,(n+1)}$ equal to 0 for all i and the map $\iota_n : R_n \to R_{n+1}$ the inclusion. The technical assumption (∗) from that section holds for the same reason as in Example 4.2.

Theorem 6.1. *For every fixed $k \in \mathbb{N}$, there exists an $n_k \in \mathbb{N}$ such that for all $n \geq n_k$ the variety $S_{kn}(K) \subseteq K^{\binom{n+1}{2}}$ is cut out set-theoretically by the polynomials in $\mathrm{Sym}(n)I_{n_k}$.*

Proof. By Proposition 4.1 and the discussion following its proof we need only prove that $S_{k\infty}(K)$ is the zero set of $\mathrm{Sym}(\infty)I_{n_k}$, for suitable n_k. Let $J_{k\infty}$ denote the ideal generated by all $(k+1) \times (k+1)$-minors of the symmetric $\mathbb{N} \times \mathbb{N}$-matrix y that do not involve diagonal entries of y, and set $S'_{k\infty} := R_\infty/J_{k\infty}$. Then surely $J_{k\infty}$ is contained in $I_{k\infty}$, so that, dually, $S'_{k\infty}(K)$ contains $S_{k\infty}(K)$.

We claim that $S'_{k\infty}(K)$ is a $\mathrm{Sym}(\infty)$-Noetherian topological space, and to prove this claim we proceed by induction. For $k = 0$ the equations in $J_{k\infty} = J_{0\infty}$ force all off-diagonal entries of the matrix y to be zero, so that $S'_{0\infty}(K)$ is just the set of K-points of $K[y_{11}, y_{22}, \ldots]$, with $\mathrm{Sym}(\infty)$ permuting the coordinates. The latter ring is $\mathrm{Sym}(\infty)$-Noetherian by Theorem 2.3, and hence its topological space of K-points is certainly $\mathrm{Sym}(\infty)$-Noetherian.

Next, assume that $S'_{k-1,\infty}(K)$ is $\mathrm{Sym}(\infty)$-Noetherian, and note that $S'_{k-1,\infty}(K)$ is a (closed) subset of $S'_{k,\infty}(K)$. On any point outside this closed subset at least one of the $k \times k$-determinants in $J_{k-1,\infty}$ is non-zero. Up to signs, these determinants

Fig. 5 The set T labelling symmetric matrix entries (in *light gray*), and the $(k+1) \times (k+1)$-determinant expressing y_{ij} as a rational function in the T-labelled variables (in *dark gray*)

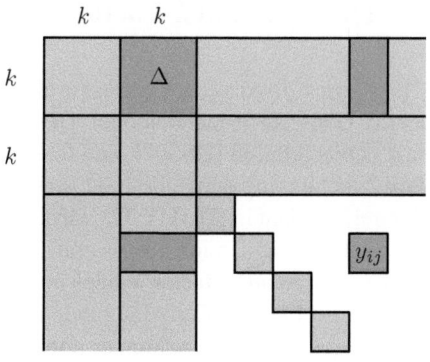

form a single orbit under $\mathrm{Sym}(\infty)$, so if we set $\Delta := \det y[[k], [2k] \setminus [k]]$, then we have

$$S'_{k,\infty}(K) = S'_{k-1,\infty}(K) \cup Z\,\mathrm{Sym}(\infty),\ \text{where}$$
$$Z = \{y \in S'_{k,\infty}(K) \mid \Delta \neq 0\}.$$

The union of two $\mathrm{Sym}(\infty)$-Noetherian topological spaces is $\mathrm{Sym}(\infty)$-Noetherian, so it suffices to prove that $Z\,\mathrm{Sym}(\infty)$ is $\mathrm{Sym}(\infty)$-Noetherian. Observe that Z itself is stable under the subgroup $H := \{\pi \in \mathrm{Sym}(\infty) \mid \pi|_{[2k]} = 1|_{[2k]}\}$. Hence by Proposition 3.5 it suffices to prove that Z is H-Noetherian. To this end, define $T \subseteq \mathbb{N} \times \mathbb{N}$ by

$$T := \{(i,j) \in \mathbb{N} \times \mathbb{N} \mid i = j \text{ or } (i < j \text{ and } i \in [2k])\}.$$

Let Q be the open subset of K^T where the $[k] \times ([2k] \setminus [k])$-submatrix has non-zero determinant. The coordinate ring of Q is H-Noetherian by Theorem 2.3 and the fact that adding finitely many H-fixed variables preserves H-Noetherianity. As a consequence, Q is an H-Noetherian space. We claim that the projection $\mathrm{pr}: Z \to Q$ that maps a matrix y to its T-labelled entries is a closed, H-equivariant embedding. Equivariance is immediate. To see that pr is injective observe that, for $y \in Z$, any matrix entry y_{ij} with $i < j$ (since we work with symmetric matrices) and $i \notin [2k]$ satisfies an equation (Fig. 5)

$$0 = \det y\,[[k] \cup \{i\}, ([2k] \setminus [k]) \cup \{j\}] = \Delta \cdot y_{ij} - E,$$

where E is an expression involving only variables in T. Since Δ is non-zero, we find that y_{ij} is determined by $\mathrm{pr}(y)$. This shows injectivity. That $\mathrm{pr}: Z \to Q$ is, in fact, a closed embedding follows by showing that the dual map $K[Q] \to K[Z]$ is surjective: the regular function $\frac{E}{\det \Delta}$ maps onto $y_{ij}|_Z$. Since Q is H-Noetherian, so is Z, and as mentioned before this concludes the induction step.

As $S'_{k,\infty}(K)$ is Noetherian, we find that in particular, the Zariski closure $S_{k,\infty}(K)$ is cut out from $S'_{k,\infty}(K)$ by finitely many $\mathrm{Sym}(\mathbb{N})$-orbits of equations. Representatives of these orbits already lie in $S'_{k,n_k}(K)$ for suitable n_k. □

7 Tensors and Δ-Varieties

This section deals with finiteness results for a wide class of varieties of tensors, introduced by Snowden [Sno13] and called Δ-*varieties*. The proof of this chapter's theorem is more involved than earlier proofs, and we have therefore decided to break the section up into more digestible sections.

Δ-*Varieties*

We work over a ground field K, which we assume to be infinite to avoid anomalies with the Zariski topology. For any tuple $(V_1, \ldots, V_n)$ of finite-dimensional vector spaces over K, we write $\mathbf{V}(V_1, \ldots, V_n)$ for the tensor product $V_1^* \otimes \cdots \otimes V_n^*$. These spaces have three types of interesting maps between them. First, given linear maps $f_i : V_i \to W_i$ there is a natural linear map $\mathbf{V}(W_1, \ldots, W_n) \to \mathbf{V}(V_1, \ldots, V_n)$, namely, the tensor product $\otimes_i f_i^*$ of the dual maps f_i^*. Second, given any $\sigma \in \mathrm{Sym}([n])$, there is a canonical map $\sigma : \mathbf{V}(V_{\sigma(1)}, \ldots, V_{\sigma(n)}) \to \mathbf{V}(V_1, \ldots, V_n)$. Third, there is a canonical *flattening* map $\mathbf{V}(V_1, \ldots, V_n, V_{n+1}) \to \mathbf{V}(V_1, \ldots, V_n \otimes V_{n+1})$, which is called like this because, in coordinates, it takes an $(n+1)$-way table of numbers and transforms it into an n-way table; see Fig. 6.

A Δ-variety is not a single variety, but rather a rule $\mathbf{X}$ that takes as input a finite sequence $(V_j)_{j \in [n]}$ of finite-dimensional vector spaces over K, and assigns to it a subvariety $\mathbf{X}(V_1, \ldots, V_n)$ of $\mathbf{V}(V_1, \ldots, V_n)$. To be a Δ-variety, each of the three types of maps above must preserve $\mathbf{X}$, i.e., $\otimes_i f_i^*$ must map $\mathbf{X}(W_1, \ldots, W_n)$ into $\mathbf{X}(V_1, \ldots, V_n)$, and σ must map $\mathbf{X}(V_{\sigma(1)}, \ldots, V_{\sigma(n)})$ into $\mathbf{X}(V_1, \ldots, V_n)$, and the flattening map must map $\mathbf{X}(V_1, \ldots, V_{n+1})$ into $\mathbf{X}(V_1, \ldots, V_n \otimes V_{n+1})$.

The Δ-varieties that we shall study will have a fourth, additional property, namely, that the inverse to the isomorphism $\mathbf{V}(V_1, \ldots, V_n, K) \to \mathbf{V}(V_1, \ldots, V_n \otimes K) = \mathbf{V}(V_1, \ldots, V_n)$ maps $\mathbf{X}(V_1, \ldots, V_n)$ into $\mathbf{X}(V_1, \ldots, V_n, K)$; we call such Δ-varieties *good*. Taking any linear function f from an additional vector space V_{n+1} to K we then find that $\mathbf{X}(V_1, \ldots, V_n) \otimes f$, being the image of $\mathbf{X}(V_1, \ldots, V_n)$ under the map above followed by $1_{V_1}^* \otimes \cdots \otimes 1_{V_n}^* \otimes f^*$, is contained in $\mathbf{X}(V_1, \ldots, V_n, V_{n+1})$. A (boring) example of a Δ-variety that is not good is that for which $\mathbf{X}(V_1, \ldots, V_n)$ equals $\mathbf{V}(V_1, \ldots, V_n)$ if $n < 10$ and the empty set otherwise.

A typical example of a good Δ-variety is **Seg**, the cone over Segre, which maps a tuple of vector spaces to the variety of pure tensors $v_1 \otimes \cdots \otimes v_n$ in the tensor product of the duals. In fact, any non-empty good Δ-variety contains **Seg**—but the class of good Δ-varieties is much larger. For instance, it is closed under taking joins:

Fig. 6 Flattening an element of $K^4 \otimes K^2 \otimes K^3$ to an element of $K^4 \otimes (K^2 \otimes K^3)$

if **X** and **Y** are Δ-varieties, then the rule **X** + **Y** that assigns to $(V_i)_{i \in [n]}$ the Zariski closure of $\mathbf{X}(V_1, \ldots, V_n) + \mathbf{Y}(V_1, \ldots, V_n)$ is also a Δ-variety, and good if both **X** and **Y** are. Similarly, (good) Δ-varieties are closed under taking tangential varieties, unions, and intersections.

Given some equations for an instance of a Δ-variety **X**, one obtains equations for other instances of **X** by pulling back along sequences of the maps appearing in the definition of a Δ-variety. For instance, start with the 2×2-determinant defining $\mathbf{Seg}(K^2, K^2)$ inside $\mathbf{V}(K^2, K^2)$. Then we obtain generators for the ideal of $\mathbf{Seg}(K^m, K^n)$ by pulling the determinant back along duals of linear maps $f_1 : K^2 \to K^m$ and $f_2 : K^2 \to K^n$.[1] And then, using the remaining two axioms, we also find equations for the variety of pure tensors in, say, $\mathbf{V}(K^2, K^2, K^2, K^2)$ through the flattening maps into $\mathbf{V}(K^2 \otimes K^2, K^2 \otimes K^2) \cong \mathbf{V}(K^4, K^4)$, $\mathbf{V}(K^2, K^2 \otimes K^2 \otimes K^2) \cong \mathbf{V}(K^2, K^8)$, etc. Indeed, one readily shows that one obtains generators of the ideals of all instances of **Seg** in this manner. The result in this section is that a similar result holds for any sufficiently small good Δ-variety, at least at a topological level.

Theorem 7.1. *Let* **X** *be a good Δ-variety which is* bounded *in the sense that there exist finite-dimensional vector spaces* W_1, W_2 *such that* $\mathbf{X}(W_1, W_2)$ *is not all of* $\mathbf{V}(W_1, W_2)$. *Then there exist an* $n_\mathbf{X} \in \mathbb{N}$ *and vector spaces* $U_1, \ldots, U_{n_\mathbf{X}}$ *such that* **X** *equals the inclusion-wise largest Δ-variety* **Y** *with* $\mathbf{Y}(U_1, \ldots, U_{n_\mathbf{X}}) = \mathbf{X}(U_1, \ldots, U_{n_\mathbf{X}})$.

This means, in more concrete terms, that the equations for $\mathbf{X}(U_1, \ldots, U_{n_\mathbf{X}})$, pulled back along all four types of linear maps from the definition of a good Δ-variety, yield equations that cut out all instances of **X** from their ambient spaces. In particular, there is a universal degree bound, depending only on **X** but not on n or $V_1, \ldots, V_n$, on equations needed to define $\mathbf{X}(V_1, \ldots, V_n)$ set-theoretically within $\mathbf{V}(V_1, \ldots, V_n)$.

Since $\mathrm{GL}(W_1) \times \mathrm{GL}(W_2)$ acts with a dense orbit on $\mathbf{V}(W_1, W_2)$—namely, the two-tensors (or matrices) of full rank—the boundedness condition on X implies that all two-tensors in instances of the form $\mathbf{X}(V_1, V_2)$ have uniformly bounded rank. This readily implies that the boundedness condition on **X** is also preserved under joins (by adding the rank bounds), tangential varieties, intersections, and unions, so that the theorem applies to a wide class of Δ-varieties of interest in applications.

---

[1] Snowden chose the notion of Δ-varieties contravariant in the linear maps f_i so as to make defining ideals and more general Δ-*modules* [Sno13] depend covariantly on them.

Related Literature

The boundedness condition on Δ-varieties was first formulated, at an ideal-theoretic level, in [Sno13]. There it is conjectured that a generalisation of Theorem 7.1 should hold, for bounded Δ-varieties, on the ideal-theoretic level; and not only for equations of instances of **X**, but also for their q-syzygies for any fixed $q \geq 1$. This general statement is proved for **Seg** in [Sno13]. The special case where $q = 1$, i.e., finiteness of equations, is known to hold for the tangential variety to **Seg** [OR11], confirming a conjecture from [LW07]; and for the variety **Seg** + **Seg** = 2**Seg** of tensors of border rank at most 2 [Rai12], confirming the GSS-conjecture from [GSS05] (a set-theoretic version of which was first proved in [LM04]). The set-theoretic theorem above was first proved in [DK13] for k**Seg**, i.e., for any fixed secant variety of **Seg**; and a discussion with Snowden led to the insight that our proof generalises to bounded, good, Δ-varieties as in the theorem. Further recent keywords closely related to the topic of this section are GL_∞-algebras [SS12], twisted commutative algebras [SS12b], FI-modules [CEF12], and cactus varieties [BB13].

From a Δ-Variety to an Infinite-Dimensional Variety

We prove the theorem by embedding all relevant instances of **X** into a single, infinite-dimensional variety given by determinantal equations, and showing that this variety is Noetherian up to symmetry preserving **X**. By the boundedness assumption, we can choose a number p that is strictly greater than the ranks of all two-tensors in instances $\mathbf{X}(V_1, V_2)$, independently of V_1 and V_2. Set $V := K^{[p]}$ and $X_n := \mathbf{X}(V, \ldots, V) \subseteq \mathbf{V}_n := \mathbf{V}(V, \ldots, V)$, where the number of Vs equals n. We first argue that the equations for all (infinitely many) $X_n, n \in \mathbb{N}$ pull back to equations defining all instances of **X**.

Indeed, let $V_1, \ldots, V_n$ be vector spaces, and let $\omega \in \mathbf{V}(V_1, \ldots, V_n)$ be a tensor. Then we claim that ω lies in $\mathbf{X}(V_1, \ldots, V_n)$ if and only if for all linear maps $f_i : V \to V_i$ the image of ω under $\otimes_{i \in [n]} f_i^*$ lies in X_n. The "only if" claim follows from the first axiom for Δ-varieties. For the "if" claim, note that if $(\otimes_{i \in [n]} f_i^*)\omega$ lies in X_n for all tuples of f_i, then for each $j \in [n]$, the linear map that ω induces from $\otimes_{i \neq j} V_i$ into V_j^*, being a flattening of ω in $\mathbf{V}(\otimes_{i \neq j} V_i, V_j)$, has image $U_j \subseteq V_j^*$ of dimension strictly smaller than p. Now take $f_j : V \to V_j$ such that f_j^* restricts to an injection $U_j \to V^*$, and let $g_j : V_j \to V$ be such that $g_j^* \circ f_j^*$ restricts to the identity on U_j. Then the tensor $\omega' := \otimes_j f_j^* \omega$ lies in $X_n = \mathbf{X}(V, \ldots, V)$ by assumption. But then, by the first axiom, the tensor $\otimes_j g_j^* \omega' = \omega$ lies in $\mathbf{X}(V_1, \ldots, V_n)$, as claimed. This argument actually also works ideal-theoretically; only later shall we need to work purely topologically.

We now cast the chain of varieties $(X_n)_{n \in \mathbb{N}}$ into the framework of Sect. 4. To this end, let R_n denote the symmetric K-algebra generated by $V^{\otimes [n]}$, which is the coordinate ring of $\mathbf{V}_n$. Pick a non-zero element $x_0 \in V$ and let $\iota_n : R_n \to R_{n+1}$ be the homomorphism of K-algebras determined by the linear map $V^{\otimes[n]} \to V^{\otimes[n+1]}$, $x \mapsto x \otimes x_0$. The group $\mathrm{GL}(V)^{[n]}$ acts on $V^{\otimes[n]}$ in the natural manner, and this extends to an action by algebra automorphisms on R_n. Similarly, the group $\mathrm{Sym}([n])$ acts on $V^{\otimes[n]}$ by permuting tensor factors. The embedding ι_n is equivariant for the group $G_n := \mathrm{Sym}([n]) \ltimes \mathrm{GL}(V)^{[n]}$ if we embed $\mathrm{Sym}([n])$ into $\mathrm{Sym}([n+1])$ by fixing n and $\mathrm{GL}(V)^{[n]}$ into $\mathrm{GL}(V)^{[n+1]}$ by adding 1_V in the last component.

The linear map $\iota^* : \mathbf{V}_{n+1} \to \mathbf{V}_n$ maps X_{n+1} into X_n, and X_n is preserved by G_n. Letting S_n be the coordinate ring of R_n, we have all the arrows in the diagram of Fig. 3 except for the arrows to the right. To obtain these, we use that $\mathbf{X}$ is good, as follows. Given any $e_0 \in V^*$ such that $e_0(x_0) = 1$, the map $\pi_n^* : \mathbf{V}_n \to \mathbf{V}_{n+1}$, $\omega \mapsto \omega \otimes e_0$ maps X_n into X_{n+1}. The dual to this linear map, extended to an algebra homomorphism, is the required map $\pi : R_{n+1} \to R_n$. This completes the diagram. The technical condition $(*)$ from page 47 is also satisfied, i.e., for all indices l, i_0, i_1 with $l \geq i_0, i_1$ and for all $g \in G_l$ there exist an index $j \leq i_0, i_1$ and group elements $g_0 \in G_{i_0}, g_1 \in G_{i_1}$ such that

$$(\pi_{li_1} \, g \, \iota_{i_0 l})^* = (g_1 \, \iota_{ji_1} \, \pi_{i_0 j} \, g_0)^*.$$

Indeed, the left-hand side is the composition of the map $\mathbf{V}_{i_1} \to \mathbf{V}_l$ tensoring with $e_0^{\otimes l - i_1}$, followed by g, followed by the map $\mathbf{V}_l \to \mathbf{V}_{i_0}$ contracting with $x_0^{\otimes l - i_0}$ in the last $l - i_0$ factors. Let $i_1 - j$ be the number of factors V^* in $\mathbf{V}_{i_1}$ that are moved, by g, into the last $l - i_0$ positions and hence end up being contracted in the last step. This means that $j \leq i_0, i_1$ is the number of factors V^* in $\mathbf{V}_{i_1}$ that are not contracted. Hence the composition can also be obtained by first applying a $g_1 \in G_{i_1}$, ensuring that the $i_1 - j$ factors V^* in $\mathbf{V}_{i_1}$ that need to be contracted are in the last $i_1 - j$ positions; then contracting by a pure tensor in those positions, which for suitable choice of g_1 may be chosen $x_0^{\otimes i_1 - j}$; then tensoring with $i_0 - j$ copies of e_0; and finally applying a suitable element $g_0 \in G_{i_0}$.

The upshot of this is that if we can prove that the projective limit $X_\infty := \lim_{\leftarrow n} X_n$ is defined by finitely many $G_\infty := \bigcup_n G_n$-orbits of equations within $\mathbf{V}_\infty := \lim_{\leftarrow n} \mathbf{V}_n$, then there exists an n_X such that for all $n \geq n_X$ the variety X_n is defined by the G_n-orbits of equations for X_{n_X}. This implies the theorem.

Flattening Varieties

To prove this finiteness result, then, we show that X_∞ is contained in a G_∞-Noetherian subvariety Y_∞ of $\mathbf{V}_\infty$, which we call a flattening variety, and that Y_∞ itself is defined by finitely many G_∞-orbits of equations. To define Y_∞, let $\mathbf{Y}^{(k)}$ denote the largest Δ-variety for which $\mathbf{Y}^{(k)}(V_1, V_2)$ consists of two-tensors of rank at most k. Then $\mathbf{Y}^{(k)}(V_1, \ldots, V_n)$ is defined by the vanishing of all $(k+1) \times (k+1)$-minors of the flattenings $\mathbf{V}(V_1, \ldots, V_n) \to \mathbf{V}(\bigotimes_{i \in A} V_i, \bigotimes_{i \in B} V_i)$ for all partitions of $[n]$ into disjoint subsets A and B. Set $Y_n^{(k)} := \mathbf{Y}^{(k)}(V, \ldots, V) \subseteq \mathbf{V}_n$, and $Y_\infty^{(k)} := \lim_{\leftarrow n} Y_n^{(k)} \subseteq \lim_{\leftarrow n} \mathbf{V}_n$. By the boundedness assumption on $\mathbf{X}$, $Y_\infty^{(p-1)}$ contains X_∞.

We first prove that each $Y_\infty^{(k)}, k \in \mathbb{N}$ is defined by finitely many G_∞-orbits of equations. Unwinding the definitions, this statement boils down to the statement that if $\omega \in \mathbf{V}_n$ with $n \gg 0$ does not lie in $Y_n^{(k)}$, then there exists an $i \in [n]$ and an $x \in V$ such that contracting ω with x in the i-th position yields a tensor $\omega' \in \mathbf{V}_{n-1}$ that does not lie in $Y_{n-1}^{(k)}$. In fact, we shall see that $n > 2k$ suffices. The condition that ω does not lie in $Y_n^{(k)}$ means that there is a partition $[n] = A \cup B$ such that ω, regarded as a linear map $V^{\otimes A} \to (V^*)^{\otimes B}$, has rank strictly larger than k. Using that $n > 2k$ and after swapping A and B if necessary we may assume that $|B| > k$.

Let $U \subseteq (V^*)^{\otimes B}$ be a $(k+1)$-dimensional subspace of the image of this linear map ω. We claim that since $|B|$ is larger than k, there exists a position $i \in B$ and an $x \in V$ such that the image of U under contraction with x in the i-th position still has dimension $k + 1$. Indeed, otherwise U would be a point in the projective variety

$$Q := \{W \in \operatorname{Gr}_{k+1}(V^*)^{\otimes B} \mid \text{contracting } U \text{ with any } x \text{ in any position}$$
$$\text{decreases the dimension}\}.$$

We claim that this variety is empty. To prove this, extend the distinguished vector $x_0 \in V$ from the definition of $\mathbf{V}_\infty$ to a basis $x_0, \ldots, x_{p-1}$, where the distinguished $e_0 \in V^*$ vanishes on $x_1, \ldots, x_{p-1}$. Then the basis of V^* dual to $x_0, \ldots, x_{p-1}$ starts with e_0; denote it $e_0, \ldots, e_{p-1}$.[2]

If Q is not empty, then by Borel's fixed point theorem [Bor91] Q contains a T^B-fixed point W, where T is the maximal torus in $\operatorname{GL}(V)$ consisting of invertible linear maps whose matrices with respect to $x_0, \ldots, x_{p-1}$ are diagonal. This means that W has a basis of common eigenvectors $e_\alpha := \otimes_{i \in B} e_{\alpha_i}$, where α runs through some set $J \subseteq \{0, \ldots, p-1\}^B$ of cardinality $k+1$. Think of the $\alpha \in J$ as B-labelled words over the alphabet $\{0, \ldots, p-1\}$. Contracting e_α with $x_0 + x_1 + \ldots + x_{p-1}$ at position $i \in B$ yields $e_{\alpha'}$, where α' is the word obtained from α by deleting the i-th letter. By assumption, the resulting words $e_{\alpha'}, \alpha \in J$ are linearly dependent, which means that at least two of them must coincide.

Summing up, J consists of $k + 1$ distinct words of length $|B| \geq k + 1$ with the property that for each $i \in B$ the collection J contains two words that differ only at position i. By induction on k we show that this is impossible, i.e., that for $k + 1$ distinct words of length $\geq k + 1$ over any alphabet there exist k positions restricted to which all words are distinct. For $k = 0$ this is immediate: restricting a single word to zero positions yields a single (empty) word. Assume that it is true for $k - 1$, and consider $k + 1$ words of length $\geq k + 1$. Set one word α apart. Then there exist $k - 1$ positions restricted to which the remaining k words are distinct. Restricted to those $k - 1$ positions α equals at most one word α' of the remaining words. So by

---

[2] The reason for labelling with $\{0, \ldots, p-1\}$ rather than $[p]$ will become apparent soon.

adding to the $k-1$ positions a position where α and α' differ we obtain k positions restricted to which all words are distinct.

This contradiction shows that there exists an $i \in B$ and an $x \in V$ such that the contraction of U with x at position i still has dimension $k+1$. As a consequence, contracting ω with x at position i yields a tensor outside $Y_{n-1}^{(k)}$, as claimed. Thus $Y_\infty^{(k)}$ is defined by finitely many G_∞-orbits of equations.

The variety $Y_\infty^{(k)}$ is defined by the vanishing of $(k+1) \times (k+1)$-determinants. For what follows, it will be convenient to understand these explicitly in terms of coordinates. The basis $x_0, \ldots, x_{p-1}$ gives rise to a basis $x_w, w \in \{0, \ldots, p-1\}^{[n]}$ of $V^{\otimes [n]}$. The ring R_n is the polynomial ring in these variables. Under the embedding $\iota_n : R_n \to R_{n+1}$ the variable x_w is mapped to x_{w0}. Hence R_∞ is the polynomial ring in variables x_w where w runs over all infinite words in $\{0, \ldots, p-1\}^{\mathbb{N}}$ of finite support $\mathrm{supp}(w) := \{j \in \mathbb{N} \mid w_j \neq 0\}$; let us call these finitary words. In these coordinates, a determinantal equation for $Y_\infty^{(k)}$ looks as follows. Fix $k+1$ finitary words $w_i, i \in [k+1]$ and $k+1$ further finitary words $w'_j, j \in [k+1]$ with the requirement that $\mathrm{supp}(w_i) \cap \mathrm{supp}(w'_j) = \emptyset$ for all i, j. Then form the square matrix

$$x[(w_i)_i, (w'_j)_j] := (x_{w_i + w'_j})_{i,j \in [k+1]}$$

and its determinant

$$\Delta[(w_i)_i, (w'_j)_j] := \det x[(w_i)_i, (w'_j)_j].$$

All determinants defining $Y_\infty^{(k)}$ have this form.[3]

Noetherianity of Flattening Varieties

Using this explicit understanding of the defining equations for $Y_\infty^{(k)}$, we prove that $Y_\infty^{(k)}$, with its Zariski-topology, is G_∞-Noetherian. The proof is similar to that in Sect. 6 for the Gaussian k-factor model. In particular, we proceed by induction on k. For $k = 0$ the variety $Y_\infty^{(0)}$ consists of a single point, namely, 0, and is certainly Noetherian. Now assume that $Y_\infty^{(k-1)}$ is G_∞-Noetherian. By the discussion of flattening varieties, $Y_\infty^{(k-1)}$ is defined by the orbits of finitely many $k \times k$-determinants of flattenings, say q of them. Let $\Delta_a, a \in [q]$ be those determinants.

---

[3]The convenient fact that the sum of two finitary words is again finitary explains our choice of labelling $x_0, \ldots, x_{p-1}$.

Then we may write

$$Y_\infty^{(k)} := Y_\infty^{(k-1)} \cup \bigcup_{a \in [q]} Z_a G_\infty, \text{ where}$$

$$Z_a := \{\omega \in Y_\infty^{(k)} \mid \Delta_a(\omega) \neq 0\}.$$

As in the case of the k-factor model, it suffices to show that Z_a is Noetherian under a suitable subgroup of G_∞ stabilising it. To this end, write $\Delta_a = \Delta[(w_i)_i, (w'_j)_j]$ for k finitary words w_i and k finitary words w'_j with $\mathrm{supp}(w_i) \cap \mathrm{supp}(w'_j) = \emptyset$ for all i, j. Set $n := \max\left(\bigcup_i \mathrm{supp}(w_i) \cup \bigcup_j \mathrm{supp}(w'_j)\right)$ and observe that Δ_a is fixed by $H := \{\pi \in \mathrm{Sym}(\infty) \mid \pi|_{[n]} = 1_{[n]}\}$, and hence Z_a is stabilised by H. We claim that Z_a is H-Noetherian. To prove this, let J be the set of finitary words w with $|\mathrm{supp}(w) \setminus [n]| \leq 1$. In particular, all variables appearing in Δ_a are in J. Let Q be the open subset of K^J where Δ_a is non-zero. By Theorem 2.3 and the fact that adding finitely many H-fixed variables to an H-Noetherian ring preserves H-Noetherianity, the coordinate ring of Q is H-Noetherian—here the crucial point is that "only one index runs off to infinity". We claim that the projection $\mathrm{pr} : Z_a \to Q$ mapping a point to its coordinates labelled by J is an H-equivariant, closed embedding. Equivariance is immediate. To see that pr is injective, we prove that on Z_a any variable x_w has an expression in terms of the variables labelled by J. We proceed by induction on the cardinality of $\mathrm{supp}(w) \setminus [n]$. For cardinality 0 and 1 the word w lies in J and we are done. So assume that the cardinality is at least 2 and that the statement is true for all smaller cardinalities. Then we can split w as $u + u'$ where $\mathrm{supp}(u) \cap \mathrm{supp}(u')$, $\mathrm{supp}(u) \cap \mathrm{supp}(w'_j)$, and $\mathrm{supp}(w_i) \cap \mathrm{supp}(u')$ are all empty and where both $\mathrm{supp}(u)$ and $\mathrm{supp}(u')$ contain at least one element of $\mathbb{N} \setminus [n]$. Then on Z_a we have

$$0 = \Delta[(w_1, \ldots, w_k, u), (w'_1, \ldots, w'_k, u')] = \Delta_a \cdot x_{u+u'} - E$$

where E is an expression involving only variables whose supports contain fewer elements of $\mathbb{N} \setminus [n]$ than $\mathrm{supp}(w)$ does. By the induction hypothesis, these may be expressed in the variables labelled by J, and as Δ_a is non-zero on Z_a, so can $x_{u+u'} = x_w$. To show that pr is a closed embedding we note that the map $K[Q] \to K[Z]$ is surjective: there is an expression for $E|_Z$ involving only J-labelled variables, and dividing by Δ_a yields such an expression for x_w.

We conclude that Z has the topology of a closed H-stable subspace of K^Q and is hence H-Noetherian. By Proposition 3.5 and the fact that finite unions of equivariantly Noetherian spaces are equivariantly Noetherian, we find that $Y_\infty^{(k)}$ is G_∞-Noetherian. This concludes the proof of the theorem.

An important final remark is in order here: our proof of the theorem shows that X_∞ is defined by finitely many G_∞-orbits of equations, which is stronger than the theorem claims. In particular, this stronger statement can be used to show that for each fixed Δ-variety there is a polynomial-time membership test. On the other hand, it is typically *not* true that the ideal of X_∞ is generated by finitely many G_∞-orbits of polynomials; indeed, this statement is already false for the cone over Segre. How to reconcile this with the aforementioned conjecture [Sno13] that an ideal-theoretic version of theorem should hold? Well, by pulling back equations along elements of G_∞, we are implicitly pulling back equations along tensor products of linear maps, and along permutations of tensor factors, and along contractions, and along tensoring with e_0, but not along flattening maps (though we did use, in the proof, that **X** was closed under flattening). This additional source of linear maps along which to pull back equations may allow for an ideal-theoretic version of the theorem. For details see [Sno13, DK13].

Acknowledgement The author was supported by a Vidi grant from the Netherlands Organisation for Scientific Research (NWO).

References

[AT03] S. Aoki, A. Takemura, Minimal basis for connected Markov chain over $3 \times 3 \times k$ contingency tables with fixed two dimensional marginals. Aust. N. Z. J. Stat. **45**, 229–249 (2003)

[AH07] M. Aschenbrenner, C.J. Hillar, Finite generation of symmetric ideals. Trans. Am. Math. Soc. **359**(11), 5171–5192 (2007)

[AH08] M. Aschenbrenner, C.J. Hillar, An algorithm for finding symmetric Gröbner bases in infinite dimensional rings (2008). Preprint available from http://arxiv.org/abs/0801.4439

[Bor91] A. Borel, *Linear Algebraic Groups* (Springer, New York, 1991)

[BD11] A.E. Brouwer, J. Draisma, Equivariant Gröbner bases and the two-factor model. Math. Comput. **80**, 1123–1133 (2011)

[BB13] W. Buczyńska, J. Buczyński, Secant varieties to high degree Veronese reembeddings, cataleticant matrices and smoothable Gorenstein schemes. J. Algebr. Geom. **23**, 63–90 (2014)

[CEF12] T. Church, J.S. Ellenberg, B. Farb, FI-modules: a new approach to stability for S_n-representations (2012). Preprint, available from http://arxiv.org/abs/1204.4533

[Coh67] D.E. Cohen, On the laws of a metabelian variety. J. Algebra **5**, 267–273 (1967)

[Coh87] D.E. Cohen, Closure relations, Buchberger's algorithm, and polynomials in infinitely many variables, in *Computation Theory and Logic*. Lecture Notes in Computer Science, vol. 270 (Springer, Berlin, 1987), pp. 78–87. 68Q40 (13B99)

[dLST95] J.A. de Loera, B. Sturmfels, R.R. Thomas, Gröbner bases and triangulations of the second hypersimplex. Combinatorica **15**, 409–424 (1995)

[DS98] P. Diaconis, B. Sturmfels, Algebraic algorithms for sampling from conditional distributions. Ann. Stat. **26**(1), 363–397 (1998)

[Dra10] J. Draisma, Finiteness for the k-factor model and chirality varieties. Adv. Math. **223**, 243–256 (2010)

[DK13] J. Draisma, J. Kuttler, Bounded-rank tensors are defined in bounded degree. Duke Math. J. **163**(1), 35–63 (2014)

[DEKL13] J. Draisma, R.H. Eggermont, R. Krone, A. Leykin, Noetherianity for infinite-dimensional toric varieties (2013). Preprint available from http://arxiv.org/abs/1306.0828

[DS06] V. Drensky, R. La Scala, Gröbner bases of ideals invariant under endomorphisms. J. Symb. Comput. **41**(7), 835–846 (2006)

[DX10] M. Drton, H. Xiao, Finiteness of small factor analysis models. Ann. Inst. Stat. Math. **62**(4), 775–783 (2010)

[DSS07] M. Drton, B. Sturmfels, S. Sullivant, Algebraic factor analysis: tetrads, pentads and beyond. Probab. Theory Relat. Fields **138**(3–4), 463–493 (2007)

[GSS05] L.D. Garcia, M. Stillman, B. Sturmfels, Algebraic geometry of Bayesian networks. J. Symb. Comput. **39**(3–4), 331–355 (2005)

[GW09] R. Goodman, N.R. Wallach, in *Symmetry, Representations, and Invariants*. Graduate Texts in Mathematics, vol. 255 (Springer, New York, 2009)

[Hig52] G. Higman, Ordering by divisibility in abstract algebras. Proc. Lond. Math. Soc. III. Ser. **2**, 326–336 (1952)

[HMdC13] C.J. Hillar, A.M. del Campo, Finiteness theorems and algorithms for permutation invariant chains of Laurent lattice ideals. J. Symb. Comput. **50**, 314–334 (2013)

[HS12] C.J. Hillar, S. Sullivant, Finite Gröbner bases in infinite dimensional polynomial rings and applications. Adv. Math. **221**, 1–25 (2012)

[SHS07] S. Hoşten, S. Sullivant, A finiteness theorem for Markov bases of hierarchical models. J. Comb. Theory Ser. A **114**(2), 311–321 (2007)

[Kru60] J.B. Kruskal, Well-quasi ordering, the tree theorem, and Vazsonyi's conjecture. Trans. Am. Math. Soc. **95**, 210–225 (1960)

[LM04] J.M. Landsberg, L. Manivel, On the ideals of secant varieties of Segre varieties. Found. Comput. Math. **4**(4), 397–422 (2004)

[LW07] J.M. Landsberg, J. Weyman, On tangential varieties of rational homogeneous varieties. J. Lond. Math. Soc. (2) **76**(2), 513–530 (2007)

[Lan65] S. Lang, *Algebra* (Addison-Wesley, Reading, 1965)

[LSL09] R. La Scala, V. Levandovskyy, Letterplace ideals and non-commutative Gröbner bases. J. Symb. Comp. **44**(10), 1374–1393 (2009)

[NW63] C.St.J.A. Nash-Williams, On well-quasi-ordering finite trees. Proc. Camb. Philos. Soc. **59**, 833–835 (1963)

[OR11] L. Oeding, C. Raicu, Tangential varieties of segre-veronese varieties (2011). Preprint, avaibable from http://arxiv.org/abs/1111.6202

[Rai12] C. Raicu, Secant varieties of Segre–Veronese varieties. Algebra Number Theory **6**(8), 1817–1868 (2012)

[SS12] S.V. Sam, A. Snowden, GL-equivariant modules over polynomial rings in infinitely many variables (2012). Preprint, available from http://arxiv.org/abs/1206.2233

[SS12b] S.V. Sam, A. Snowden, Introduction to twisted commutative algebras. Preprint (2012), available from http://arxiv.org/abs/1209.5122

[Sno13] A. Snowden, Syzygies of Segre embeddings and Δ-modules. Duke Math. J. **162**(2), 225–277 (2013)

Likelihood Geometry

June Huh and Bernd Sturmfels

Introduction

Maximum likelihood estimation (MLE) is a fundamental computational problem in statistics, and it has recently been studied with some success from the perspective of algebraic geometry. In these notes we give an introduction to the geometry behind MLE for algebraic statistical models for discrete data. As is customary in algebraic statistics [LiAS], we shall identify such models with certain algebraic subvarieties of high-dimensional complex projective spaces.

The article is organized into four sections. The first three sections correspond to the three lectures given at Levico Terme. The last section will contain proofs of new results.

In Sect. 1, we start out with plane curves, and we explain how to identify the relevant punctured Riemann surfaces. We next present the definitions and basic results for likelihood geometry in $\mathbb{P}^n$. Theorems 1.6 and 1.7 are concerned with the likelihood correspondence, the sheaf of differential one-forms with logarithmic poles, and the topological Euler characteristic. The ML degree of generic complete intersections is given in Theorem 1.10. Theorem 1.15 shows that the likelihood fibration behaves well over strictly positive data. Examples of Grassmannians and Segre varieties are discussed in detail. Our treatment of linear spaces in Theorem 1.20 will appeal to readers interested in matroids and hyperplane arrangements.

Section 2 begins leisurely, with the question *Does watching soccer on TV cause hair loss?* [MSS]. This leads us to conditional independence and low rank matrices.

J. Huh (✉)
Department of Mathematics, University of Michigan, Ann Arbor, MI 48109, USA
e-mail: junehuh@umich.edu

B. Sturmfels
Department of Mathematics, University of California, Berkeley, CA 94720, USA
e-mail: bernd@berkeley.edu

We study likelihood geometry of determinantal varieties, culminating in the duality theorem of Draisma and Rodriguez [DR]. The ML degrees in Theorems 2.2 and 2.6 were computed using the software Bertini [Bertini], underscoring the benefits of using numerical algebraic geometry for MLE. After a discussion of mixture models, highlighting the distinction between rank and nonnegative rank, we end Sect. 2 with a review of recent results in [ARSZ] on tensors of nonnegative rank 2.

Section 3 starts out with toric models [PS, §1.22] and geometric programming [BoydVan, §4.5]. Theorem 3.2 identifies the ML degree of a toric variety with the Euler characteristic of the complement of a hypersurface in a torus. Theorem 3.7 furnishes the ML degree of a variety parametrized by generic polynomials. Theorem 3.10 characterizes varieties of ML degree 1 and it reveals a beautiful connection to the A-discriminant of [GKZ]. We introduce the ML bidegree and the sectional ML degree of an arbitrary projective variety in $\mathbb{P}^n$, and we explain how these two are related. Section 3 ends with a study of the operations of intersection, projection, and restriction in likelihood geometry. This concerns the algebro-geometric meaning of the distinction between sampling zeros and structural zeros in statistical modeling.

In Sect. 4 we offer precise definitions and technical explanations of more advanced concepts from algebraic geometry, including logarithmic differential forms, Chern–Schwartz–MacPherson classes, and schön very affine varieties. This enables us to present complete proofs of various results, both old and new, that are stated in the earlier sections.

We close the introduction with a disclaimer regarding our overly ambitious title. There are many important topics in the statistical study of likelihood inference that should belong to "Likelihood Geometry" but are not covered in this article. Such topics include Watanabe's theory of singular Bayesian integrals [Wat], differential geometry of likelihood in information geometry [AN], and real algebraic geometry of Gaussian models [Uhl]. We regret not being able to talk about these topics and many others. Our presentation here is restricted to the setting of [LiAS, §2.2], namely statistical models for discrete data viewed as projective varieties in $\mathbb{P}^n$.

1 First Lecture

Let us begin our discussion with likelihood on algebraic curves in the complex projective plane $\mathbb{P}^2$. We fix a system of homogeneous coordinates p_0, p_1, p_2 on $\mathbb{P}^2$. The set of real points in $\mathbb{P}^2$ with $\text{sign}(p_0) = \text{sign}(p_1) = \text{sign}(p_2)$ is identified with the open triangle

$$\Delta_2 = \{(p_0, p_1, p_2) \in \mathbb{R}^3 : p_0, p_1, p_2 > 0 \text{ and } p_0 + p_1 + p_2 = 1\}.$$

Given three positive integers u_0, u_1, u_2, the corresponding likelihood function is

$$\ell_{u_0,u_1,u_2}(p_0, p_1, p_2) = \frac{p_0^{u_0} p_1^{u_1} p_2^{u_2}}{(p_0 + p_1 + p_2)^{u_0+u_1+u_2}}.$$

Likelihood Geometry

This defines a rational function on $\mathbb{P}^2$, and it restricts to a regular function on $\mathbb{P}^2 \backslash \mathcal{H}$, where

$$\mathcal{H} = \{(p_0 : p_1 : p_2) \in \mathbb{P}^2 : p_0 p_1 p_2 (p_0 + p_1 + p_2) = 0\}$$

is our arrangement of four distinguished lines. The likelihood function ℓ_{u_0, u_1, u_2} is positive on the triangle Δ_2, it is zero on the boundary of Δ_2, and it attains its maximum at the point

$$(\hat{p}_0, \hat{p}_1, \hat{p}_2) = \frac{1}{u_0 + u_1 + u_2}(u_0, u_1, u_2). \tag{1}$$

The corresponding point $(\hat{p}_0 : \hat{p}_1 : \hat{p}_2)$ is the only critical point of the function ℓ_{u_0, u_1, u_2} on the four-dimensional real manifold $\mathbb{P}^2 \backslash \mathcal{H}$. To see this, we consider the *logarithmic derivative*

$$\mathrm{dlog}(\ell_{u_0, u_1, u_2}) = \left(\frac{u_0}{p_0} - \frac{u_0 + u_1 + u_2}{p_0 + p_1 + p_2}, \frac{u_1}{p_1} - \frac{u_0 + u_1 + u_2}{p_0 + p_1 + p_2}, \frac{u_2}{p_2} - \frac{u_0 + u_1 + u_2}{p_0 + p_1 + p_2} \right).$$

We note that this equals $(0, 0, 0)$ if and only if $(p_0 : p_1 : p_2)$ is the point $(\hat{p}_0 : \hat{p}_1 : \hat{p}_2)$ in (1).

Let X be a smooth curve in $\mathbb{P}^2$ defined by a homogeneous polynomial $f(p_0, p_1, p_2)$. This curve plays the role of a statistical model, and our task is to maximize the likelihood function ℓ_{u_0, u_1, u_2} over its set $X \cap \Delta_2$ of positive real points. To compute that maximum algebraically, we examine the set of all critical points of ℓ_{u_0, u_1, u_2} on the complex curve $X \backslash \mathcal{H}$. That set of critical points is the *likelihood locus*. Using Lagrange Multipliers from Calculus, we see that it consists of all points of $X \backslash \mathcal{H}$ such that $\mathrm{dlog}(\ell_{u_0, u_1, u_2})$ lies in the plane spanned by $\mathrm{d}f$ and $(1, 1, 1)$ in $\mathbb{C}^3$. Thus, our task is to study the solutions in $\mathbb{P}^2 \backslash \mathcal{H}$ of the equations

$$f(p_0, p_1, p_2) = 0 \quad \text{and} \quad \det \begin{pmatrix} 1 & 1 & 1 \\ \frac{u_0}{p_0} & \frac{u_1}{p_1} & \frac{u_2}{p_2} \\ \frac{\partial f}{\partial p_0} & \frac{\partial f}{\partial p_1} & \frac{\partial f}{\partial p_2} \end{pmatrix} = 0. \tag{2}$$

Suppose that X has degree d. Then, after clearing denominators, the second equation has degree $d + 1$. By Bézout's Theorem, we expect the likelihood locus to consist of $d(d + 1)$ points in $\mathbb{P}^2 \backslash \mathcal{H}$. This is indeed what happens when f is a generic polynomial of degree d.

We define the *maximum likelihood degree* (or *ML degree*) of our curve X to be the cardinality of the likelihood locus for generic choices of u_0, u_1, u_2. Thus a general plane curve of degree d has ML degree $d(d + 1)$. However, for special curves, the ML degree can be smaller.

Theorem 1.1. *Let X be a smooth curve of degree d in $\mathbb{P}^2$, and $a = \#(X \cap \mathcal{H})$ the number of its points on the distinguished arrangement. Then the ML degree of X equals $d^2 - 3d + a$.*

This is a very special case of Theorem 1.7 which identifies the ML degree with the signed Euler characteristic of $X \backslash \mathcal{H}$. For a general curve of degree d in $\mathbb{P}^2$, we have $a = 4d$, and so $d^2 - 3d + a = d(d+1)$ as predicted. However, the number a of points in $X \cap \mathcal{H}$ can drop:

Example 1.2. Consider the case $d = 1$ of lines. A generic line has ML degree 2. The line $X = V(p_0 + cp_1)$ has ML degree 1 provided $c \notin \{0, 1\}$. The special line $X = V(p_0 + p_1)$ has ML degree 0: (2) has no solutions on $X \backslash \mathcal{H}$ unless $u_0 + u_1 = 0$. In the three cases, $X \backslash \mathcal{H}$ is the Riemann sphere $\mathbb{P}^1$ with four, three, or two points removed. ◇

Example 1.3. Consider the case $d = 2$ of quadrics. A general quadric has ML degree 6. The *Hardy–Weinberg curve*, which plays a fundamental role in population genetics, is given by

$$f(p_0, p_1, p_2) = \det \begin{pmatrix} 2p_0 & p_1 \\ p_1 & 2p_2 \end{pmatrix} = 4p_0 p_2 - p_1^2.$$

The curve has only three points on the distinguished arrangement:

$$X \cap \mathcal{H} = \{(1:0:0), (0:0:1), (1:-2:1)\}.$$

Hence the ML degree of the Hardy–Weinberg curve equals 1. This means that the maximum likelihood estimate (MLE) is a rational function of the data. Explicitly, the MLE equals

$$(\hat{p}_0, \hat{p}_1, \hat{p}_2) = \frac{1}{4(u_0 + u_1 + u_2)^2} \left((2u_0 + u_1)^2, \ 2(2u_0 + u_1)(u_1 + 2u_2), \ (u_1 + 2u_2)^2 \right). \tag{3}$$

In applications, the Hardy–Weinberg curve arises via its parametric representation

$$\begin{aligned} p_0(s) &= s^2 \\ p_1(s) &= 2s(1-s) \\ p_2(s) &= (1-s)^2 \end{aligned} \tag{4}$$

Here the parameter s is the probability that a biased coin lands on tails. If we toss that same biased coin twice, then the above formulas represent the following probabilities:

$$\begin{aligned} p_0(s) &= \text{probability of 0 heads} \\ p_1(s) &= \text{probability of 1 head} \\ p_2(s) &= \text{probability of 2 heads} \end{aligned}$$

Suppose now that the experiment of tossing the coin twice is repeated N times. We record the following counts, where $N = u_0 + u_1 + u_2$ is the sample size of our repeated experiment:

$$u_0 = \text{number of times 0 heads were observed}$$
$$u_1 = \text{number of times 1 head was observed}$$
$$u_2 = \text{number of times 2 heads were observed}$$

The MLE problem is to estimate the unknown parameter s by maximizing

$$\ell_{u_0,u_1,u_2} = p_0(s)^{u_0} p_1(s)^{u_1} p_2(s)^{u_2} = 2^{u_1} s^{2u_0+u_1} (1-s)^{u_1+2u_2}.$$

The unique solution to this optimization problem is

$$\hat{s} = \frac{2u_0 + u_1}{2u_0 + 2u_1 + 2u_2}.$$

Substituting this expression into (4) gives the estimator $(p_0(\hat{s}), p_1(\hat{s}), p_2(\hat{s}))$ for the three probabilities in our model. The resulting rational function coincides with (3). ◇

The ML degree is also defined when the given curve $X \subset \mathbb{P}^2$ is not smooth, but it counts critical points of ℓ_u only in the regular locus of X. Here is an example to illustrate this.

Example 1.4. A general cubic curve X in $\mathbb{P}^2$ has ML degree 12. Suppose now that X is a cubic which meets $\mathcal{H}$ transversally but has one isolated singular point in $\mathbb{P}^2 \backslash \mathcal{H}$. If the singular point is a *node* then the ML degree of X is 10, and if the singular point is a *cusp* then the ML degree of X is 9. The ML degrees are found by saturating the equations in (2) with respect to the homogenous ideal of the singular point. ◇

Moving beyond likelihood geometry in the plane, we shall introduce our objects in any dimension. We fix the complex projective space $\mathbb{P}^n$ with coordinates $p_0, p_1, \ldots, p_n$, representing probabilities. We summarize the observed data in a vector $u = (u_0, u_1, \ldots, u_n) \in \mathbb{N}^{n+1}$, where u_i is the number of samples in state i. The likelihood function on $\mathbb{P}^n$ given by u equals

$$\ell_u = \frac{p_0^{u_0} p_1^{u_1} \cdots p_n^{u_n}}{(p_0 + p_1 + \cdots + p_n)^{u_0 + u_1 + \cdots + u_n}}.$$

The unique critical point of this rational function on $\mathbb{P}^n$ is the data point itself:

$$(u_0 : u_1 : \cdots : u_n).$$

Moreover, this point is the global maximum of the likelihood function ℓ_u on the probability simplex Δ_n. Throughout, we identify Δ_n with the set of all positive real points in $\mathbb{P}^n$.

The linear forms in ℓ_u define an arrangement $\mathcal{H}$ of $n+2$ distinguished hyperplanes in $\mathbb{P}^n$. The differential of the logarithm of the likelihood function is the vector of rational functions

$$\mathrm{dlog}(\ell_u) = \left(\frac{u_0}{p_0}, \frac{u_1}{p_1}, \ldots, \frac{u_n}{p_n}\right) - \frac{u_+}{p_+} \cdot (1, 1, \ldots, 1). \tag{5}$$

Here $p_+ = \sum_{i=0}^n p_i$ and $u_+ = \sum_{i=0}^n u_i$. The vector (5) represents a section of the sheaf of differential one-forms on $\mathbb{P}^n$ that have logarithmic singularities along $\mathcal{H}$. This sheaf is denoted

$$\Omega^1_{\mathbb{P}^n}(\log(\mathcal{H})).$$

Our aim is to study the restriction of ℓ_u to a closed subvariety $X \subseteq \mathbb{P}^n$. We will assume that X is defined over the real numbers, irreducible, and not contained in $\mathcal{H}$. Let X_{sing} denote the singular locus of X, and X_{reg} denote $X \backslash X_{\mathrm{sing}}$. When X serves as a statistical model, the goal is to maximize the rational function ℓ_u on the semialgebraic set $X \cap \Delta_n$. To solve this problem algebraically, we determine all critical points of the log-likelihood function $\log(\ell_u)$ on the complex variety X. Here we must exclude points that are singular or lie in $\mathcal{H}$.

Definition 1.5. The *maximum likelihood degree* of X is the number of complex critical points of the function ℓ_u on $X_{\mathrm{reg}} \backslash \mathcal{H}$, for generic data u. The *likelihood correspondence* $\mathcal{L}_X$ is the universal family of these critical points. To be precise, $\mathcal{L}_X$ is the closure in $\mathbb{P}^n \times \mathbb{P}^n$ of

$$\{(p, u) : p \in X_{\mathrm{reg}} \backslash \mathcal{H} \text{ and } \mathrm{dlog}(\ell_u) \text{ vanishes at } p\}.$$

We sometimes write $\mathbb{P}^n_p \times \mathbb{P}^n_u$ for $\mathbb{P}^n \times \mathbb{P}^n$ to highlight that the first factor is the probability space, with coordinates p, while the second factor is the data space, with coordinates u. The first part of the following result appears in [Huh1, §2]. A precursor was [HKS, Proposition 3].

Theorem 1.6. *The likelihood correspondence $\mathcal{L}_X$ of any irreducible subvariety X in $\mathbb{P}^n_p$ is an irreducible variety of dimension n in the product $\mathbb{P}^n_p \times \mathbb{P}^n_u$. The map $\mathrm{pr}_1 : \mathcal{L}_X \to \mathbb{P}^n_p$ is a projective bundle over $X_{\mathrm{reg}} \backslash \mathcal{H}$, and the map $\mathrm{pr}_2 : \mathcal{L}_X \to \mathbb{P}^n_u$ is generically finite-to-one.*

See Sect. 4 for a proof. The degree of the map $\mathrm{pr}_2 : \mathcal{L}_X \to \mathbb{P}^n_u$ to data space is the ML degree of X. This number has a topological interpretation as an Euler characteristic, provided suitable assumptions on X are being made. The relationship between the homology of a manifold and critical points of a suitable function on it is the topic of Morse theory.

The study of ML degrees was started in [CHKS, §2] by developing the connection to the sheaf $\Omega^1_X(\log(\mathcal{H}))$ of differential one-forms on X with logarithmic poles along $\mathcal{H}$. It was shown in [CHKS, Theorem 20] that the ML degree of X equals the signed topological Euler characteristic

$$(-1)^{\dim X} \cdot \chi(X \backslash \mathcal{H}),$$

provided X is smooth and the intersection $\mathcal{H} \cap X$ defines a normal crossing divisor in $X \subseteq \mathbb{P}^n$. A major drawback of that early result was that the hypotheses are so restrictive that they essentially never hold for varieties X that arise from statistical models used in practice. From a theoretical point view, this issue can be addressed by passing to a resolution of singularities. However, in spite of existing algorithms for resolution in characteristic zero, these algorithms do not scale to problems of the sizes of interest in algebraic statistics. Thus, whatever computations we wish to do should not be based on resolution of singularities.

The following result due to [Huh1] gives the same topological interpretation of the ML degree. The hypotheses here are much more realistic and inclusive than those in [CHKS, Theorem 20].

Theorem 1.7. *If the very affine variety $X \backslash \mathcal{H}$ is smooth of dimension d, then the ML degree of X equals the signed topological Euler characteristic of $(-1)^d \cdot \chi(X \backslash \mathcal{H})$.*

The term *very affine variety* refers to a closed subvariety of some algebraic torus $(\mathbb{C}^*)^m$. Our ambient space $\mathbb{P}^n \backslash \mathcal{H}$ is a very affine variety because it has a closed embedding

$$\mathbb{P}^n \backslash \mathcal{H} \longrightarrow (\mathbb{C}^*)^{n+1}, \qquad (p_0 : \cdots : p_n) \longmapsto \left(\frac{p_0}{p_+}, \ldots, \frac{p_n}{p_+}\right).$$

The study of such varieties is foundational for *tropical geometry*. The special case when $X \backslash \mathcal{H}$ is a Riemann surface with a punctures, arising from a curve in $\mathbb{P}^2$, was seen in Theorem 1.1. We remark that Theorem 1.7 can be deduced from works of Gabber–Loeser [Gabber-Loeser] and Franecki–Kapranov [Franecki-Kapranov] on perverse sheaves on algebraic tori.

The smoothness hypothesis is essential for Theorem 1.7 to hold. If X is singular then, generally, neither $X \backslash \mathcal{H}$ nor $X_{\text{reg}} \backslash \mathcal{H}$ has its signed Euler characteristic equal to the ML degree of X. Varieties X that demonstrate this are the two singular cubic curves in Example 1.4.

Conjecture 1.8. For any projective variety $X \subseteq \mathbb{P}^n$ of dimension d, not contained in $\mathcal{H}$,

$$(-1)^d \cdot \chi(X \backslash \mathcal{H}) \geq \text{MLdegree}(X).$$

In particular, the signed topological Euler characteristic $(-1)^d \cdot \chi(X \backslash \mathcal{H})$ is nonnegative.

Analogous conjectures can be made in the slightly more general setting of [Huh1]. In particular, we conjecture that the inequality

$$(-1)^d \cdot \chi(V) \geq 0$$

holds for any closed d-dimensional subvariety $V \subseteq (\mathbb{C}^*)^m$.

Remark 1.9. We saw in Example 1.2 that the ML degree of a projective variety X can be 0. In all situations of statistical interest, the variety $X \subset \mathbb{P}^n$ intersects the open simplex Δ_n in a subset that is Zariski dense in X. If that intersection is smooth then MLdegree$(X) \geq 1$. In fact, arguing as in [CHKS, Proposition 11], it can be shown that for smooth X,

$$\text{MLdegree}(X) \geq \#(\text{bounded regions of } X_\mathbb{R} \backslash \mathcal{H}).$$

Here a *bounded region* is a connected component of the semialgebraic set $X_\mathbb{R} \backslash \mathcal{H}$ whose classical closure is disjoint from the distinguished hyperplane $V(p_+)$ in $\mathbb{P}^n$.

If X is singular then the number of bounded regions of $X_\mathbb{R} \backslash \mathcal{H}$ can exceed MLdegree(X). For instance, let $X \subset \mathbb{P}^2$ be the cuspidal cubic curve defined by

$$(p_0+p_1+p_2)(7p_0-9p_1-2p_2)^2 = (3p_0+5p_1+4p_2)^3.$$

The real part $X_\mathbb{R} \backslash \mathcal{H}$ consists of 8 bounded and 2 unbounded regions, but the ML degree of X is 7. The bounded region that contains the cusp $(13 : 17 : -31)$ has no other critical points for ℓ_u. ◇

In what follows we present instances that illustrate the computation of the ML degree. We begin with the case of *generic complete intersections*. Suppose that $X \subset \mathbb{P}^n$ is a complete intersection defined by r generic homogeneous polynomials $g_1, \ldots, g_r$ of degrees $d_1, d_2, \ldots, d_r$.

Theorem 1.10. *The ML degree of X equals $Dd_1 d_2 \cdots d_r$, where*

$$D = \sum_{i_1+i_2+\cdots+i_r \leq n-r} d_1^{i_1} d_2^{i_2} \cdots d_r^{i_r}. \tag{6}$$

Proof. By Bertini's Theorem, the generic complete intersection X is smooth in $\mathbb{P}^n$. All critical points of the likelihood function ℓ_u on X lie in the dense open subset $X \backslash \mathcal{H}$. Consider the following $(r+2) \times (n+1)$-matrix with entries in the polynomial ring $\mathbb{R}[p_0, p_1, \ldots, p_n]$:

$$\begin{bmatrix} u \\ \tilde{J}(p) \end{bmatrix} = \begin{bmatrix} u_0 & u_1 & \cdots & u_n \\ p_0 & p_1 & \cdots & p_n \\ p_0 \frac{\partial g_1}{\partial p_0} & p_1 \frac{\partial g_1}{\partial p_1} & \cdots & p_n \frac{\partial g_1}{\partial p_n} \\ p_0 \frac{\partial g_2}{\partial p_0} & p_1 \frac{\partial g_2}{\partial p_1} & \cdots & p_n \frac{\partial g_2}{\partial p_n} \\ \vdots & \vdots & \ddots & \vdots \\ p_0 \frac{\partial g_r}{\partial p_0} & p_1 \frac{\partial g_r}{\partial p_1} & \cdots & p_n \frac{\partial g_r}{\partial p_n} \end{bmatrix}. \tag{7}$$

Likelihood Geometry

Let Y denote the determinantal variety in $\mathbb{P}^n$ given by the vanishing of its $(r+2) \times (r+2)$ minors. The codimension of Y is at most $n-r$, which is a general upper bound for ideals of maximal minors, and hence the dimension of Y is at least r. Our genericity assumptions ensure that the matrix $\tilde{J}(p)$ has maximal row rank $r+1$ for all $p \in X$. Hence a point $p \in X$ lies in Y if and only if the vector u is in the row span of $\tilde{J}(p)$. Moreover, by Theorem 1.6,

$$(X_{\text{reg}} \backslash \mathcal{H}) \cap Y \;=\; X \cap Y$$

is a finite subset of $\mathbb{P}^n$, and its cardinality is the desired ML degree of X.

Since X has dimension $n-r$, we conclude that Y has the maximum possible codimension, namely $n-r$, and that the intersection of X with the determinantal variety Y is proper. We note that Y is Cohen–Macaulay, since Y has maximal codimension $n-r$, and ideals of minors of generic matrices are Cohen–Macaulay. Bézout's Theorem implies

$$\text{MLdegree}(X) \;=\; \text{degree}(X) \cdot \text{degree}(Y) \;=\; d_1 \cdots d_r \cdot \text{degree}(Y).$$

The degree of the determinantal variety Y equals the degree of the determinantal variety given by generic forms of the same row degrees. By the Thom–Porteous–Giambelli formula, this degree is the complete homogeneous symmetric function of degree $\text{codim}(Y) = n-r$ evaluated at the row degrees of the matrix. Here, the row degrees are $0, 1, d_1, \ldots, d_r$, and the value of that symmetric function is precisely D. We conclude that $\text{degree}(Y) = D$. Hence the ML degree of the generic complete intersection $X = \mathcal{V}(g_1, \ldots, g_r)$ equals $D \cdot d_1 d_2 \cdots d_n$. □

Example 1.11 ($r = 1$). A generic hypersurface of degree d in $\mathbb{P}^n$ has ML degree

$$d \cdot D = d + d^2 + d^3 + \cdots + d^n.$$

Example 1.12 ($r = 2, n = 3$). A space curve that is the generic intersection of two surfaces of degree d and e in $\mathbb{P}^3$ has ML degree $de + d^2 e + de^2$. ◇

Remark 1.13. It was shown in [HKS, Theorem 5] that (6) is an upper bound for the ML degree of any variety X *of codimension r* that is defined by polynomials of degree $d_1, \ldots, d_r$. In fact, the same is true under the weaker hypothesis that X is cut out by polynomials of degrees $d_1 \geq \cdots \geq d_r \geq d_{r+1} \geq \cdots \geq d_s$, so X need not be a complete intersection. However, the hypothesis $\text{codim}(X) = r$ is essential in order for $\text{MLdegree}(X) \leq (6)$ to hold. That codimension hypothesis was forgotten when this upper bound was cited in [LiAS, Theorem 2.2.6] and in [PS, Theorem 3.31]. Hence these two book references are not correct as stated.

Here is a simple counterexample. Let $n = 3$ and $d_1 = d_2 = d_3 = 2$. Then the bound (6) is the Bézout number 8, and this is also the correct ML degree for a general complete intersection of three quadrics in $\mathbb{P}^3$. Now let X be a general rational normal curve in $\mathbb{P}^3$. The curve X is defined by three quadrics, namely, the 2×2-minors of a 2×3-matrix filled with general linear forms in p_0, p_1, p_2, p_3. Since X is a Riemann sphere with 15 punctures, Theorem 1.7 tells us that MLdegree$(X) = 13$, and this exceeds the bound of 8. ◇

We now come to a variety that is ubiquitous in statistics, namely the model of *independence* for two binary random variables [LiAS, §1.1]. This model is represented by Segre's quadric surface X in $\mathbb{P}^3$. By this we mean the surface defined by the 2×2-determinant:

$$X = V(p_{00}p_{11} - p_{01}p_{10}) \subset \mathbb{P}^3.$$

The surface X is isomorphic to $\mathbb{P}^1 \times \mathbb{P}^1$, so it is smooth, and we can apply Theorem 1.7 to find the ML degree. In other words, we seek to determine the Euler characteristic of the open complex surface $X \setminus \mathcal{H}$ where

$$\mathcal{H} = \{ p \in \mathbb{P}^3 : p_{00}p_{01}p_{10}p_{11}(p_{00}+p_{01}+p_{10}+p_{11}) = 0 \}.$$

To this end, we write $X = \mathbb{P}^1 \times \mathbb{P}^1$ with coordinates $((x_0 : x_1), (y_0 : y_1))$. Our surface is parametrized by $p_{ij} = x_i y_j$, and hence

$$\begin{aligned} X \setminus \mathcal{H} &= (\mathbb{P}^1 \times \mathbb{P}^1) \setminus \{x_0 x_1 y_0 y_1 (x_0 + x_1)(y_0 + y_1) = 0\} \\ &= (\mathbb{P}^1 \setminus \{x_0 x_1 (x_0 + x_1) = 0\}) \times (\mathbb{P}^1 \setminus \{y_0 y_1 (y_0 + y_1) = 0\}) \\ &= (\text{2-sphere} \setminus \{\text{three points}\}) \times (\text{2-sphere} \setminus \{\text{three points}\}). \end{aligned}$$

Since the Euler characteristic is additive and multiplicative,

$$\chi(X \setminus \mathcal{H}) = (-1) \cdot (-1) = 1.$$

This means that the map $u \mapsto \hat{p}$ from the data to the MLE is a rational function in each coordinate. The following "word problem for freshmen" is aimed at finding that function.

Example 1.14. Do this exercise: A biologist friend of yours wishes to test whether two binary random variables are independent. She collects data and records the matrix of counts

$$u = \begin{pmatrix} u_{00} & u_{01} \\ u_{10} & u_{11} \end{pmatrix}.$$

How to ascertain whether u lies close to the independence model

$$X = V(p_{00}p_{11} - p_{01}p_{10})\,?$$

A statistician who recently started working in her lab explains that, as the first step in the analysis of her data, the biologist should calculate the maximum likelihood estimate (MLE)

$$\hat{p} = \begin{pmatrix} \hat{p}_{00} & \hat{p}_{01} \\ \hat{p}_{10} & \hat{p}_{11} \end{pmatrix}.$$

Can you help your friend by supplying the formula for $\hat{p}$ as a rational function in u? The solution to this word problem is as follows. The MLE is the rank 1 matrix

$$\hat{p} = \frac{1}{(u_{++})^2} \begin{pmatrix} u_{0+} \\ u_{1+} \end{pmatrix} \cdot \begin{pmatrix} u_{+0} & u_{+1} \end{pmatrix}. \tag{8}$$

We illustrate the concepts introduced above by deriving this well-known formula. The likelihood correspondence $\mathcal{L}_X$ of $X = V(p_{00}p_{11} - p_{01}p_{10})$ is the subvariety of $X \times \mathbb{P}^3$ defined by

$$U \cdot (p_{00}, p_{01}, p_{10}, p_{11})^T = 0, \tag{9}$$

where U is the matrix

$$U = \begin{pmatrix} 0 & -u_{10} - u_{11} & 0 & u_{00} + u_{01} \\ u_{11} + u_{01} & -u_{00} - u_{10} & 0 & 0 \\ u_{11} + u_{10} & 0 & -u_{01} - u_{00} & 0 \\ 0 & 0 & -u_{01} - u_{11} & u_{00} + u_{10} \end{pmatrix}.$$

We urge the reader to derive (9) from Definition 1.5 using a computer algebra system.

Note that the determinant of U vanishes identically. In fact, for generic u_{ij}, the matrix U has rank 3, so its kernel is spanned by a single vector. The coordinates of that vector are given by Cramer's rule, and we find them to be equal to the rational functions in (8).

The locus where the function $u \mapsto \hat{p}$ is undefined consists of those u where the matrix rank of U drops below 3. A computation shows that the rank of U drops to 2 on the variety

$$V(u_{00} + u_{10}, u_{01} + u_{11}) \cup V(u_{00} + u_{01}, u_{10} + u_{11}),$$

and it drops to 0 on the point $V(u_{00} + u_{01}, u_{10} + u_{11}, u_{01} + u_{11})$. In particular, the likelihood function ℓ_u given by that point u has infinitely many critical points in the quadric X. ◇

We note that all coefficients of the linear forms that define the exceptional loci in $\mathbb{P}^3_u$ for the independence model are positive. This means that data points u with all coordinates positive can never be exceptional. We will prove in Sect. 4 that this usually holds. Let $\mathrm{pr}_1 : \mathcal{L}_X \to \mathbb{P}^n_p$ and $\mathrm{pr}_2 : \mathcal{L}_X \to \mathbb{P}^n_u$ be the projections from the likelihood correspondence to p-space and u-space respectively. We are interested in the fibers of pr_2 over positive points u.

Theorem 1.15. *Let $u \in \mathbb{R}^{n+1}_{>0}$, and let $X \subset \mathbb{P}^n$ be an irreducible variety such that no singular points of any intersection $X \cap \{p_i = 0\}$ lies in the hyperplane at infinity $\{p_+ = 0\}$. Then*

(1) the likelihood function ℓ_u on X has only finitely many critical points in $X_{\text{reg}} \backslash \mathcal{H}$;
(2) if the fiber $\text{pr}_2^{-1}(u)$ is contained in X_{reg}, then its length equals the ML degree of X.

The hypothesis concerning "no singular point" will be satisfied for essentially all statistical models of interest. Here is an example which shows that this hypothesis is necessary.

Example 1.16. We consider the smooth cubic curve X in $\mathbb{P}^2$ that is defined by

$$f = (p_0 + p_1 + p_2)^3 + p_0 p_1 p_2.$$

The ML degree of the curve X is 3. Each intersection $X \cap \{p_i = 0\}$ is a triple point that lies on the line at infinity $\{p_+ = 0\}$. The fiber $\text{pr}_2^{-1}(u)$ of the likelihood fibration over the positive point $u = (1:1:1)$ is the entire curve X. ◇

If u is not positive in Theorem 1.15, then the fiber of pr_2 over u may have positive dimension. We saw an instance of this at the end of Example 1.14. Such *resonance loci* have been studied extensively when X is a linear subspace of $\mathbb{P}^n$. See [CDFV] and references therein.

The following cautionary example shows that the length of the scheme-theoretic fiber of $\mathcal{L}_X \to \mathbb{P}^n_u$ over special points u in the open simplex Δ_n may exceed the ML degree of X.

Example 1.17. Let X be the curve in $\mathbb{P}^2$ defined by the ternary cubic

$$f = p_2(p_1 - p_2)^2 + (p_0 - p_2)^3.$$

This curve intersects $\mathcal{H}$ in eight points, has ML degree 5, and has a cuspidal singularity at

$$P := (1:1:1).$$

The prime ideal in $\mathbb{R}[p_0, p_1, p_2, u_0, u_1, u_2]$ for the likelihood correspondence $\mathcal{L}_X$ is minimally generated by five polynomials, having degrees $(3, 0), (2, 2), (3, 1), (3, 1), (3, 1)$. They are obtained by saturating the two equations in (2) with respect to $\langle p_0 p_2 \rangle \cap \langle p_0 - p_1, p_2 - p_1 \rangle$.

The scheme-theoretic fiber of pr_1 over a general point of X is a reduced line in the u-plane, while the fiber of pr_1 over P is the double line

$$L := \{(u_0 : u_1 : u_2) \in \mathbb{P}^2 : (2u_0 - u_1 - u_2)^2 = 0\}.$$

Likelihood Geometry

The reader is invited to verify the following assertions using a computer algebra system:

(a) If u is a general point of $\mathbb{P}_u^2$, then $\text{pr}_2^{-1}(u)$ consists of 5 reduced points in $X_{\text{reg}} \backslash \mathcal{H}$.
(b) If u is a general point on the line L, then the locus of critical points $\text{pr}_2^{-1}(u)$ consists of four reduced points in $X_{\text{reg}} \backslash \mathcal{H}$ and the reduced point P.
(c) If u is the point $(1 : 1 : 1) \in L$, then $\text{pr}_2^{-1}(u)$ is a zero-dimensional scheme of length 6. This scheme consists of three reduced points in $X_{\text{reg}} \backslash \mathcal{H}$ and P counted with multiplicity 3.

In particular, the fiber in (c) is not algebraically equivalent to the general fiber (a). This example illustrates one of the difficulties classical geometers had to face when formulating the "principle of conservation of numbers". See [Fulton, Chap. 10] for a modern treatment. ◇

It is instructive to examine classical varieties from projective geometry from the likelihood perspective. For instance, we may study the Grassmannian in its Plücker embedding. Grassmannians are a nice test case because they are smooth, so that Theorem 1.7 applies.

Example 1.18. Let $X = G(2, 4)$ denote the Grassmannian of lines in $\mathbb{P}^3$. In its Plücker embedding in $\mathbb{P}^5$, this Grassmannian is the quadric hypersurface defined by

$$p_{12}p_{34} - p_{13}p_{24} + p_{14}p_{23} = 0. \qquad (10)$$

As in (7), the critical equations for the likelihood function ℓ_u are the 3×3-minors of

$$\begin{bmatrix} u_{12} & u_{13} & u_{14} & u_{23} & u_{24} & u_{34} \\ p_{12} & p_{13} & p_{14} & p_{23} & p_{24} & p_{34} \\ p_{12}p_{34} & -p_{13}p_{24} & p_{14}p_{23} & p_{14}p_{23} & -p_{13}p_{24} & p_{12}p_{34} \end{bmatrix}. \qquad (11)$$

By Theorem 1.6, the likelihood correspondence $\mathcal{L}_X$ is a five-dimensional subvariety of $\mathbb{P}^5 \times \mathbb{P}^5$. The cohomology class of this subvariety can be represented by the bidegree of its ideal:

$$B_X(p, u) = 4p^5 + 6p^4u + 6p^3u^2 + 6p^2u^3 + 2pu^4. \qquad (12)$$

This is the *multidegree*, in the sense of [ch3:MS, §8.5], of $\mathcal{L}_X$ with respect to the natural $\mathbb{Z}^2$-grading on the polynomial ring $\mathbb{R}[p, u]$. We can use [ch3:MS, Proposition 8.49] to compute the bidegree from the prime ideal of $\mathcal{L}_X$. Its leading coefficient 4 is the ML degree of X. Its trailing coefficient 2 is the degree of X. The polynomials $B_X(p, u)$ will be studied in Sect. 3.

The prime ideal of $\mathcal{L}_X$ is computed from the equations in (10) and (11) by saturation with respect to $\mathcal{H}$. It is minimally generated by the following eight polynomials in $\mathbb{R}[p, u]$:

(a) one polynomial of degree $(2,0)$, namely the Plücker quadric,
(b) six polynomials of degree $(1,1)$, given by 2×2-minors of

$$\begin{pmatrix} p_{12} - p_{34} & p_{13} - p_{24} & p_{14} - p_{23} \\ u_{12} - u_{34} & u_{13} - u_{24} & u_{14} - u_{23} \end{pmatrix} \quad \text{and}$$

$$\begin{pmatrix} p_{12}+p_{13}+p_{23} & p_{12}+p_{14}+p_{24} & p_{13}+p_{14}+p_{34} & p_{23}+p_{24}+p_{34} \\ u_{12}+u_{13}+u_{23} & u_{12}+u_{14}+u_{24} & u_{13}+u_{14}+u_{34} & u_{23}+u_{24}+u_{34} \end{pmatrix},$$

(c) one polynomial of degree $(2,1)$, for instance

$$2u_{24}p_{12}p_{34} + 2u_{34}p_{13}p_{24} + (u_{23} + u_{24} + u_{34})p_{14}p_{24}$$
$$- (u_{13} + u_{14} + u_{34})p_{24}^2 - (u_{12} + 2u_{13} + u_{14} - u_{24})p_{24}p_{34}.$$

For a fixed positive data vector $u > 0$, these six polynomials in (b) reduce to three linear equations, and these cut out a plane $\mathbb{P}^2$ inside $\mathbb{P}^5$. To find the four critical points of ℓ_u on $X = G(2,4)$, we must then intersect the two conics (a) and (c) in that plane $\mathbb{P}^2$.

The ML degree of the Grassmannian $G(r,m)$ in $\mathbb{P}^{\binom{m}{r}-1}$ is the signed Euler characteristic of the manifold $G(r,m)\backslash \mathcal{H}$ obtained by removing $\binom{m}{r} + 1$ distinguished hyperplane sections. It would be very interesting to find a general formula for this ML degree. At present, we only know that the ML degree of $G(2,5)$ is 26, and that the ML degree of $G(2,6)$ is 156. By Theorem 1.7, these numbers give the Euler characteristic of $G(2,m)\backslash \mathcal{H}$ for $m \leq 6$. ◇

We end this lecture with a discussion of the delightful case when X is a linear subspace of $\mathbb{P}^n$, and the open variety $X\backslash \mathcal{H}$ is the complement of a *hyperplane arrangement*. In this context, following Varchenko [Varchenko], the likelihood function ℓ_u is known as the *master function*, and the statement of Theorem 1.7 was first proved by Orlik and Terao in [Orlik-Terao]. We assume that X has dimension d, is defined over $\mathbb{R}$, and does not contain the vector $\mathbf{1} = (1,1,\ldots,1)$. We can regard X as a $(d+1)$-dimensional linear subspace of $\mathbb{R}^{n+1}$. The orthogonal complement $X^\perp$ with respect to the standard dot product is a linear space of dimension $n-d$ in $\mathbb{R}^{n+1}$. The linear space $X^\perp + \mathbf{1}$ spanned by $X^\perp$ and the vector $\mathbf{1}$ has dimension $n-d+1$ in $\mathbb{R}^{n+1}$, and hence can be viewed as subspace of codimension d in $\mathbb{P}_u^n$. In our next formula, the operation $\star$ is the Hadamard product or coordinatewise product.

Proposition 1.19. *The likelihood correspondence* $\mathcal{L}_X$ *in* $\mathbb{P}^n \times \mathbb{P}^n$ *is defined by*

$$p \in X \quad \text{and} \quad u \in p \star (X^\perp + \mathbf{1}). \tag{13}$$

The prime ideal of $\mathcal{L}_X$ *is obtained from these constraints by saturation with respect to* $\mathcal{H}$.

Likelihood Geometry

Proof. If all p_i are non-zero then $u \in p \star (X^\perp + 1)$ says that

$$u/p := \left(\frac{u_0}{p_0}, \frac{u_1}{p_1}, \ldots, \frac{u_n}{p_n}\right)$$

lies in the subspace $X^\perp + 1$. Equivalently, the vector obtained by adding a multiple of $(1, 1, \ldots, 1)$ to u/p is perpendicular to X. We can take that vector to be the differential (5). Hence (13) expresses the condition that p is a critical point of ℓ_u on X. □

The intersection $X \cap \mathcal{H}$ is an arrangement of $n + 2$ hyperplanes in $X \simeq \mathbb{P}^d$. For special choices of the subspace X, it may happen that two or more hyperplanes coincide. Taking $\{p_+ = 0\}$ as the hyperplane at infinity, we view $X \cap \mathcal{H}$ as an arrangement of $n + 1$ hyperplanes in the affine space $\mathbb{R}^d$. A region of this arrangement is *bounded* if it is disjoint from $\{p_+ = 0\}$.

Theorem 1.20. *The ML degree of X is the number of bounded regions of the real affine hyperplane arrangement $X \cap \mathcal{H}$ in $\mathbb{R}^d$. The bidegree of the likelihood correspondence $\mathcal{L}_X$ is the h-polynomial of the broken circuit complex of the rank $d+1$ matroid associated with $X \cap \mathcal{H}$.*

We need to explain the second assertion. The hyperplane arrangement $X \cap \mathcal{H}$ consists of the intersections of the $n + 2$ hyperplanes in $\mathcal{H}$ with $X \simeq \mathbb{P}^d$. We regard these as hyperplanes through the origin in $\mathbb{R}^{d+1}$. They define a matroid M of rank $d + 1$ on $n + 2$ elements. We identify these elements with the variables $x_1, x_2, \ldots, x_{n+2}$. For each circuit C of M let $m_C = (\prod_{i \in C} x_i)/x_j$ where j is the smallest index such that $x_j \in C$. The *broken circuit complex* of M is the simplicial complex with Stanley–Reisner ring $\mathbb{R}[x_1, \ldots, x_{n+2}]/\langle m_C : C$ circuit of $M \rangle$. See [ch3:MS, §1.1] for Stanley–Reisner basics. The Hilbert series of this graded ring has the form

$$\frac{h_0 + h_1 z + \cdots + h_d z^d}{(1-z)^{d+1}}.$$

What is being claimed in Theorem 1.20 is that the bidegree of $\mathcal{L}_X$ equals

$$B_X(p, u) = (h_0 u^d + h_1 p u^{d-1} + h_2 p^2 u^{d-2} + \cdots + h_d p^d) \cdot p^{n-d} \qquad (14)$$

Equivalently, this is the class of $\mathcal{L}_X$ in the cohomology ring

$$H^*(\mathbb{P}^n \times \mathbb{P}^n; \mathbb{Z}) = \mathbb{Z}[p, u]/\langle p^{n+1}, u^{n+1}\rangle.$$

There are several (purely combinatorial) definitions of the invariants h_i of the matroid M. For instance, they are coefficients of the following specialization of the *characteristic polynomial*:

$$\chi_M(q+1) = q \cdot \left(h_0 q^d - h_{d-1} q^{d-1} + \cdots + (-1)^{d-1} h_1 q + (-1)^d h_0\right). \qquad (15)$$

Theorem 1.20 was used in [Huh0] to prove a conjecture of Dawson, stating that the sequence $h_0, h_1, \ldots, h_d$ is log-concave, when M is representable over a field of characteristic zero.

The first assertion in Theorem 1.20 was proved by Varchenko in [Varchenko]. For definitions and characterizations of the characteristic polynomial χ, and many pointers to matroid basics, we refer to [OTBook]. A proof of the second assertion was given by Denham et al. in a slightly different setting [Denham-Garrousian-Schulze, Theorem 1]. We give a proof in Sect. 4 following [Huh1, §3]. The ramification locus of the likelihood fibration $\mathrm{pr}_2 : \mathcal{L}_X \to \mathbb{P}_u^n$ is known as the *entropic discriminant* [SSV].

Example 1.21. Let $d = 2$ and $n = 4$, so X is a plane in $\mathbb{P}^4$, defined by two linear forms

$$c_{10}p_0 + c_{11}p_1 + c_{12}p_2 + c_{13}p_3 + c_{14}p_4 = 0, \\ c_{20}p_0 + c_{21}p_1 + c_{22}p_2 + c_{23}p_3 + c_{24}p_4 = 0. \quad (16)$$

Following Theorem 1.20, we view $X \cap \mathcal{H}$ as an arrangement of five lines in the affine plane

$$\{p \in X : p_0 + p_1 + p_2 + p_3 + p_4 \neq 0\} \simeq \mathbb{C}^2.$$

Hence, for generic c_{ij}, the ML degree of X is equal to 6, the number of bounded regions of this arrangement. The condition $u \in p \star (X^\perp + 1)$ in Proposition 1.19 translates into

$$\mathrm{rank} \begin{bmatrix} u_0 & u_1 & u_2 & u_3 & u_4 \\ p_0 & p_1 & p_2 & p_3 & p_4 \\ c_{10}p_0 & c_{11}p_1 & c_{12}p_2 & c_{13}p_3 & c_{14}p_4 \\ c_{20}p_0 & c_{21}p_1 & c_{22}p_2 & c_{23}p_3 & c_{24}p_4 \end{bmatrix} \leq 3. \quad (17)$$

The 4×4-minors of this 4×5-matrix, together with the two linear forms defining X, form a system of equations that has six solutions in $\mathbb{P}^4$, for generic c_{ij}. All solutions have real coordinates. In fact, there is one solution in each bounded region of $X \backslash \mathcal{H}$. The likelihood correspondence $\mathcal{L}_X$ is the fourfold in $\mathbb{P}^4 \times \mathbb{P}^4$ given by the Eqs. (16) and (17).

We now illustrate the second statement in Theorem 1.20. Suppose that the real numbers c_{ij} are generic, so M is the uniform matroid of rank three on six elements. The Stanley–Reisner ring of the broken circuit complex of M equals

$$\mathbb{R}[x_1, x_2, x_3, x_4, x_5, x_6]/\langle x_2 x_3 x_4, x_2 x_3 x_5, x_2 x_3 x_6, \ldots, x_4 x_5 x_6 \rangle.$$

The Hilbert series of this graded algebra is

$$\frac{h_0 + h_1 z + h_2 z^2}{(1-z)^3} = \frac{1 + 3z + 6z^2}{(1-z)^3}.$$

Likelihood Geometry

We conclude that the bidegree (14) of the likelihood correspondence $\mathcal{L}_X$ equals

$$B_X(p, u) \;\; = \;\; 6p^4 + 3p^3 u + p^2 u^2.$$

For special choices of the coefficients c_{ij} in (16), some triples of lines in the arrangement $X \cap \mathcal{H}$ may meet in a point. For such matroids, the ML degree drops from 6 to some integer between 0 and 5. We recommend it as an exercise to the reader to explore these cases. For instance, can you find explicit c_{ij} so that the ML degree of X equals 3? What are the prime ideal and the bidegree of $\mathcal{L}_X$ in that case? How can the ML degree of X be 0 or 1? ◇

It would be interesting to know which statistical model X in $\mathbb{P}^n$ defines the likelihood correspondence $\mathcal{L}_X$ which is a complete intersection in $\mathbb{P}^n \times \mathbb{P}^n$. When X is a linear subspace of $\mathbb{P}^n$, this question is closely related to the concept of *freeness* of a hyperplane arrangement.

Proposition 1.22. *If the hyperplane arrangement $X \cap \mathcal{H}$ in X is free, then the likelihood correspondence $\mathcal{L}_X$ is an ideal-theoretic complete intersection in $\mathbb{P}^n \times \mathbb{P}^n$.*

Proof. For the definition of freeness see §1 in the paper [CDFV] by Cohen, Denman, Falk and Varchenko. The proposition is implied by their [CDFV, Theorem 2.13] and [CDFV, Corollary 3.8]. □

Using Theorem 1.20, this provides a likelihood geometry proof of Terao's theorem that the characteristic polynomial of a free arrangement factors into integral linear forms [Terao].

2 Second Lecture

In our newspaper we frequently read about studies aimed at proving that a behavior or food causes a certain medical condition. We begin the second lecture with an introduction to statistical issues arising in such studies. The "medical question" we wish to address is *Does Watching Soccer on TV Cause Hair Loss?* We learned this amusing example from [MSS, §1].

In a fictional study, 296 British subjects aged between 40 and 50 were interviewed about their hair length and how many hours per week they watch soccer (a.k.a. "football") on TV. Their responses are summarized in the following *contingency table* of format 3×3:

$$U \;\;=\;\; \begin{array}{c} \leq 2\text{h} \\ 2\text{-}6\text{h} \\ \geq 6\text{h} \end{array} \!\!\left(\begin{array}{ccc} \text{lots of hair} & \text{medium hair} & \text{little hair} \\ 51 & 45 & 33 \\ 28 & 30 & 29 \\ 15 & 27 & 38 \end{array} \right)$$

For instance, 29 respondents reported having little hair and watching between 2 and 6 h of soccer on TV per week. Based on these data, are these two random variables independent, or are we inclined to believe that watching soccer on TV and hair loss are correlated?

On first glance, the latter seems to be the case. Indeed, being independent means that the data matrix U should be close to a rank 1 matrix. However, all 2×2-minors of U are strictly positive, indeed by quite a margin, and this suggests a positive correlation.

However, this interpretation is deceptive. A much better explanation of our data can be given by identifying a certain hidden random variable. That hidden variable is *gender*. Indeed, suppose that among the respondents 126 were males and 170 were females. Our data matrix U is then the sum of the male table and the female table, maybe as follows:

$$U = \begin{pmatrix} 3 & 9 & 15 \\ 4 & 12 & 20 \\ 7 & 21 & 35 \end{pmatrix} + \begin{pmatrix} 48 & 36 & 18 \\ 24 & 18 & 9 \\ 8 & 6 & 3 \end{pmatrix}. \tag{18}$$

Both of these tables have rank 1, hence U has rank 2. Hence, the appropriate null hypothesis H_0 for analyzing our situation is not independence but it is *conditional independence*:

H_0 : *Soccer on TV and Hair Loss are Independent given Gender.*

And, based on the data U, we most definitely do not reject that null hypothesis.

The key feature of the matrix U above was that it has rank 2. We now define low rank matrix models in general. Consider two discrete random variables X and Y having m and n states respectively. Their *joint probability distribution* is written as an $m \times n$-matrix

$$P = \begin{pmatrix} p_{11} & p_{12} & \cdots & p_{1n} \\ p_{21} & p_{22} & \cdots & p_{2n} \\ \vdots & \vdots & \ddots & \vdots \\ p_{m1} & p_{m2} & \cdots & p_{mn} \end{pmatrix}$$

whose entries are nonnegative and sum to 1. Here p_{ij} represents the probability that X is in state i and Y is in state j. The of all probability distributions is the standard simplex Δ_{mn-1} of dimension $mn - 1$. We write $\mathcal{M}_r$ for the manifold of rank r matrices in Δ_{mn-1}.

The matrices P in $\mathcal{M}_1$ represent independent distributions. Mixtures of r independent distributions correspond to matrices in $\mathcal{M}_r$. As always in applied algebraic geometry, we can make any problem that involves semi-algebraic sets progressively easier by three steps:

Likelihood Geometry

- disregard inequalities,
- replace real numbers with complex numbers,
- replace affine space by projective space.

In our situation, this leads us to replacing $\mathcal{M}_r$ with its Zariski closure in complex projective space $\mathbb{P}^{mn-1}$. This Zariski closure is the projective variety $\mathcal{V}_r$ of complex $m \times n$ matrices of rank $\leq r$. Note that $\mathcal{V}_r$ is singular along $\mathcal{V}_{r-1}$. The codimension of $\mathcal{V}_r$ is $(m-r)(n-r)$. It is a non-trivial exercise to write the degree of $\mathcal{V}_r$ in terms of m, n, r. Hint: [ch3:MS, Example 15.2].

Suppose now that i.i.d. samples are drawn from an unknown joint distribution on our two random variables X and Y. We summarize the resulting data in a contingency table

$$U = \begin{pmatrix} u_{11} & u_{12} & \cdots & u_{1n} \\ u_{21} & u_{22} & \cdots & u_{2n} \\ \vdots & \vdots & \ddots & \vdots \\ u_{m1} & u_{m2} & \cdots & u_{mn} \end{pmatrix}.$$

The entries of the matrix U are nonnegative integers whose sum is u_{++}.

The *likelihood function* for the contingency table U is the following function on Δ_{mn-1}:

$$P \mapsto \binom{u_{++}}{u_{11} u_{12} \cdots u_{mn}} \prod_{i=1}^{m} \prod_{j=1}^{n} p_{ij}^{u_{ij}}.$$

Assuming fixed sample size, this is the likelihood of observing the data U given an unknown probability distribution P in Δ_{mn-1}. In what follows we suppress the multinomial coefficient. Furthermore, we regard the likelihood function as a rational function on $\mathbb{P}^{mn-1}$, so we write

$$\ell_U = \frac{\prod_{i=1}^{m} \prod_{j=1}^{n} p_{ij}^{u_{ij}}}{p_{++}^{u_{++}}}.$$

We wish to find a low rank probability matrix P that best explains the data U. Maximum likelihood estimation means solving the following optimization problem:

$$\text{Maximize } \ell_U(P) \text{ subject to } P \in \mathcal{M}_r. \tag{19}$$

The optimal solution $\hat{P}$ is a rank r matrix. This is the *maximum likelihood estimate* for U.

For $r = 1$, the independence model, the maximum likelihood estimate $\hat{P}$ is obtained from the data matrix U by the following formula, already seen for $m = n = 2$ in (8). Multiply the vector of row sums with the vector of column sums and divide by the sample size:

$$\hat{P} = \frac{1}{(u_{++})^2} \cdot \begin{pmatrix} u_{1+} \\ u_{2+} \\ \vdots \\ u_{m+} \end{pmatrix} \cdot (u_{+1}\ u_{+2}\ \cdots\ u_{+n}). \tag{20}$$

Statisticians, scientists and engineers refer to such a formula as an "analytic solution". In our view, it would be more appropriate to call this an "algebraic solution". After all, we are here using algebra not analysis. Our algebraic solution for $r = 1$ reveals the following points:

- The MLE $\hat{P}$ is a *rational function* of the data U.
- The function $U \mapsto \hat{P}$ is an algebraic function of degree 1.
- The ML degree of the independence model $\mathcal{V}_1$ equals 1.

We next discuss the smallest case when the ML degree is larger than 1.

Example 2.1. Let $m = n = 3$ and $r = 2$. Our MLE problem is to maximize

$$\ell_U = (p_{11}^{u_{11}} p_{12}^{u_{12}} p_{13}^{u_{13}} p_{21}^{u_{21}} p_{22}^{u_{22}} p_{23}^{u_{23}} p_{31}^{u_{31}} p_{32}^{u_{32}} p_{33}^{u_{33}})/p_{++}^{u_{++}}$$

subject to the constraints $P \geq 0$ and $\text{rank}(P) = 2$, where $P = (p_{ij})$ is a 3×3-matrix of unknowns. The equations that characterize the critical points of this optimization problem are

$$\det(P) = \begin{matrix} p_{11}p_{22}p_{33} - p_{11}p_{23}p_{32} - p_{12}p_{21}p_{33} \\ +p_{12}p_{23}p_{31} + p_{13}p_{21}p_{32} - p_{13}p_{22}p_{31} \end{matrix} = 0$$

and the vanishing of the 3×3-minors of the following 3×9-matrix:

$$\begin{bmatrix} u_{11} & u_{12} & u_{13} & u_{21} & u_{22} & u_{23} & u_{31} & u_{32} & u_{33} \\ p_{11} & p_{12} & p_{13} & p_{21} & p_{22} & p_{23} & p_{31} & p_{32} & p_{33} \\ p_{11}a_{11} & p_{12}a_{12} & p_{13}a_{13} & p_{21}a_{21} & p_{22}a_{22} & p_{33}a_{33} & p_{31}a_{31} & p_{32}a_{32} & p_{33}a_{33} \end{bmatrix}$$

where $a_{ij} = \frac{\partial \det(P)}{\partial p_{ij}}$ is the cofactor of p_{ij} in P. For random positive data u_{ij}, these equations have ten solutions with $\text{rank}(P) = 2$ in $\mathbb{P}^8 \backslash \mathcal{H}$. Hence the ML degree of $\mathcal{V}_2$ is 10. If we regard the u_{ij} as unknowns, then saturating the above determinantal equations with respect to $\mathcal{H} \cup \mathcal{V}_1$ yields the prime ideal of the likelihood correspondence $\mathcal{L}_{\mathcal{V}_2} \subset \mathbb{P}^8 \times \mathbb{P}^8$. See Example 4.8 for the bidegree and other enumerative invariants of the eight-dimensional variety $\mathcal{L}_{\mathcal{V}_2}$. ◇

Recall from Definition 1.5 that the *ML degree* of a statistical model (or a projective variety) is the number of critical points of the likelihood function for generic data.

Theorem 2.2. *The known values for the ML degrees of the determinantal varieties $\mathcal{V}_r$ are*

$(m,n) =$	(3,3)	(3,4)	(3,5)	(4,4)	(4,5)	(4,6)	(5,5)
$r = 1$	1	1	1	1	1	1	1
$r = 2$	10	26	58	**191**	**843**	**3119**	**6776**
$r = 3$	1	1	1	191	843	3119	**61326**
$r = 4$				1	1	1	6776
$r = 5$							1

The numbers 10 and 26 were computed back in 2004 using the symbolic software Singular, and they were reported in [HKS, §5]. The bold face numbers were found in 2012 in [HRS] using the numerical software Bertini. In what follows we shall describe some of the details.

Remark 2.3. Each determinantal variety $\mathcal{V}_r$ is singular along the smaller variety $\mathcal{V}_{r-1}$. Hence, the very affine variety $\mathcal{V}_r \backslash \mathcal{H}$ is singular for $r \geq 2$, so Theorem 1.7 does not apply. Here, $\mathcal{H} = \{p_{++} \prod p_{ij} = 0\}$. According to Conjecture 1.8, the ML degree above provides a lower bound for the signed topological Euler characteristic of $\mathcal{V}_r \backslash \mathcal{H}$. The difference between the two numbers reflect the nature of the singular locus $\mathcal{V}_{r-1} \backslash \mathcal{H}$ inside $\mathcal{V}_r \backslash \mathcal{H}$. For plane curves that have nodes and cusps, we encountered this issue in Examples 1.4 and 1.17.

We begin with a geometric description of the likelihood correspondence. An $m \times n$-matrix P is a regular point in $\mathcal{V}_r$ if and only if $\mathrm{rank}(P) = r$. The tangent space T_P is a subspace of dimension $rn + rm - r^2$ in $\mathbb{C}^{m \times n}$. Its orthogonal complement $T_P^\perp$ has dimension $(m-r)(n-r)$.

The partial derivatives of the log-likelihood function $\log(\ell_U)$ on $\mathbb{P}^{mn-1}$ are

$$\frac{\partial \log(\ell_U)}{\partial p_{ij}} = \frac{u_{ij}}{p_{ij}} - \frac{u_{++}}{p_{++}}.$$

Proposition 2.4. *An $m \times n$-matrix P of rank r is a critical point for $\log(\ell_U)$ on $\mathcal{V}_r$ if and only if the linear subspace $T_P^\perp$ contains the matrix*

$$\left[\frac{u_{ij}}{p_{ij}} - \frac{u_{++}}{p_{++}} \right]_{\substack{i=1,\ldots,m \\ j=1,\ldots,n}}$$

In order to get to the numbers in Theorem 2.2, the geometric formulation was replaced in [HRS] with a parametric representation of the rank constraints. The following linear algebra formulation worked well for non-trivial computations. Assume $m \leq n$. Let P_1, R_1, L_1 and Λ be matrices of unknowns of formats $r \times r$, $r \times (n-r)$, $(m-r) \times r$, and $(n-r) \times (m-r)$. Set

$$L = \begin{pmatrix} L_1 & -I_{m-r} \end{pmatrix}, \quad P = \begin{pmatrix} P_1 & P_1 R_1 \\ L_1 P_1 & L_1 P_1 R_1 \end{pmatrix}, \quad \text{and} \quad R = \begin{pmatrix} R_1 \\ -I_{n-r} \end{pmatrix},$$

where I_{m-r} and I_{n-r} are identity matrices. In the next statement we use the symbol $\star$ for the Hadamard (entrywise) product of two matrices that have the same format.

Proposition 2.5. *Fix a general $m \times n$ data matrix U. The polynomial system*

$$P \star (R \cdot \Lambda \cdot L)^T + u_{++} \cdot P = U$$

consists of mn equations in mn unknowns. For generic U, it has finitely many complex solutions (P_1, L_1, R_1, Λ). The $m\times n$-matrices P resulting from these solutions are precisely the critical points of the likelihood function ℓ_U on the determinantal variety $\mathcal{V}_r$.

We next present the analogue to Theorem 2.2 for symmetric matrices

$$P = \begin{pmatrix} 2p_{11} & p_{12} & p_{13} & \cdots & p_{1n} \\ p_{12} & 2p_{22} & p_{23} & \cdots & p_{2n} \\ p_{13} & p_{23} & 2p_{33} & \cdots & p_{3n} \\ \vdots & \vdots & \vdots & \ddots & \vdots \\ p_{1n} & p_{2n} & p_{3n} & \cdots & 2p_{nn} \end{pmatrix}.$$

Such matrices, with nonnegative coordinates p_{ij} that sum to 1, represent joint probability distributions for two identically distributed random variables with n states. The case $n = 2$ and $r = 1$ is the Hardy–Weinberg curve, which we discussed in detail in Example 1.3.

Theorem 2.6. *The known values for ML degrees of symmetric matrices of rank at most r (mixtures of r independent identically distributed random variables) are*

$n =$	2	3	4	5	6
$r = 1$	1	1	1	1	1
$r = 2$		1	6	37	270	2341
$r = 3$			1	37	1394	?
$r = 4$				1	270	?
$r = 5$					1	2341

At present we do not know the common value of the ML degree for $n = 6$ and $r = 3, 4$. In what follows we take a closer look at the model for symmetric 3×3-matrices of rank 2.

Example 2.7. Let $n = 3$ and $r = 2$, so X is a cubic hypersurface in $\mathbb{P}^5$. The likelihood correspondence $\mathcal{L}_X$ is a five-dimensional subvariety of $\mathbb{P}^5 \times \mathbb{P}^5$ having bidegree

$$B_X(p, u) = 6p^5 + 12p^4u + 15p^3u^2 + 12p^2u^3 + 3pu^4.$$

Likelihood Geometry

The bihomogeneous prime ideal of $\mathcal{L}_X$ is minimally generated by 23 polynomials, namely:

- One polynomial of bidegree $(3, 0)$; this is the determinant of P.
- Three polynomials of degree $(1, 1)$. These come from the underlying toric model $\{\mathrm{rank}(P) = 1\}$. As suggested in Proposition 3.5, they are the 2×2-minors of

$$\begin{pmatrix} 2p_0 + p_1 + p_2 & p_1 + 2p_3 + p_4 & p_2 + p_4 + 2p_5 \\ 2u_0 + u_1 + u_2 & u_1 + 2u_3 + u_4 & u_2 + u_4 + 2u_5 \end{pmatrix}.$$

- One polynomial of degree $(2, 1)$,
- three polynomial of degree $(2, 2)$,
- nine polynomials of degree $(3, 1)$,
- six polynomials of degree $(3, 2)$.

It turns out that this ideal represents an expression for the MLE $\hat{P}$ in terms of radicals in U.

We shall work this out for one numerical example. Consider the data matrix U with

$$u_{11} = 10,\ u_{12} = 9,\ u_{13} = 1,\ u_{22} = 21,\ u_{23} = 3,\ u_{33} = 7.$$

For this choice, all six critical points of the likelihood function are real and positive:

p_{11}	p_{12}	p_{13}	p_{22}	p_{23}	p_{33}	$\log \ell_U(p)$
0.1037	0.3623	0.0186	0.3179	0.0607	0.1368	−82.18102
0.1084	0.2092	0.1623	0.3997	0.0503	0.0702	−84.94446
0.0945	0.2554	0.1438	0.3781	0.4712	0.0810	−84.99184
0.1794	0.2152	0.0142	0.3052	0.2333	0.0528	−85.14678
0.1565	0.2627	0.0125	0.2887	0.2186	0.0609	−85.19415
0.1636	0.1517	0.1093	0.3629	0.1811	0.0312	−87.95759

The first three points are local maxima in Δ_5 and the last three points are local minima. These six points define an algebraic field extension of degree 6 over $\mathbb{Q}$. One might expect that the Galois group of these six points over $\mathbb{Q}$ is the full symmetric group S_6. If this were the case then the above coordinates could not be written in radicals. However, that expectation is wrong. The Galois group of the likelihood fibration $\mathrm{pr}_2 : \mathcal{L}_X \to \mathbb{P}_U^5$ given by the 3×3 symmetric problem is a subgroup of S_6 isomorphic to the solvable group S_4.

To be concrete, for the data above, the minimal polynomial for the MLE $\hat{p}_{33}$ equals

$$9528773052286944 p_{33}^6 - 4125267629399052 p_{33}^5 + 713452955656677 p_{33}^4$$
$$- 63349419858182 p_{33}^3 + 3049564842009 p_{33}^2 - 75369770028 p_{33}$$
$$+ 744139872 = 0.$$

We solve this equation in radicals as follows:

$$p_{33} = \frac{16427}{227664} + \frac{1}{12}(\zeta - \zeta^2)\omega_2 - \frac{66004846384302}{19221271018849}\omega_2^2 + \left(\frac{14779904193}{211433981207339}\zeta^2 - \frac{14779904193}{211433981207339}\zeta\right)\omega_1\omega_2^2 + \frac{1}{2}\omega_3,$$

where ζ is a primitive third root of unity, $\omega_1^2 = 94834811/3$, and

$$\omega_2^3 = \left(\frac{5992589425361}{150972770845322208}\zeta - \frac{5992589425361}{150972770845322208}\zeta^2\right) + \frac{97163}{40083040181952}\omega_1,$$

$$\omega_3^2 = \frac{5006721709}{1248260766912} + \left(\frac{212309132509}{4242035935404}\zeta - \frac{212309132509}{4242035935404}\zeta^2\right)\omega_2 - \frac{2409}{20272573168}\omega_1\omega_2$$
$$- \frac{158808750548335}{76885084075396}\omega_2^2 + \left(\frac{17063004159}{422867962414678}\zeta^2 - \frac{17063004159}{422867962414678}\zeta\right)\omega_1\omega_2^2.$$

The explanation for the extra symmetry stems from the duality theorem below. It furnishes an involution on the set of six critical points that allows us to express them in radicals. ◇

The tables in Theorems 2.2 and 2.6 suggest that the columns will always be symmetric. This fact was conjectured in [HRS] and subsequently proved by Draisma and Rodriguez in [DR].

Theorem 2.8. *Fix $m \leq n$ and consider the determinantal varieties $\mathcal{V}_i$ for either general or symmetric matrices. Then the ML degrees for rank r and for rank $m-r+1$ coincide.*

In fact, the main result in [DR] establishes the following more precise statement. Given a data matrix U of format $m \times n$, we write Ω_U for the $m \times n$-matrix whose (i, j) entry equals

$$\frac{u_{ij} \cdot u_{i+} \cdot u_{+j}}{(u_{++})^3}.$$

Theorem 2.9. *Fix $m \leq n$ and U an $m \times n$-matrix with strictly positive integer entries. There exists a bijection between the complex critical points $P_1, P_2, \ldots, P_s$ of the likelihood function ℓ_U on $\mathcal{V}_r$ and the complex critical points $Q_1, Q_2, \ldots, Q_s$ of ℓ_U on $\mathcal{V}_{m-r+1}$ such that*

$$P_1 \star Q_1 = P_2 \star Q_2 = \cdots = P_s \star Q_s = \Omega_U.$$

Thus, this bijection preserves reality, positivity, and rationality.

The key to computing the ML degree tables and to formulating the duality conjectures in [HRS], was the use of numerical algebraic geometry. The software Bertini allowed for the computation of thousands of instances in which the formula of Theorem 2.9 was confirmed.

Bertini is numerical software, based on homotopy continuation, for finding all complex solutions to a system of polynomial equations (and much more). The software is available at [Bertini]. The developers, Daniel Bates, Jonathan

Hauenstein, Andrew Sommese, Charles Wampler, have just completed a new textbook [BHSW] on the mathematics behind Bertini.

For the past two decades, algebraic geometers have increasingly employed computational methods as a tool for their research. However, these computations have almost always been symbolic (and hence exact). They relied on Gröbner-based software such as Singular or Macaulay2. Algebraists often feel a certain discomfort when asked to trust a numerical computation. We encourage discussion about this issue, by raising the following question.

Example 2.10. In the rightmost column of Theorem 2.6, it is asserted that the solution to a certain enumerative geometry problem is **2341**. Which of these would **you** trust most:

- the output of a symbolic computation?
- the output of a numerical computation?
- a proof written by an algebraic geometer?

In the authors' view, it always pays off to be critical and double-check all computations, regardless of how they were carried out. And, this applies to all three of the above. ◇

One of the big advantages of numerical algebraic geometry over Gröbner bases when it comes to MLE is the separation between *Preprocessing* and *Solving*. For any particular variety $X \subset \mathbb{P}^n$, such as $X = \mathcal{V}_r$, we preprocess by solving the likelihood equations once, for a generic data set U_0 chosen by us. The coordinates of U_0 may be complex (rather than real) numbers. We can chose them with stable numerics in mind, so as to compute all critical points up to high accuracy. This step can take a long time, but the output is highly reliable.

After solving the equations once, for that generic U_0, all subsequent computations for any other data set U are very fast. In particular, the computation is fully parallelizable. If we have m processors at our disposal, where $m = $ MLdegree(X), then each processor can track one of the paths. To be precise, homotopy continuation starts from the critical points of ℓ_{U_0} and transform them into the critical points of ℓ_U. Geometrically speaking, for fixed X, the homotopy amounts to walking on the sheets of the likelihood fibration pr$_2 : \mathcal{L}_X \to \mathbb{P}^n_u$.

To illustrate this point, here are the timings (in seconds) that were reported in [HRS] for the determinantal varieties $X = \mathcal{V}_r$. Those computations were carried out in Bertini on a 64-bit Linux cluster with 160 processors. The first row is the preprocessing time for solving the equations once. The second row is the time needed to solve any subsequent instance:

(m,n,r)	(4, 4, 2)	(4, 4, 3)	(4, 5, 2)	(4, 5, 3)	(5, 5, 2)	(5, 5, 4)
Preprocessing	257	427	1938	2902	348555	146952
Solving	4	4	20	20	83	83

This table suggests that combining numerical algebraic geometry with existing tools from computational statistics might lead to a viable tool for certifiably solving MLE problems.

We are now at the point where it is essential to offer a disclaimer. The low rank model $\mathcal{M}_r$ does not correctly represent the notion of conditional independence. The model we should have used instead is the *mixture model* Mix$_r$. By definition, Mix$_r$ is the set of probability distributions P in Δ_{mn-1} that are convex combinations of r independent distributions, each taken from $\mathcal{M}_1$. Equivalently, the mixture model Mix$_r$ consists of all matrices

$$P = A \cdot \Lambda \cdot B, \tag{21}$$

where A is a nonnegative $m \times r$-matrix whose rows sum to 1, Λ is a nonnegative $r \times r$ diagonal matrix whose entries sum to 1, and B is a nonnegative $r \times n$-matrix whose columns sum to 1. The formula (21) expresses Mix$_r$ as the image of a trilinear map between polytopes:

$$\phi : (\Delta_{m-1})^r \times \Delta_{r-1} \times (\Delta_{n-1})^r \to \Delta_{mn-1}, \quad (A, \Lambda, B) \mapsto P.$$

The following result is well-known; see e.g. [LiAS, Example 4.1.2].

Proposition 2.11. *Our low rank model $\mathcal{M}_r$ is the Zariski closure of the mixture model* Mix$_r$ *in the probability simplex Δ_{mn-1}. If $r \leq 2$ then* Mix$_r = \mathcal{M}_r$. *If $r \geq 3$ then* Mix$_r \subsetneq \mathcal{M}_r$.

The point here is the distinction between the rank and the nonnegative rank of a nonnegative matrix. Matrices in $\mathcal{M}_r$ have rank $\leq r$ and matrices in Mix$_r$ have nonnegative rank $\leq r$. Thus elements of $\mathcal{M}_r \backslash$ Mix$_r$ are matrices whose nonnegative rank exceeds its rank.

Example 2.12. The following 4×4-matrix has rank 3 but nonnegative rank 4:

$$P = \frac{1}{8} \cdot \begin{pmatrix} 1 & 1 & 0 & 0 \\ 0 & 1 & 1 & 0 \\ 0 & 0 & 1 & 1 \\ 1 & 0 & 0 & 1 \end{pmatrix}$$

This is the slack matrix of a regular square. It is an element of $\mathcal{M}_3 \backslash$ Mix$_3$. ◇

Engineers and scientists care more about Mix$_r$ than $\mathcal{M}_r$. In many applications, nonnegative rank is more relevant than rank. The reason can be seen in (18). In such a low-rank decomposition, we do not want the female table or the male table to have a negative entry.

This raises the following important questions: How to maximize the likelihood function ℓ_U over Mix$_r$? What are the algebraic degrees associated with that optimization problem?

Statisticians seek to maximize the likelihood function ℓ_U on Mix_r by using the *expectation-maximization* (EM) algorithm in the space $(\Delta_{m-1})^r \times \Delta_{r-1} \times (\Delta_{n-1})^r$ of parameters (A, Λ, B). In each iteration, the EM algorithm strictly decreases the *Kullback–Leibler divergence* from the current model point $P = \phi(A, \Lambda, B)$ to the empirical distribution $\frac{1}{u_{++}} \cdot U$. The hope in running the EM algorithm for given data U is that it converges to the global maximum $\hat{P}$ on Mix_r. For a presentation of the EM algorithm for discrete algebraic models see [PS, §1.3]. A study of the geometry of this algorithm for the mixture model Mix_r is undertaken in [KRS].

If the EM algorithm converges to a point that lies in the interior of the parameter polytope, and is non-singular with respect to ϕ, then that point will be among the critical points on $\mathcal{M}_r$. These are characterized by Proposition 2.4. However, since Mix_r is properly contained in $\mathcal{M}_r$, it frequently happens that the true MLE $\hat{P}$ lies on the boundary of Mix_r. In that case, $\hat{P}$ is not a critical point of ℓ_U on $\mathcal{M}_r$, meaning that $(\hat{P}, U)$ is not in the likelihood correspondence on $\mathcal{V}_r$. Such points will never be found by the method described above.

In order to address this issue, we need to identify the divisors in the variety $\mathcal{V}_r \subset \mathbb{P}^{mn-1}$ that appear in the algebraic boundary of Mix_r. By this we mean the irreducible components $W_1, W_2, \ldots, W_s$ of the Zariski closure of ∂Mix_r. Each of these W_i has codimension 1 in $\mathcal{V}_r$. Once the W_i are identified, one would need to examine their ML degree, and also the ML degree of the various strata $W_{i_1} \cap \cdots \cap W_{i_s}$ in which ℓ_U might attain its maximum. At present we do not have this information even in the smallest non-trivial case $m = n = 4$ and $r = 3$.

Example 2.13. We illustrate this issue by describing one of the components W of the algebraic boundary for the mixture model Mix_3 when $m = n = 4$. Consider the equation

$$\begin{pmatrix} p_{11} & p_{12} & p_{13} & p_{14} \\ p_{21} & p_{22} & p_{23} & p_{24} \\ p_{31} & p_{32} & p_{33} & p_{34} \\ p_{41} & p_{42} & p_{43} & p_{44} \end{pmatrix} = \begin{pmatrix} 0 & a_{12} & a_{13} \\ 0 & a_{22} & a_{23} \\ a_{31} & 0 & a_{33} \\ a_{41} & a_{42} & 0 \end{pmatrix} \cdot \begin{pmatrix} 0 & b_{12} & b_{13} & b_{14} \\ b_{21} & 0 & b_{23} & b_{24} \\ b_{31} & b_{32} & b_{33} & 0 \end{pmatrix}$$

This parametrizes a 13-dimensional subvariety W of the hypersurface $\mathcal{V}_3 = \{\det(P) = 0\}$ in $\mathbb{P}^{15}$. The variety W is a component in the algebraic boundary of Mix_3. To see this, we choose the a_{ij} and b_{ij} to be positive, and we note that P lies outside Mix_3 when precisely one of the 0 entries gets replaced by $-\epsilon$. The prime ideal of W in $\mathbb{Q}[p_{11}, \ldots, p_{44}]$ is obtained by eliminating the 17 unknowns a_{ij} and b_{ij} from the 16 scalar equations. A direct computation with `Macaulay 2` shows that the variety W is Cohen–Macaulay of codimension-2. By the Hilbert–Burch Theorem, it is defined by the 4×4-minors of the 4×5-matrix. This following specific matrix representation was suggested to us by Aldo Conca and Matteo Varbaro:

$$\begin{pmatrix} p_{11} & p_{12} & p_{13} & p_{14} & 0 \\ p_{21} & p_{22} & p_{23} & p_{24} & 0 \\ p_{31} & p_{32} & p_{33} & p_{34} & p_{34}(p_{11}p_{22} - p_{12}p_{21}) \\ p_{41} & p_{42} & p_{43} & p_{44} & p_{41}(p_{12}p_{24} - p_{14}p_{22}) + p_{44}(p_{11}p_{22} - p_{12}p_{21}) \end{pmatrix}.$$

Tte algebraic boundary of Mix_3 consists of precisely 304 irreducible components, namely the 16 coordinate hyperplanes and 288 hypersurfaces that are all isomorphic to W. This is proved in [KRS]. In that paper, it is also shown that the ML degree of W equals 633. ◇

The definition of rank varieties and mixture models extends to m-dimensional tensors P of arbitrary format $d_1 \times d_2 \times \cdots \times d_m$. We refer to Landsberg's book [Land] for an introduction to tensors and their rank. Now, $\mathcal{V}_r$ is the variety of tensors of borderrank $\leq r$, the model $\mathcal{M}_r$ is the set of all probability distributions in $\mathcal{V}_r$, and the model Mix_r is the subset of tensors of nonnegative rank $\leq r$. Unlike in the matrix case $m = 2$, the mixture model for borderrank $r = 2$ is already quite interesting when $m \geq 3$. We state two theorems that characterize our objects. The set-theoretic version of Theorem 2.14 is due to Landsberg and Manivel [LM]. The ideal-theoretic statement was proved more recently by Raicu [Rai].

Theorem 2.14. *The variety $\mathcal{V}_2$ is defined by the 3×3-minors of all flattenings of P.*

Here, *flattening* means picking any subset A of $[n] = \{1, 2, \ldots, n\}$ with $1 \leq |A| \leq n - 1$ and writing the tensor P as an ordinary matrix with $\prod_{i \in A} d_i$ rows and $\prod_{j \notin A} d_j$ columns.

Theorem 2.15. *The mixture model Mix_2 is the subset of supermodular distributions in $\mathcal{M}_2$.*

This theorem was proved in [ARSZ]. Being *supermodular* means that P satisfies a natural family of quadratic binomial inequalities. We explain these for $m = 3, d_1 = d_2 = d_3 = 2$.

Example 2.16. We consider $2 \times 2 \times 2$ tensors. Since secant lines of the Segre variety $\mathbb{P}^1 \times \mathbb{P}^1 \times \mathbb{P}^1$ fill all of $\mathbb{P}^7$, we have that $\mathcal{V}_2 = \mathbb{P}^7$ and $\mathcal{M}_2 = \Delta_7$. The mixture model Mix_2 is an interesting, full-dimensional, closed, semi-algebraic subset of Δ_7. By definition, Mix_2 is the image of a 2-to-1 map $\phi : (\Delta_1)^7 \to \Delta_7$ analogous to (21). The branch locus is the $2 \times 2 \times 2$-hyperdeterminant, which is a hypersurface in $\mathbb{P}^7$ of degree 4 and ML degree 13.

The analysis in [ARSZ, §2] represents the model Mix_2 as the union of four *toric cells*. One of these toric cells is the set of tensors satisfying

$$\begin{array}{lll} p_{111}p_{222} \geq p_{112}p_{221} & p_{111}p_{222} \geq p_{121}p_{212} & p_{111}p_{222} \geq p_{211}p_{122} \\ p_{112}p_{222} \geq p_{122}p_{212} & p_{121}p_{222} \geq p_{122}p_{221} & p_{211}p_{222} \geq p_{212}p_{221} \\ p_{111}p_{122} \geq p_{112}p_{121} & p_{111}p_{212} \geq p_{112}p_{211} & p_{111}p_{221} \geq p_{121}p_{211} \end{array} \quad (22)$$

A nonnegative $2 \times 2 \times 2$-tensor P in Δ_7 is *supermodular* if it satisfies these inequalities, possibly after label swapping $1 \leftrightarrow 2$. We visualize Mix_2 by restricting to the three-dimensional subspace H given by $p_{111} = p_{222}, p_{112} = p_{221}, p_{121} = p_{212}$ and $p_{211} = p_{122}$. The intersection $H \cap \Delta_7$ is a tetrahedron, and we consider $H \cap \text{Mix}_2$ inside that tetrahedron. The restricted model $H \cap \text{Mix}_2$ is shown on the left in Fig. 1. It consists of four toric cells as shown on the right side. The boundary

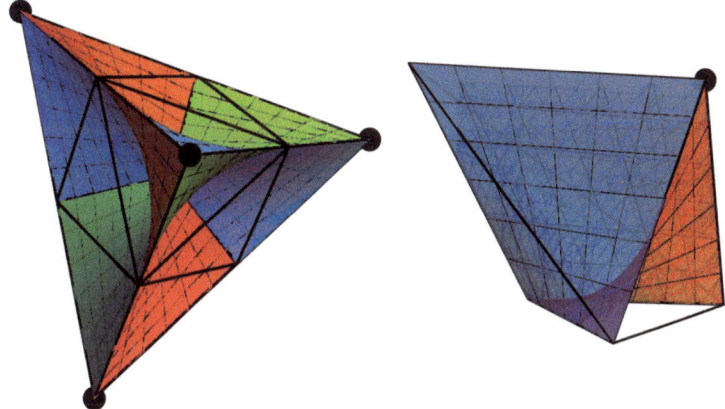

Fig. 1 A three-dimensional slice of the seven-dimensional model of 2×2×2 tensors of nonnegative rank ≤ 2. Each toric cell is bounded by 3 quadrics and contains a vertex of the tetrahedron

is given by three quadratic surfaces, shown in red, green and blue, and which are obtained from either the first or the second row in (22) by restriction to H.

The boundary analysis suggested in Example 2.13 turns out to be quite simple in the present example. All boundary strata of the model Mix_2 are varieties of ML degree 1.

One such boundary stratum for Mix_2 is the five-dimensional toric variety

$$X \;=\; V(p_{112}p_{222} - p_{122}p_{212},\, p_{111}p_{122} - p_{112}p_{121},\, p_{111}p_{222} - p_{121}p_{212}) \;\subset\; \mathbb{P}^7.$$

As a preview for what is to come, we report its ML bidegree and its sectional ML degree:

$$\begin{aligned}
B_X(p,u) &= p^7 + 2p^6u + 3p^5u^2 + 3p^4u^3 + 3p^3u^4 + 3p^2u^5, \\
S_X(p,u) &= p^7 + 14p^6u + 30p^5u^2 + 30p^4u^3 + 15p^3u^4 + 3p^2u^5.
\end{aligned} \qquad (23)$$

In the next section, we shall study the class of toric varieties and the class of varieties having ML degree 1. Our variety X lies in the intersection of these two important classes. ◇

3 Third Lecture

In our third lecture we start out with the likelihood geometry of embedded toric varieties. Fix a $(d+1) \times (n+1)$ integer matrix $A = (a_0, a_1, \ldots, a_n)$ of rank $d+1$ that has $(1, 1, \ldots, 1)$ as its last row. This matrix defines an effective action of the torus $(\mathbb{C}^*)^d$ on projective space $\mathbb{P}^n$:

$$(\mathbb{C}^*)^d \times \mathbb{P}^n \longrightarrow \mathbb{P}^n, \qquad t \times (p_0 : p_1 : \cdots : p_n) \longmapsto (t^{\tilde{a}_0} \cdot p_0 : t^{\tilde{a}_1} \cdot p_1 : \cdots : t^{\tilde{a}_n} \cdot p_n).$$

Here $\tilde{a}_i$ is the column vector a_i with the last entry 1 removed. We also fix

$$c = (c_0, c_1, \ldots, c_n) \in (\mathbb{C}^*)^{n+1},$$

viewed as a point in $\mathbb{P}^n$. Let X_c be the closure in $\mathbb{P}^n$ of the orbit $(\mathbb{C}^*)^d \cdot c$. This is a projective toric variety of dimension d, defined by the pair (A, c). The ideal that defines X_c is the familiar *toric ideal* I_A as in [LiAS, §1.3], but with $p = (p_0, \ldots, p_n)$ replaced by

$$p/c = \left(\frac{p_0}{c_0}, \frac{p_1}{c_1}, \ldots, \frac{p_n}{c_n} \right). \tag{24}$$

Example 3.1. Fix $d = 2$ and $n = 3$. The matrix

$$A = \begin{pmatrix} 0 & 3 & 0 & 1 \\ 0 & 0 & 3 & 1 \\ 1 & 1 & 1 & 1 \end{pmatrix}$$

specifies the following family of toric surfaces of degree three in $\mathbb{P}^3$:

$$X_c = \overline{\{(c_0 : c_1 x_1^3 : c_2 x_2^3 : c_3 x_1 x_2) : (x_1, x_2) \in (\mathbb{C}^*)^2\}} = V(c_3^3 \cdot p_0 p_1 p_2 - c_0 c_1 c_2 \cdot p_3^3).$$

Of course, the prime ideal of any particular surface X_c is the principal ideal generated by

$$\frac{p_0 \, p_1 \, p_2}{c_0 \, c_1 \, c_2} - \left(\frac{p_3}{c_3} \right)^3.$$

How does the ML degree of X_c depend on the parameter $c = (c_0, c_1, c_2, c_3) \in (\mathbb{C}^*)^4$? ◇

We shall express the ML degree of the toric variety X_c in terms of the complement of a hypersurface in the torus $(\mathbb{C}^*)^d$. The pair (A, c) define the sparse Laurent polynomial

$$f(x) = c_0 \cdot x^{\tilde{a}_0} + c_1 \cdot x^{\tilde{a}_1} + \cdots + c_n \cdot x^{\tilde{a}_n}.$$

Theorem 3.2. *The ML degree of the d-dimensional toric variety $X_c \subset \mathbb{P}^n$ is equal to $(-1)^d$ times the Euler characteristic of the very affine variety*

$$X_c \backslash \mathcal{H} \;\simeq\; \{x \in (\mathbb{C}^*)^d \,:\, f(x) \neq 0\}. \tag{25}$$

For generic c, the ML degree agrees with the degree of X_c, which is the normalized volume of the d-dimensional lattice polytope $\mathrm{conv}(A)$ obtained as the convex hull of the columns of A.

Proof. We first argue that the identification (25) holds. The map

$$x \longmapsto p = (c_0 \cdot x^{\tilde{a}_0} : c_1 \cdot x^{\tilde{a}_1} : \cdots : c_n \cdot x^{\tilde{a}_n})$$

defines an injective group homomorphism from $(\mathbb{C}^*)^d$ into the dense torus of $\mathbb{P}^n$. Its image is equal to the dense torus of X_c, so we have an isomorphism between $(\mathbb{C}^*)^d$ and the dense torus of X_c. Under this isomorphism, the affine open set $\{f \neq 0\}$ in $(\mathbb{C}^*)^d$ is identified with the affine open set $\{p_0 + \cdots + p_n \neq 0\}$ in the dense torus of X_c. The latter is precisely $X_c \backslash \mathcal{H}$. Since $(\mathbb{C}^*)^d$ is smooth, we see that $X_c \backslash \mathcal{H}$ is smooth, so our first assertion follows from Theorem 1.7. The second assertion is a consequence of the description of the likelihood correspondence $\mathcal{L}_{X_c}$ via linear sections of X_c that is given in Proposition 3.5 below. □

Example 3.3. We return to the cubic surface X_c in Example 3.1. For a general parameter vector c, the ML degree of X_c is 3. For instance, the surface $V(p_0 p_1 p_2 - p_3^3) \subset \mathbb{P}^3$ has ML degree 3. However, the ML degree of X_c drops to 2 whenever the plane curve defined by

$$f(x_1, x_2) = c_0 + c_1 x_1^3 + c_2 x_2^3 + c_3 x_1 x_2$$

has a singularity in $(\mathbb{C}^*)^2$. For instance, this happens for $c = (1 : 1 : 1 : -3)$. The corresponding surface $V(27 p_0 p_1 p_2 + p_3^3) \subset \mathbb{P}^3$ has ML degree 2. ◇

The isomorphism (25) has a nice interpretation in terms of Convex Optimization. Namely, it implies that maximum likelihood estimation for toric varieties is equivalent to global minimization of *posynomials*, and hence to the most fundamental case of *Geometric Programming*. We refer to [BoydVan, §4.5] for an introduction to posynomials and geometric programming.

We write $|\cdot|$ for the one-norm on $\mathbb{R}^{n+1}$, we set $b = Au$, and we assume that $c = (c_0, c_1, \ldots, c_n)$ is in $\mathbb{R}_{>0}^{n+1}$. Maximum likelihood estimation for toric models is the problem

$$\text{Maximize } \frac{p^u}{|p|^{|u|}} \text{ subject to } p \in X_c \cap \Delta_n. \tag{26}$$

Setting $p_i = c_i \cdot x^{\tilde{a}_i}$ as above, this problem becomes equivalent to the geometric program

$$\text{Minimize } \frac{f(x)^{|u|}}{x^b} \text{ subject to } x \in \mathbb{R}_{>0}^d. \tag{27}$$

By construction, $f(x)^{|u|}/x^b$ is a posynomial whose Newton polytope contains the origin. Such a posynomial attains a unique global minimum on the open orthant $\mathbb{R}_{>0}^d$. This can be seen by convexifying as in [BoydVan, §4.5.3]. This global minimum of (27) corresponds to the solution of (26), which exists and is unique by Birch's Theorem [PS, Theorem 1.10].

Example 3.4. Consider the geometric program for the surfaces in Example 3.1, with

$$A = \begin{pmatrix} 0 & 3 & 0 & 1 \\ 0 & 0 & 3 & 1 \\ 1 & 1 & 1 & 1 \end{pmatrix} \quad \text{and} \quad u = (0, 0, 0, 1).$$

The problem (27) is to find the global minimum, over all positive $x = (x_1, x_2)$, of the function

$$\frac{f(x_1, x_2)}{x_1 x_2} = c_0 x_1^{-1} x_2^{-1} + c_1 x_1^2 x_2^{-1} + c_2 x_1^{-1} x_2^2 + c_3.$$

This is equivalent to maximizing p_3/p_+ subject to $p \in V(c_3^3 \cdot p_0 p_1 p_2 - c_0 c_1 c_2 \cdot p_3^3) \cap \Delta_3$. ◇

We now describe the toric likelihood correspondence $\mathcal{L}_{X_c}$ in $\mathbb{P}^n \times \mathbb{P}^n$ associated with the pair (A, c). This is the likelihood correspondence of the toric variety $X_c \subset \mathbb{P}^n$ defined above.

Proposition 3.5. *On the open subset $(X_c \backslash \mathcal{H}) \times \mathbb{P}^n$, the toric likelihood correspondence $\mathcal{L}_{X_c}$ is defined by the 2×2-minors of the $2 \times (d+1)$-matrix*

$$\begin{pmatrix} p/c \cdot A^T \\ u/c \cdot A^T \end{pmatrix}. \tag{28}$$

Here the notation p/c is as in (24). In particular, for any fixed data vector u, the critical points of ℓ_u are characterized by a linear system of equations in p restricted to X_c.

Proof. This is an immediate consequence of Birch's Theorem [PS, Theorem 1.10]. □

Example 3.6. The *Hardy–Weinberg curve* of Example 1.3 is the subvariety $X_c = V(p_1^2 - 4p_0 p_2)$ in the projective plane $\mathbb{P}^2$. As a toric variety, this plane curve is given by

$$A = \begin{pmatrix} 0 & 1 & 2 \\ 2 & 1 & 0 \end{pmatrix} \quad \text{and} \quad c = (1, 2, 1).$$

The likelihood correspondence of X_c is the surface in $\mathbb{P}^2 \times \mathbb{P}^2$ given by

$$\det \begin{pmatrix} 2p_0 & p_1 \\ p_1 & 2p_2 \end{pmatrix} = \det \begin{pmatrix} p_1 + 2p_2 & 2p_0 + p_1 \\ u_1 + 2u_2 & 2u_0 + u_1 \end{pmatrix} = 0. \tag{29}$$

Note that the second determinant equals the determinant of the 2×2-matrix (28) times 4. Saturating (29) with respect to $p_0 + p_1 + p_2$ reveals two further equations of degree (1, 1):

Likelihood Geometry

$$2(u_1 + 2u_2)p_0 = (2u_0 + u_1)p_1 \quad \text{and} \quad (u_1 + 2u_2)p_1 = 2(2u_0 + u_1)p_2.$$

For fixed u, these equations have a unique solution in $\mathbb{P}^2$, given by the formula in (3). ◇

Toric varieties are rational varieties that are parametrized by monomials. We now examine those varieties that are parametrized by generic polynomials. Understanding these is useful for statistics since many widely used models for discrete data are given in the form

$$f : \Theta \to \Delta_n,$$

where Θ is a d-dimensional polytope and f is a polynomial map. The coordinates $f_0, f_1, \ldots, f_n$ are polynomial functions in the parameters $\theta = (\theta_1, \ldots, \theta_d)$ satisfying $f_0 + f_1 + \cdots + f_n = 1$. Such models include the mixture models in Proposition 2.11, phylogenetic models, Bayesian networks, hidden Markov models, and many others arising in computational biology [PS].

The model specified by the polynomials $f_0, \ldots, f_n$ is the semialgebraic set $f(\Theta) \subset \Delta_n$. We study its Zariski closure $X = \overline{f(\Theta)}$ in $\mathbb{P}^n$. Finding its equations is hard and interesting.

Theorem 3.7. *Let $f_0, f_1, \ldots, f_n$ be polynomials of degrees $b_0, b_1, \ldots, b_n$ satisfying $\sum f_i = 1$. The ML degree of the variety X is at most the coefficient of z^d in the generating function*

$$\frac{(1-z)^d}{(1-zb_0)(1-zb_1)\cdots(1-zb_n)}.$$

Equality holds when the coefficients of $f_0, f_1, \ldots, f_n$ are generic relative to $\sum f_i = 1$.

Proof. This is the content of [CHKS, Theorem 1]. □

Example 3.8. We examine the case of quartic surfaces in $\mathbb{P}^3$. Let $d = 2, n = 3$, pick random affine quadrics f_1, f_2, f_3 in two unknowns and set $f_0 = 1 - f_1 - f_2 - f_3$. This defines a map

$$f : \mathbb{C}^2 \to \mathbb{C}^3 \subset \mathbb{P}^3.$$

The ML degree of the image surface $X = \overline{f(\mathbb{C}^2)}$ in $\mathbb{P}^3$ is equal to 25 since

$$\frac{(1-z)^2}{(1-2z)^4} = 1 + 6z + \mathbf{25}z^2 + 88z^3 + \cdots$$

The rational surface X is a Steiner surface (or Roman surface). Its singular locus consists of three lines that meet in a point P. To understand the graph of f, we observe that the linear span of $\{f_0, f_1, f_2, f_3\}$ in $\mathbb{C}[x,y]$ has a basis $\{1, L^2, M^2, N^2\}$ where L, M, N represent lines in $\mathbb{C}^2$. Let l denote the line through $M \cap N$ parallel to L, m the line through $L \cap N$ parallel to M, and n the line through $L \cap M$ parallel to N. The map $\mathbb{C}^2 \to X$ is a bijection outside these three lines, and it maps each line 2-to-1 onto one of the lines in X_{sing}. The fiber over the special point P on X consists of three points, namely, $l \cap m$, $l \cap n$ and $m \cap n$. If the quadric f_0 were also picked at random, rather than as $1 - f_1 - f_2 - f_3$, then we would still get a Steiner surface $X \subset \mathbb{P}^3$. However, now the ML degree of X increases to 33.

On the other hand, if we take X to be a general quartic surface in $\mathbb{P}^3$, so X is a smooth K3 surface of Picard rank 1, then X has ML degree 84. This is the formula in Example 1.11 evaluated at $n = 3$ and $d = 4$. Here $X \backslash \mathcal{H}$ is the generic quartic surface in $\mathbb{P}^3$ with five plane sections removed. The number 84 is the Euler characteristic of that open K3 surface.

In the first case, $X \backslash \mathcal{H}$ is singular, so we cannot apply Theorem 1.7 directly to our Steiner surface X in $\mathbb{P}^3$. However, we can work in the parameter space and consider the smooth very affine surface $\mathbb{C}^2 \backslash V(f_0 f_1 f_2 f_3)$. The number 25 is the Euler characteristic of that surface.

It is instructive to verify Conjecture 1.8 for our three quartic surfaces in $\mathbb{P}^3$. We found

$$\chi(X \backslash \mathcal{H}) = 38 > 25 = \text{MLdegree}(X),$$
$$\chi(X \backslash \mathcal{H}) = 49 > 33 = \text{MLdegree}(X),$$
$$\chi(X \backslash \mathcal{H}) = 84 = 84 = \text{MLdegree}(X).$$

The Euler characteristics of the three surfaces were computed using Aluffi's method [AluJSC]. ◇

We now turn to the following question: *which projective varieties X have ML degree one?* This question is important for likelihood inference because a model having ML degree one means that the MLE $\hat{p}$ is a rational function in the data u. It is known that Bayesian networks and decomposable graphical models enjoy this property, and it is natural to wonder which other statistical models are in this class. The answer to this question was given by the first author in [Huh2]. We shall here present the result of [Huh2] from a slightly different angle.

Our point of departure is the notion of the *A-discriminant*, as introduced and studied by Gel'fand, Kapranov and Zelevinsky in [GKZ]. We fix an $r \times m$ integer matrix $A = (a_1, a_2, \ldots, a_m)$ of rank r which has $(1, 1, \ldots, 1)$ in its row space. The Zariski closure of

$$\{(t^{a_1} : t^{a_2} : \cdots : t^{a_m}) \in \mathbb{P}^{m-1} \; : \; t \in (\mathbb{C}^*)^r\}$$

is an $(r-1)$-dimensional toric variety Y_A in $\mathbb{P}^{m-1}$. We here intentionally changed the notation relative to that used for toric varieties at the beginning of this section. The reason is that d and n are always reserved for the dimension and embedding dimension of a statistical model.

Likelihood Geometry

The *dual variety* Y_A^* is an irreducible variety in the dual projective space $(\mathbb{P}^{m-1})^\vee$ whose coordinates are $x = (x_1 : x_2 : \cdots : x_m)$. We identify points x in $(\mathbb{P}^{m-1})^\vee$ with hypersurfaces

$$\{t \in (\mathbb{C}^*)^r \;:\; x_1 \cdot t^{a_1} + x_2 \cdot t^{a_2} + \cdots + x_m \cdot t^{a_m} = 0\}. \tag{30}$$

The dual variety Y_A^* is the Zariski closure in $(\mathbb{P}^{m-1})^\vee$ of the locus of all hypersurfaces (30) that are singular. Typically, Y_A^* is a hypersurface. In that case, Y_A^* is defined by a unique (up to sign) irreducible polynomial $\Delta_A \in \mathbb{Z}[x_1, x_2, \ldots, x_m]$. The homogeneous polynomial Δ_A is called the *A-discriminant*. Many classical discriminants and resultants are instances of Δ_A. So are determinants and hyperdeterminants. This is the punch line of the book [GKZ].

Example 3.9. Let $m = 4, r = 2$, and $A = \begin{pmatrix} 3 & 2 & 1 & 0 \\ 0 & 1 & 2 & 3 \end{pmatrix}$. The associated toric variety is the twisted cubic curve

$$Y_A = \overline{\{(1 : t : t^2 : t^3) \,|\, t \in \mathbb{C}\}} \subset \mathbb{P}^3.$$

The variety Y_A^* that is dual to the curve Y_A is a surface in $(\mathbb{P}^3)^\vee$. The surface Y_A^* parametrizes all planes that are tangent to the curve Y_A. These represent univariate cubics

$$x_1 + x_2 t + x_3 t^2 + x_4 t^3$$

that have a double root. Here the A-discriminant is the classical discriminant

$$\Delta_A = 27 x_1^2 x_4^2 - 18 x_1 x_2 x_3 x_4 + 4 x_1 x_3^3 + 4 x_2^3 x_4 - x_2^2 x_3^2.$$

The surface Y_A^* in $\mathbb{P}^3$ defined by this equation is the discriminant of the univariate cubic. $\diamond$

Theorem 3.10. *Let $X \subseteq \mathbb{P}^n$ be a projective variety of ML degree 1. Each coordinate $\hat{p}_i$ of the rational function $u \mapsto \hat{p}$ is an alternating product of linear forms in $u_0, u_1, \ldots, u_n$.*

The paper [Huh2] gives an explicit construction of the map $u \mapsto \hat{p}$ as a *Horn uniformization*. A precursor was [Kapranov]. We explain this construction. The point of departure is a matrix A as above. We now take Δ_A to be any non-zero homogenous polynomial that vanishes on the dual variety Y_A^* of the toric variety Y_A. If Y_A^* is a hypersurface then Δ_A is the A-discriminant.

First, we write Δ_A as a Laurent polynomial by dividing it by one of its monomials:

$$\frac{1}{\text{monomial}} \cdot \Delta_A = 1 - c_0 \cdot x^{b_0} - c_1 \cdot x^{b_1} - \cdots - c_n \cdot x^{b_n}. \tag{31}$$

This expression defines an $m \times (n+1)$ integer matrix $B = (b_0, \ldots, b_n)$ satisfying $AB = 0$. Second, we define X to be the rational subvariety of $\mathbb{P}^n$ that is given parametrically by

$$\frac{p_i}{p_0 + p_1 + \cdots + p_n} = c_i \cdot x^{b_i} \qquad \text{for } i = 0, 1, \ldots, n. \tag{32}$$

The defining ideal of X is obtained by eliminating $x_1, \ldots, x_m$ from the equations above. Then X has ML degree 1, and, by Huh [Huh2], every variety of ML degree 1 arises in this manner.

Example 3.11. The following curve in $\mathbb{P}^3$ happens to be a variety of ML degree 1:

$$X = V\left(9p_1p_2 - 8p_0p_3,\ p_0^2 - 12(p_0+p_1+p_2+p_3)p_3\right).$$

This curve comes from the discriminant of the univariate cubic in Example 3.9:

$$\frac{1}{\text{monomial}} \cdot \Delta_A = 1 - \left(\frac{2}{3}\frac{x_2 x_3}{x_1 x_4}\right) - \left(-\frac{4}{27}\frac{x_2^3}{x_1^2 x_4}\right) - \left(-\frac{4}{27}\frac{x_3^3}{x_1 x_4^2}\right) - \left(\frac{1}{27}\frac{x_2^2 x_3^2}{x_1^2 x_4^2}\right).$$

We derived the curve X from the four parenthesized monomials via the formula (32). The maximum likelihood estimate for this model is given by the products of linear forms

$$\hat{p}_0 = \frac{2}{3}\frac{x_2 x_3}{x_1 x_4} \qquad \hat{p}_1 = -\frac{4}{27}\frac{x_2^3}{x_1^2 x_4} \qquad \hat{p}_2 = -\frac{4}{27}\frac{x_3^3}{x_1 x_4^2} \qquad \hat{p}_3 = \frac{1}{27}\frac{x_2^2 x_3^2}{x_1^2 x_4^2}$$

where

$$\begin{aligned} x_1 &= -u_0 - u_1 - 2u_2 - 2u_3 & x_2 &= u_0 + 3u_2 + 2u_3 \\ x_3 &= u_0 + 3u_1 + 2u_3 & x_4 &= -u_0 - 2u_1 - u_2 - 2u_3 \end{aligned}$$

These expressions are the alternating products of linear forms promised in Theorem 3.10. ◇

We now give the formula for $\hat{p}_i$ in general. This is the *Horn uniformization* of [GKZ, §9.3].

Corollary 3.12. *Let $X \subset \mathbb{P}^n$ be the variety of ML degree 1 with parametrization (32) derived from a scaled A-discriminant (31). The coordinates of the MLE function $u \mapsto \hat{p}$ are*

$$\hat{p}_k = c_k \cdot \prod_{j=1}^{m} \left(\sum_{i=0}^{n} b_{ij} u_i\right)^{b_{kj}}.$$

Likelihood Geometry

It is not obvious (but true) that $\hat{p}_0 + \hat{p}_1 + \cdots + \hat{p}_n = 1$ holds in the formula above. In light of its monomial parametrization, our variety X is toric in $\mathbb{P}^n \setminus \mathcal{H}$. In general, it is not toric in $\mathbb{P}^n$, due to appearances of the factor $(p_0 + p_1 + \cdots + p_n)$ in equations for X. Interestingly, there are numerous instances when this factor does not appear and X is toric also in $\mathbb{P}^n$.

One toric instance is the independence model $X = V(p_{00}p_{11} - p_{01}p_{10})$, whose MLE was derived in Example 1.14. What is the matrix A in this case? We shall answer this question for a slightly larger example, which serves as an illustration for *decomposable graphical models*.

Example 3.13. Consider the conditional independence model for three binary variables given by the graph •—•—•. We claim that this graphical model is derived from

$$A = \begin{array}{c} \\ x \\ y \\ z \\ w \end{array} \begin{pmatrix} a_{00} & a_{10} & a_{01} & a_{11} & b_{00} & b_{01} & b_{10} & b_{11} & c_0 & c_1 & d \\ 1 & 1 & 1 & 1 & 1 & 1 & 1 & 1 & 1 & 1 & 1 \\ 1 & 1 & 0 & 0 & 0 & 0 & 0 & 0 & 1 & 0 & 0 \\ 0 & 0 & 1 & 1 & 0 & 0 & 0 & 0 & 0 & 1 & 0 \\ 0 & 0 & 0 & 0 & 1 & 1 & 0 & 0 & 1 & 0 & 0 \\ 0 & 0 & 0 & 0 & 0 & 0 & 1 & 1 & 0 & 1 & 0 \end{pmatrix}.$$

The discriminant of the corresponding family of hypersurfaces

$$\{(x, y, z, w) \in (\mathbb{C}^*)^4 \mid (a_{00} + a_{10})x + (a_{01} + a_{11})y + (b_{00} + b_{01})z$$
$$+ (b_{10} + b_{11})w + c_0 xz + c_1 yw + d = 0\}$$

equals

$$\Delta_A = c_0 c_1 d - a_{01} b_{10} c_0 - a_{11} b_{10} c_0 - a_{01} b_{11} c_0 - a_{11} b_{11} c_0$$
$$- a_{00} b_{00} c_1 - a_{10} b_{00} c_1 - a_{00} b_{01} c_1 - a_{10} b_{01} c_1.$$

We divide this A-discriminant by its first term $c_0 c_1 d$ to rewrite it in the form (31) with $n = 7$. The parametrization of $X \subset \mathbb{P}^7$ given by (32) can be expressed as

$$p_{ijk} = \frac{a_{ij} \cdot b_{jk}}{c_j \cdot d} \qquad \text{for } i, j, k \in \{0, 1\}. \tag{33}$$

This is indeed the desired graphical model •—•—• with implicit representation

$$X = V\bigl(p_{000}p_{101} - p_{001}p_{100},\ p_{010}p_{111} - p_{011}p_{110}\bigr) \subset \mathbb{P}^7.$$

The linear forms used in the Horn uniformization of Corollary 3.12 are

$$a_{ij} = u_{ij+} \qquad b_{jk} = u_{+jk} \qquad c_j = u_{+j+} \qquad d = u_{+++}$$

Substituting these expressions into (33), we obtain

$$\hat{p}_{ijk} = \frac{u_{ij+} \cdot u_{+jk}}{u_{+j+} \cdot u_{+++}} \qquad \text{for } i, j, k \in \{0, 1\}.$$

This is the formula in Lauritzen's book [Lau] for MLE of decomposable graphical models. ◇

We now return to the likelihood geometry of an arbitrary d-dimensional projective variety X in $\mathbb{P}^n$, as always defined over $\mathbb{R}$ and not contained in $\mathcal{H}$. We define the *ML bidegree* of X to be the bidegree of its likelihood correspondence $\mathcal{L}_X \subset \mathbb{P}^n \times \mathbb{P}^n$. This is a binary form

$$B_X(p, u) = (b_0 \cdot p^d + b_1 \cdot p^{d-1} u + \cdots + b_d \cdot u^d) \cdot p^{n-d},$$

where $b_0, b_1, \ldots, b_d$ are certain positive integers. By definition, $B_X(p, u)$ is the multidegree [ch3:MS, §8.5] of the prime ideal of $\mathcal{L}_X$, with respect to the natural $\mathbb{Z}^2$-grading on the polynomial ring $\mathbb{R}[p, u] = \mathbb{R}[p_0, \ldots, p_n, u_0, \ldots, u_n]$. Equivalently, the ML bidegree $B_X(p, u)$ is the class defined by $\mathcal{L}_X$ in the cohomology ring

$$H^*(\mathbb{P}^n \times \mathbb{P}^n; \mathbb{Z}) = \mathbb{Z}[p, u]/\langle p^{n+1}, u^{n+1} \rangle.$$

We already saw some examples, for the Grassmannian $G(2, 4)$ in (12), for arbitrary linear spaces in (14), and for a toric model of ML degree 1 in (23). We note that the bidegree $B_X(p, u)$ can be computed conveniently using the command `multidegree` in `Macaulay2`.

To understand the geometric meaning of the ML bidegree, we introduce a second polynomial. Let L_{n-i} be a sufficiently general linear subspace of $\mathbb{P}^n$ of codimension i, and define

$$s_i = \text{MLdegree}(X \cap L_{n-i}).$$

We define the *sectional ML degree* of X to be the polynomial

$$S_X(p, u) = (s_0 \cdot p^d + s_1 \cdot p^{d-1} u + \cdots + s_d \cdot u^d) \cdot p^{n-d},$$

Example 3.14. The sectional ML degree of the Grassmannian $G(2, 4)$ in (10) equals

$$S_X(p, u) = 4p^5 + 20p^4 u + 24p^3 u^2 + 12p^2 u^3 + 2pu^4.$$

Thus, if H_1, H_2, H_3 denote generic hyperplanes in $\mathbb{P}^5$, then the threefold $G(2, 4) \cap H_1$ has ML degree 20, the surface $G(2, 4) \cap H_1 \cap H_2$ has ML degree 24, and the curve $G(2, 4) \cap H_1 \cap H_2 \cap H_3$ has ML degree 12. Lastly, the coefficient 2 of pu^4 is simply the degree of $G(2, 4)$ in $\mathbb{P}^5$. ◇

Conjecture 3.15. The ML bidegree and the sectional ML degree of any projective variety $X \subset \mathbb{P}^n$, not lying in $\mathcal{H}$, are related by the following involution on binary forms of degree n:

$$B_X(p,u) = \frac{u \cdot S_X(p, u-p) - p \cdot S_X(p,0)}{u-p},$$

$$S_X(p,u) = \frac{u \cdot B_X(p, u+p) + p \cdot B_X(p,0)}{u+p}.$$

This conjecture is a theorem when $X \backslash \mathcal{H}$ is smooth and its boundary is *schön*. See Theorem 4.6 below. In that case, the ML bidegree is identified, by Huh [Huh1, Theorem 2], with the Chern–Schwartz–MacPherson (CSM) class of the constructible function on $\mathbb{P}^n$ that is 1 on $X \backslash \mathcal{H}$ and 0 elsewhere. Aluffi proved in [Alu, Theorem 1.1] that the CSM class of an locally closed subset of $\mathbb{P}^n$ satisfies such a *log-adjunction formula*. Our formula in Conjecture 3.15 is precisely the homogenization of Aluffi's involution. The combination of [Alu, Theorem 1.1] and [Huh1, Theorem 2] proves Conjecture 3.15 in cases such as generic complete intersections (Theorem 1.10) and arbitrary linear spaces (Theorem 1.20). In the latter case, it can also be verified using matroid theory. Conjecture 3.15 says that this holds for any X, indicating a deeper connection between likelihood correspondences and CSM classes.

We note that $B_X(p,u)$ and $S_X(p,u)$ always share the same leading term and the same trailing term, and this is compatible with our formulas. Both polynomials start and end like

$$\text{MLdegree}(X) \cdot p^n + \cdots + \text{degree}(X) \cdot p^{\text{codim}(X)} u^{\text{dim}(X)}.$$

We now illustrate Conjecture 3.15 by verifying it computationally for a few more examples.

Example 3.16. Let us examine some cubic fourfolds in $\mathbb{P}^5$. If X is a generic hypersurface of degree 4 in $\mathbb{P}^5$ then its sectional ML degree and ML bidegree satisfy the conjectured formula:

$$S_X(p,u) = 1364p^5 + 448p^4u + 136p^3u^2 + 32p^2u^3 + 3pu^4,$$
$$B_X(p,u) = 1364p^5 + 341p^4u + 81p^3u^2 + 23p^2u^3 + 3pu^4.$$

Of course, in algebraic statistics, we are more interested in special hypersurfaces that are statistically meaningful. One such instance was seen in Example 2.7. The mixture model for two identically distributed ternary random variables is the fourfold $X \subset \mathbb{P}^5$ defined by

$$\det \begin{pmatrix} 2p_{11} & p_{12} & p_{13} \\ p_{12} & 2p_{22} & p_{23} \\ p_{13} & p_{23} & 2p_{33} \end{pmatrix} = 0. \qquad (34)$$

The sectional ML degree and the ML bidegree of this determinantal fourfold are

$$S_X(p,u) = 6p^5 + 42p^4u + 48p^3u^2 + 21p^2u^3 + 3pu^4$$
$$B_X(p,u) = 6p^5 + 12p^4u + 15p^3u^2 + 12p^2u^3 + 3pu^4.$$

For the toric fourfold $X = V(p_{11}p_{22}p_{33} - p_{12}p_{13}p_{23})$, ML bidegree and sectional ML degree are

$$B_X(p,u) = \mathbf{3}p^5 + 3p^4u + 3p^3u^2 + 3p^2u^3 + 3pu^4,$$
$$S_X(p,u) = \mathbf{3}p^5 + 12p^4u + 18p^3u^2 + 12p^2u^3 + 3pu^4.$$

Now, taking $X = V(p_{11}p_{22}p_{33} + p_{12}p_{13}p_{23})$ instead, the leading coefficient 3 changes to 2. ◇

Remark 3.17. Conjecture 3.15 is true when X_c is a toric variety with c generic, as in Theorem 3.2. Here we can use Proposition 3.5 to infer that all coefficients of B_X are equal to the normalized volume of the lattice polytope conv(A). In symbols, for generic c, we have

$$B_{X_c}(p,u) = \text{degree}(X_c) \cdot \sum_{i=0}^{d} p^{n-i}u^i.$$

It is now an exercise to transform this into a formula for the sectional ML degree $S_{X_c}(p,u)$.

In general, it is hard to compute generators for the ideal of the likelihood correspondence.

Example 3.18. The following submodel of (34) was featured prominently in [HKS, §1]:

$$\det \begin{pmatrix} 12p_0 & 3p_1 & 2p_2 \\ 3p_1 & 2p_2 & 3p_3 \\ 2p_2 & 3p_3 & 12p_4 \end{pmatrix} = 0. \tag{35}$$

This cubic threefold X is the secant variety of a rational normal curve in $\mathbb{P}^4$, and it represents the mixture model for a binomial random variable (tossing a biased coin four times). It takes several hours in Macaulay2 to compute the prime ideal of the likelihood correspondence $\mathcal{L}_X \subset \mathbb{P}^4 \times \mathbb{P}^4$. That ideal has 20 minimal generators one in degree $(1,1)$, one in degree $(3,0)$, five in degree $(3,1)$, ten in degree $(4,1)$ and three in degree $(3,2)$. After passing to a Gröbner basis, we use the formula in [ch3:MS, Definition 8.45] to compute the bidegree of $\mathcal{L}_X$:

$$B_X(p,u) = 12p^4 + 15p^3u + 12p^2u^2 + 3pu^3.$$

We now intersect X with random hyperplanes in $\mathbb{P}^4$, and we compute the ML degrees of the intersections. Repeating this experiment many times reveals the sectional ML degree of X:

$$S_X(p,u) \;=\; 12p^4 + 30p^3u + 18p^2u^2 + 3pu^3.$$

The two polynomials satisfy our transformation rule, thus confirming Conjecture 3.15. We note that Conjecture 1.8 also holds for this example: using Aluffi's method [AluJSC], we find $\chi(X\backslash\mathcal{H}) = -13$. ◇

Our last topic is the operation of restriction and deletion. This is a standard tool for complements of hyperplane arrangements, as in Theorem 1.20. It was developed in [Huh1] for arbitrary very affine varieties, such as $X\backslash\mathcal{H}$. We motivate this by explaining the distinction between *structural zeros* and *sampling zeros* for contingency tables in statistics [BFH, §5.1.1].

Returning to the "hair loss due to TV soccer" example from the beginning of Sect. 2, let us consider the following questions. What is the difference between the data set

$$U \;=\; \begin{pmatrix} & \text{lots of hair} & \text{medium hair} & \text{little hair} \\ \leq 2\text{h} & 15 & 0 & 9 \\ 2\text{–}6\text{h} & 20 & 24 & 12 \\ \geq 6\text{h} & 10 & 12 & 6 \end{pmatrix}$$

and the data set

$$\tilde{U} \;=\; \begin{pmatrix} & \text{lots of hair} & \text{medium hair} & \text{little hair} \\ \leq 2\text{h} & 10 & 0 & 5 \\ 2\text{–}6\text{h} & 9 & 3 & 6 \\ \geq 6\text{h} & 7 & 9 & 8 \end{pmatrix} ?$$

How should we think about the zero entries in row 1 and column 2 of these two contingency tables? Would the rank 1 model $\mathcal{M}_1$ or the rank 2 model $\mathcal{M}_2$ be more appropriate?

The first matrix U has rank 2 and it can be completed to a rank 1 matrix by replacing the zero entry with 18. Thus, the model $\mathcal{M}_1$ fits perfectly except for the *structural zero* in row 1 and column 2. It seems that this zero is inherent in the structure of the problem: planet Earth simply has no people with medium hair length who rarely watch soccer on TV.

The second matrix $\tilde{U}$ also has rank two, but it cannot be completed to rank 1. The model $\mathcal{M}_2$ is a perfect fit. The zero entry in $\tilde{U}$ appeared to be an artifact of the particular group that was interviewed in this study. This is a *sampling zero*. It arose because, by chance, in this cohort nobody happened to have medium hair length and watch soccer on TV rarely. We refer to the book of Bishop et al. [BFH, Chap. 5] for an introduction.

We now consider an arbitrary projective variety $X \subseteq \mathbb{P}^n$, serving as our statistical model. Suppose that structural zeros or sampling zeros occur in the last coordinate u_n. Following [Rapallo, Theorem 4], we model structural zeros by the projection $\pi_n(X)$. This model is the variety in $\mathbb{P}^{n-1}$ that is the closure of the image of X under the rational map

$$\pi_n : \mathbb{P}^n \dashrightarrow \mathbb{P}^{n-1}, \qquad (p_0 : p_1 : \cdots : p_{n-1} : p_n) \mapsto (p_0 : p_1 : \cdots : p_{n-1}).$$

Which projective variety is a good representation for sampling zeros? We propose that sampling zeros be modeled by the intersection $X \cap \{p_n=0\}$. This is now to be regarded as a subvariety in $\mathbb{P}^{n-1}$. In this manner, both structural zeros and sampling zeros are modeled by closed subvarieties of $\mathbb{P}^{n-1}$. Inside that ambient $\mathbb{P}^{n-1}$, our standard arrangement $\mathcal{H}$ consists of $n+1$ hyperplanes. Usually, none of these hyperplanes contains $X \cap \{p_n=0\}$ or $\pi_n(X)$.

It would be desirable to express the (sectional) ML degree of X in terms of those of the intersection $X \cap \{p_n = 0\}$ and the projection $\pi_n(X)$. As an alternative to the ML degree of the projection $\pi_n(X)$ into $\mathbb{P}^{n-1}$, here is a quantity in $\mathbb{P}^n$ that reflects the presence of structural zeros even more accurately. We denote by

$$\text{MLdegree}(X|_{u_n=0})$$

the number of critical points $\hat{p} = (\hat{p}_0 : \hat{p}_1 : \cdots : \hat{p}_{n-1} : \hat{p}_n)$ of ℓ_u in $X_{\text{reg}} \backslash \mathcal{H}$ for those data vectors $u = (u_0, u_1, \ldots, u_{n-1}, 0)$ whose first n coordinates u_i are positive and generic.

Conjecture 3.19. The maximum likelihood degree satisfies the inductive formula

$$\text{MLdegree}(X) = \text{MLdegree}(X \cap \{p_n=0\}) + \text{MLdegree}(X|_{u_n=0}), \qquad (36)$$

provided X and $X \cap \{p_n=0\}$ are reduced, irreducible, and not contained in their respective $\mathcal{H}$.

We expect that an analogous formula will hold for the sectional ML degree $S_X(p, u)$. The intuition behind equation (36) is as follows. As the data vector u moves from a general point in $\mathbb{P}^n_u$ to a general point on the hyperplane $\{u_n = 0\}$, the corresponding fiber $\text{pr}_2^{-1}(u)$ of the likelihood fibration splits into two clusters. One cluster has size $\text{MLdegree}(X|_{u_n=0})$ and stays away from $\mathcal{H}$. The other cluster moves onto the hyperplane $\{p_n = 0\}$ in $\mathbb{P}^n_p$, where it approaches the various critical points of ℓ_u in that intersection. This degeneration is the perfect scenario for a numerical homotopy, e.g. in Bertini, as discussed in Sect. 2. These homotopies are currently being studied for determinantal varieties by Elizabeth Gross and Jose Rodriguez [GR]. The formula (36) has been verified computationally for many examples. Also, Conjecture 3.19 is known to be true in the slightly different setting of [Huh1], under a certain smoothness assumption. This is the content of [Huh1, Corollary 3.2].

Example 3.20. Fix the space $\mathbb{P}^8$ of 3×3-matrices as in Sect. 2. For the rank 2 variety $X = \mathcal{V}_2$, the formula (36) reads $10 = 5+5$. For the rank 1 variety $X = \mathcal{V}_1$, it reads $1 = 0+1$. ◇

Example 3.21. If X is a generic (d,e)-curve in $\mathbb{P}^3$, then

$$\text{MLdegree}(X) = d^2e + de^2 + de \quad \text{and} \quad X \cap \{p_3 = 0\} = (d \cdot e \text{ distinct points}).$$

Computations suggest that

$$\text{MLdegree}(X|_{u_3=0}) = d^2e + de^2 \quad \text{and} \quad \text{MLdegree}(\pi_3(X)) = d^2e + de^2.$$

To derive the second equality geometrically, one may argue as follows. Both curves $X \subset \mathbb{P}^3$ and $\pi_3(X) \subset \mathbb{P}^2$ have degree de and genus $\frac{1}{2}(d^2e + de^2) - 2de + 1$. Subtracting this from the expected genus $\frac{1}{2}(de-1)(de-2)$ of a plane curve of degree de, we find that $\pi_3(X)$ has $\frac{1}{2}d(d-1)e(e-1)$ nodes. Example 1.4 suggests that each node decreases the ML degree of a plane curve by 2. Assuming this to bet the case, we conclude

$$\text{MLdegree}(\pi_3(X)) = de(de+1) - d(d-1)e(e-1) = d^2e + de^2.$$

Here we are using that a general plane curve of degree de has ML degree $de(de+1)$. ◇

This example suggests that, in favorable circumstances, the following identity would hold:

$$\text{MLdegree}(X|_{u_n=0}) = \text{MLdegree}(\pi_n(X)). \tag{37}$$

However, this is certainly not true in general. Here is a particularly telling example:

Example 3.22. Suppose that X is a generic surface of degree d in $\mathbb{P}^3$. Then

$$\begin{aligned}
\text{MLdegree}(X) &= d + d^2 + d^3, \\
\text{MLdegree}(X \cap \{p_3 = 0\}) &= d + d^2, \\
\text{MLdegree}(X|_{u_3=0}) &= d^3, \\
\text{MLdegree}(\pi_3(X)) &= 1.
\end{aligned}$$

Indeed, for most hypersurfaces $X \subset \mathbb{P}^n$, the same will happen, since $\pi_n(X) = \mathbb{P}^{n-1}$. ◇

As a next step, one might conjecture that (37) holds when the map is birational and the center $(0 : \cdots : 0 : 1)$ of the projection does not lie on the variety X. But this also fails:

Example 3.23. Let X be the twisted cubic curve in $\mathbb{P}^3$ defined by the 2×2-minors of

$$\begin{pmatrix} p_0 + p_1 - p_2 & 2p_0 - p_2 + 9p_3 & p_0 - 6p_1 + 8p_2 \\ 2p_0 - p_2 + 9p_3 & p_0 - 6p_1 + 8p_2 & 7p_0 + p_1 + 2p_2 \end{pmatrix}.$$

The ML degree of X is $13 = 3 + 10$, and X intersects $\{p_3 = 0\}$ in three distinct points. The projection of the curve X into $\mathbb{P}^2$ is a cuspidal cubic, as in Example 1.4. We have

$$\text{MLdegree}\,(X|_{u_3=0}) = 10 \quad \text{and} \quad \text{MLdegree}\,(\pi_3(X)) = 9.$$

It is also instructive to compare the number $13 = -\chi(X \setminus \mathcal{H})$ with the number 11 one gets in Theorem 3.7 for the special twisted cubic curve with $d = 1$, $n = 3$ and $b_0 = b_1 = b_2 = b_3 = 3$. There are many mysteries still to be explored in likelihood geometry, even within $\mathbb{P}^3$. ◇

4 Characteristic Classes

We start by giving an alternative description of the likelihood correspondence which reveals its intimate connection with the theory of Chern classes on possibly noncompact varieties. An important role will be played by the Lie algebra and cotangent bundle of the algebraic torus $(\mathbb{C}^*)^{n+1}$. This section ties our discussion to the work of Aluffi [AluJSC, AluLectures, Alu] and Huh [Huh1, Huh0, Huh2]. In particular, we introduce and explain Chern–Schwartz–MacPherson (CSM) classes. And, most importantly, we present proofs for Theorems 1.6, 1.7, 1.15, and 1.20.

Let $X \subseteq \mathbb{P}^n$ be a closed and irreducible subvariety of dimension d, not contained in our distinguished arrangement of $n + 2$ hyperplanes,

$$\mathcal{H} = \{(p_0 : p_1 : \cdots : p_n) \in \mathbb{P}^n \mid p_0 \cdot p_1 \cdots p_n \cdot p_+ = 0\}, \qquad p_+ = \sum_{i=0}^{n} p_i.$$

Let φ_i denote the restriction of the rational function p_i/p_+ to $X \setminus \mathcal{H}$. The closed embedding

$$\varphi : X \setminus \mathcal{H} \longrightarrow (\mathbb{C}^*)^{n+1}, \qquad \varphi = (\varphi_0, \ldots, \varphi_n),$$

shows that the variety $X \setminus \mathcal{H}$ is *very affine*. Let x be a smooth point of $X \setminus \mathcal{H}$. We define

$$\gamma_x : T_x X \longrightarrow T_{\varphi(x)}(\mathbb{C}^*)^{n+1} \longrightarrow \mathfrak{g} := T_1(\mathbb{C}^*)^{n+1} \tag{38}$$

to be the derivative of φ at x followed by that of left-translation by $\varphi(x)^{-1}$. Here $\mathfrak{g}$ is the Lie algebra of the algebraic torus $(\mathbb{C}^*)^{n+1}$. In local coordinates $(x_1, \ldots, x_d)$ around the smooth point x, the linear map γ_x is represented by the logarithmic Jacobian matrix

Likelihood Geometry

$$\left(\frac{\partial \log \varphi_i}{\partial x_j}\right), \quad 0 \leq i \leq n, \quad 1 \leq j \leq d.$$

The linear map γ_x in (38) is injective because φ is injective. We write $q_0, \ldots, q_n$ for the coordinate functions on the torus $(\mathbb{C}^*)^{n+1}$. These functions define a $\mathbb{C}$-linear basis of the dual Lie algebra $\mathfrak{g}^\vee$ corresponding to differential forms

$$\mathrm{dlog}(q_0), \ldots, \mathrm{dlog}(q_n) \in H^0\left((\mathbb{C}^*)^{n+1}, \Omega^1_{(\mathbb{C}^*)^{n+1}}\right) \simeq \mathfrak{g}^\vee \simeq \mathbb{C}^{n+1}.$$

We fix this choice of basis of $\mathfrak{g}^\vee$, and we identify $\mathbb{P}(\mathfrak{g}^\vee)$ with the space of data vectors $\mathbb{P}^n_u$:

$$\mathfrak{g}^\vee \simeq \left\{\sum_{i=0}^n u_i \cdot \mathrm{dlog}(q_i) \mid u = (u_0, \ldots, u_n) \in \mathbb{C}^{n+1}\right\}.$$

Consider the vector bundle homomorphism defined by the pullback of differential forms

$$\gamma^\vee : \mathfrak{g}^\vee_{X_{\mathrm{reg}}\backslash\mathcal{H}} \longrightarrow \Omega^1_{X_{\mathrm{reg}}\backslash\mathcal{H}}, \quad (x, u) \longmapsto \sum_{i=0}^n u_i \cdot \mathrm{dlog}(\varphi_i)(x). \tag{39}$$

Here $\mathfrak{g}^\vee_{X_{\mathrm{reg}}\backslash\mathcal{H}}$ is the trivial vector bundle over $X_{\mathrm{reg}}\backslash\mathcal{H}$ modeled on the vector space $\mathfrak{g}^\vee$. The induced linear map $\gamma^\vee_x$ between the fibers over a smooth point x is dual to the injective linear map $\gamma_x : T_x X \longrightarrow \mathfrak{g}$. Therefore $\gamma^\vee$ is surjective and $\ker(\gamma^\vee)$ is a vector bundle over $X_{\mathrm{reg}}\backslash\mathcal{H}$. This vector bundle has positive rank $n - d + 1$, and hence its projectivization is nonempty.

Proof of Theorem 1.6. Under the identification $\mathbb{P}(\mathfrak{g}^\vee) \simeq \mathbb{P}^n_u$, the projective bundle $\mathbb{P}(\ker \gamma^\vee)$ corresponds to the following constructible subset of dimension n:

$$\mathcal{L}_X \cap \left((X_{\mathrm{reg}}\backslash\mathcal{H}) \times \mathbb{P}^n_u\right) \subseteq \mathbb{P}^n_p \times \mathbb{P}^n_u.$$

Therefore its Zariski closure $\mathcal{L}_X$ is irreducible of dimension n, and $\mathrm{pr}_1 : \mathcal{L}_X \to \mathbb{P}^n_p$ is a projective bundle over $X_{\mathrm{reg}}\backslash\mathcal{H}$. The likelihood vibration $\mathrm{pr}_2 : \mathcal{L}_X \to \mathbb{P}^n_u$ is generically finite-to-one because the domain and the range are algebraic varieties of the same dimension. $\square$

Our next aim is to prove Theorem 1.15. For this we fix a resolution of singularities

$$\begin{array}{ccc} \pi^{-1}(X_{\text{reg}}\backslash\mathcal{H}) & \longrightarrow & \tilde{X} \\ \downarrow & & \downarrow \pi \\ X_{\text{reg}}\backslash\mathcal{H} & \longrightarrow X \longrightarrow & \mathbb{P}^n, \end{array}$$

where π is an isomorphism over $X_{\text{reg}}\backslash\mathcal{H}$, the variety $\tilde{X}$ is smooth and projective, and the complement of $\pi^{-1}(X_{\text{reg}}\backslash\mathcal{H})$ is a simple normal crossing divisor in $\tilde{X}$ with irreducible components $D_1, \ldots, D_k$. Each φ_i lifts to a rational function on $\tilde{X}$ which is regular on $\pi^{-1}(X\backslash\mathcal{H})$. If $u = (u_0, \ldots, u_n)$ is an integer vector in $\mathbb{Z}^{n+1}$, then these functions satisfy

$$\operatorname{ord}_{D_j}(\ell_u) = \sum_{i=0}^{n} u_i \cdot \operatorname{ord}_{D_j}(\varphi_i). \tag{40}$$

If $u \in \mathbb{C}^{n+1}\backslash\mathbb{Z}^{n+1}$ then $\operatorname{ord}_{D_j}(\ell_u)$ is the complex number defined by the Eq. (40) for $j = 1, \ldots, k$. We write $H_i := \{p_i = 0\}$ and $H_+ := \{p_+ = 0\}$ for the $n+2$ hyperplanes in $\mathcal{H}$.

Lemma 4.1. *Suppose that $X \cap H_i$ is smooth along H_+, and let D_j be a divisor in the boundary of $\tilde{X}$ such that $\pi(D_j) \subseteq \mathcal{H}$. Then the following three statements hold:*

(1) *If $\pi(D_j) \not\subseteq H_+$ then $\operatorname{ord}_{D_j}(\varphi_i)$ is* $\begin{cases} \text{positive} & \text{if } \pi(D_j) \subseteq H_i, \\ \text{zero} & \text{if } \pi(D_j) \not\subseteq H_i. \end{cases}$

(2) *If $\pi(D_j) \subseteq H_+$ then $-\operatorname{ord}_{D_j}(\varphi_i)$ is* $\begin{cases} \text{positive} & \text{if } \pi(D_j) \not\subseteq H_i, \\ \text{nonnegative} & \text{if } \pi(D_j) \subseteq H_i. \end{cases}$

(3) *In each of the above two cases, $\operatorname{ord}_{D_j}(\varphi_i)$ is non-zero for at least one index i.*

Proof. Write H'_i and H'_+ for the pullbacks of H_i and H_+ to X respectively. Note that $\operatorname{ord}_{D_j}(\pi^*(H'_i))$ is positive if D_j is contained in $\pi^{-1}(H'_i)$ and otherwise zero. Since

$$\operatorname{ord}_{D_j}(\varphi_i) = \operatorname{ord}_{D_j}(\pi^*(H'_i)) - \operatorname{ord}_{D_j}(\pi^*(H'_+)),$$

this proves the first and second assertion, except for the case when $\pi(D_j) \subseteq H_i \cap H_+$. In this case, our assumption that H'_i is smooth along H'_+ shows that $\pi(D_j) \subseteq X_{\text{reg}}$ and the order of vanishing of H'_i along $\pi(D_j)$ is 1. Therefore

$$-\operatorname{ord}_{D_j}(\varphi_i) = \operatorname{ord}_{D_j}(\pi^*(H'_+)) - 1 \geq 0.$$

The third assertion of Lemma 4.1 is derived by the following set-theoretic reasoning:

- If $\pi(D_j) \not\subseteq H_+$, then $\pi(D_j) \subseteq H_i$ for some i because $\pi(D_j) \subseteq \mathcal{H}$ is irreducible.
- If $\pi(D_j) \subseteq H_+$, then $\pi(D_j) \not\subseteq H_i$ for some i because $\bigcap_{i=0}^{n} H_i = \emptyset$.

□

From Lemma 4.1 and Eq. (40) we deduce the following result. In Lemmas 4.2 and 4.3 we retain the hypothesis from Lemma 4.1 which coincides with that in Theorem 1.15.

Lemma 4.2. *If $\pi(D_j) \subseteq \mathcal{H}$ and $u \in \mathbb{R}_{>0}^{n+1}$ is strictly positive, then $\mathrm{ord}_{D_j}(\ell_u)$ is nonzero.*

Consider the sheaf of logarithmic differential one-forms $\Omega_{\tilde{X}}^1(\log D)$, where D is the sum of the irreducible components of $\pi^{-1}(\mathcal{H})$. If u is an integer vector, then the corresponding likelihood function ℓ_u on $\tilde{X}$ defines a global section of this sheaf:

$$\mathrm{dlog}(\ell_u) = \sum_{i=0}^{n} u_i \cdot \mathrm{dlog}(\varphi_i) \in H^0(\tilde{X}, \Omega_{\tilde{X}}^1(\log D)). \tag{41}$$

If $u \in \mathbb{C}^{n+1} \setminus \mathbb{Z}^{n+1}$ then we define the global section $\mathrm{dlog}(\ell_u)$ by the above expression (41).

Lemma 4.3. *If $u \in \mathbb{R}_{>0}^{n+1}$ is strictly positive, then $\mathrm{dlog}(\ell_u)$ does not vanish on $\pi^{-1}(\mathcal{H})$.*

Proof. Let $x \in \pi^{-1}(\mathcal{H})$ and $D_1, \ldots, D_l$ the irreducible components of D containing x, with local equations $g_1, \ldots, g_l$ on a small neighborhood G of x. Clearly, $l \geq 1$. By passing to a smaller neighborhood if necessary, we may assume that $\Omega_{\tilde{X}}^1(\log D)$ trivializes over G, and

$$\mathrm{dlog}(\ell_u) = \sum_{j=1}^{l} \mathrm{ord}_{D_j}(\ell_u) \cdot \mathrm{dlog}(g_j) + \psi,$$

where ψ is a regular 1-form. Since the $\mathrm{dlog}(g_j)$ form part of a free basis of a trivialization of $\Omega_{\tilde{X}}^1(\log D)$ over G, Lemma 4.2 implies that $\mathrm{dlog}(\ell_u)$ is nonzero on $\pi^{-1}(\mathcal{H})$ if $u \in \mathbb{R}_{>0}^{n+1}$. □

Proof of Theorem 1.7. In the notation above, the logarithmic Poincaré–Hopf theorem states

$$\int_{\tilde{X}} c_d(\Omega_{\tilde{X}}^1(\log D)) = (-1)^d \cdot \chi(\tilde{X} \setminus \pi^{-1}(\mathcal{H})).$$

See [AluLectures, Sect. 3.4] for example. If $X \setminus \mathcal{H}$ is smooth, then Lemma 4.3 shows that, for generic u, the zero-scheme of the Eq. (41) is equal to the likelihood locus

$$\{x \in X \backslash \mathcal{H} \mid \mathrm{dlog}(\ell_u)(x) = 0\}.$$

Since the likelihood locus is a zero-dimensional scheme of length equal to the ML degree of X, the logarithmic Poincaré–Hopf theorem implies Theorem 1.7. □

Proof of Theorem 1.15. Suppose that the likelihood locus $\{x \in X_{\mathrm{reg}} \backslash \mathcal{H} \mid \mathrm{dlog}(\ell_u)(x) = 0\}$ contains a curve. Let C and $\tilde{C}$ denote the closures of that curve in X and $\tilde{X}$ respectively. Let $\pi^*(\mathcal{H})$ be the pullback of the divisor $\mathcal{H} \cap X$ of X. If $u \in \mathbb{R}^{n+1}_{>0}$ then Lemma 4.3 implies that $\pi^*(\mathcal{H}) \cdot \tilde{C}$ is rationally equivalent to zero in $\tilde{X}$. It then follows from the Projection Formula that $\mathcal{H} \cdot C$ is also rationally equivalent to zero in $\mathbb{P}^n$. But this is impossible. Therefore the likelihood locus does not contain a curve. This proves the first part of Theorem 1.15.

For the second part, we first show that $\mathrm{pr}_2^{-1}(u)$ is contained in $X \backslash \mathcal{H}$ for a strictly positive vector u. This means there is no pair $(x, u) \in \mathcal{L}_X$ with $x \in \mathcal{H}$ which is a limit of the form

$$(x, u) = \lim_{t \to 0}(x_t, u_t), \qquad x_t \in X_{\mathrm{reg}} \backslash \mathcal{H}, \qquad \mathrm{dlog}(\ell_{u_t})(x_t) = 0.$$

If there is such a sequence (x_t, u_t), then we can take its limit over $\tilde{X}$ to find a point $\tilde{x} \in \tilde{X}$ such that $\mathrm{dlog}(\ell_u)(\tilde{x}) = 0$, but this would contradict Lemma 4.3.

Now suppose that the fiber $\mathrm{pr}_2^{-1}(u)$ is contained in X_{reg}, and hence in $X_{\mathrm{reg}} \backslash \mathcal{H}$. By Theorem 1.6, this fiber $\mathrm{pr}_2^{-1}(u)$ is contained the smooth variety $(\mathcal{L}_X)_{\mathrm{reg}}$. Furthermore, by the first part of Theorem 1.15, $\mathrm{pr}_2^{-1}(u)$ is a zero-dimensional subscheme of $(\mathcal{L}_X)_{\mathrm{reg}}$. The assertion on the length of the fiber now follows from a standard result on intersection theory on Cohen–Macaulay varieties. More precisely, we have

$$\mathrm{MLdegree}(X) = (U_1 \cdot \ldots \cdot U_n)_{\mathcal{L}_X} = (U_1 \cdot \ldots \cdot U_n)_{(\mathcal{L}_X)_{\mathrm{reg}}} = \deg(\mathrm{pr}_2^{-1}(u)),$$

where the U_i are pullbacks of sufficiently general hyperplanes in $\mathbb{P}^n_u$ containing u, and the two terms in the middle are the intersection numbers defined in [Fulton, Definition 2.4.2]. The fact that $(\mathcal{L}_X)_{\mathrm{reg}}$ is Cohen–Macaulay is used in the last equality [Fulton, Example 2.4.8]. □

Remark 4.4. If X is a curve, then the zero-scheme of the Eq. (41) is zero-dimensional for generic u, even if $X \backslash \mathcal{H}$ is singular. Furthermore, the length of this zero-scheme is at least as large as ML degree of X. Therefore

$$-\chi(X \backslash \mathcal{H}) \geq -\chi(\tilde{X} \backslash \pi^{-1}(\mathcal{H})) \geq \mathrm{MLdegree}(X).$$

This proves that Conjecture 1.8 holds for $d = 1$.

Next we give a brief description of the Chern–Schwartz–MacPherson (CSM) class. For a gentle introduction we refer to [AluLectures]. The group $C(X)$ of *constructible functions* on a complex algebraic variety X is a subgroup of the group

Likelihood Geometry

of integer valued functions on X. It is generated by the characteristic functions $\mathbf{1}_Z$ of all closed subvarieties Z of X. If $f : X \to Y$ is a morphism between complex algebraic varieties, then the pushforward of constructible functions is the homomorphism

$$f_* : C(X) \longrightarrow C(Y), \qquad \mathbf{1}_Z \mapsto \left(y \mapsto \chi(f^{-1}(y) \cap Z), \quad y \in Y \right).$$

If X is a compact complex manifold, then the characteristic class of X is the Chern class of the tangent bundle $c(TX) \cap [X] \in H_*(X; \mathbb{Z})$. A generalization to possibly singular or noncompact varieties is provided by the Chern–Schwartz–MacPherson class, whose existence was once a conjecture of Deligne and Grothendieck.

In the next definition, we write C for the functor of constructible functions from the category of complete complex algebraic varieties to the category of abelian groups.

Definition 4.5. The *CSM class* is the unique natural transformation

$$c_{SM} : C \longrightarrow H_*$$

such that $c_{SM}(\mathbf{1}_X) = c(TX) \cap [X] \in H_*(X; \mathbb{Z})$ when X is smooth and complete.

The uniqueness follows from the naturality, the resolution of singularities over $\mathbb{C}$, and the requirement for smooth and complete varieties. We highlight two properties of the CSM class which follow directly from Definition 4.5:

1. The CSM class satisfies the inclusion–exclusion relation

$$c_{SM}(\mathbf{1}_{U \cup U'}) = c_{SM}(\mathbf{1}_U) + c_{SM}(\mathbf{1}_{U'}) - c_{SM}(\mathbf{1}_{U \cap U'}) \in H_*(X; \mathbb{Z}). \tag{42}$$

2. The CSM class captures the topological Euler characteristic as its degree:

$$\chi(U) = \int_X c_{SM}(\mathbf{1}_U) \in \mathbb{Z}. \tag{43}$$

Here U and U' are arbitrary constructible subsets of a complete variety X.

What kind of information on a constructible subset is encoded in its CSM class? In likelihood geometry, U is a constructible subset in the complex projective space $\mathbb{P}^n$, and we identify $c_{SM}(\mathbf{1}_U)$ with its image in $H_*(\mathbb{P}^n, \mathbb{Z}) = \mathbb{Z}[p]/\langle p^{n+1} \rangle$. Thus $c_{SM}(\mathbf{1}_U)$ is a polynomial of degree $\leq n$ is one variable p. To be consistent with the earlier sections, we introduce a homogenizing variable u, and we write $c_{SM}(\mathbf{1}_U)$ as a binary form of degree n in (p, u).

The CSM class of U carries the same information as the *sectional Euler characteristic*

$$\chi_{\text{sec}}(\mathbf{1}_U) = \sum_{i=0}^{n} \chi(U \cap L_{n-i}) \cdot p^{n-i} u^i.$$

Here L_{n-i} is a generic linear subspace of codimension i in $\mathbb{P}^n$. Indeed, it was proved by Aluffi in [Alu, Theorem 1.1] that $c_{SM}(\mathbf{1}_U)$ is the transform of $\chi_{\mathrm{sec}}(\mathbf{1}_U)$ under a linear involution on binary forms of degree n in (p, u). In fact, our involution in Conjecture 3.15 is nothing but the signed version of the Aluffi's involution. This is explained by the following result.

Theorem 4.6. *Let $X \subset \mathbb{P}^n$ be closed subvariety of dimension d that is not contained in $\mathcal{H}$. If the very affine variety $X \backslash \mathcal{H}$ is schön then, up to signs, the ML bidegree equals the CSM class and the sectional ML degree equals the sectional Euler characteristic. In symbols,*

$$c_{SM}(\mathbf{1}_{X\backslash\mathcal{H}}) = (-1)^{n-d} \cdot B_X(-p, u) \quad and \quad \chi_{\mathrm{sec}}(\mathbf{1}_{X\backslash\mathcal{H}}) = (-1)^{n-d} \cdot S_X(-p, u).$$

Proof. The first identity is a special case of [Huh1, Theorem 2], here adapted to $\mathbb{P}^n$ minus $n + 2$ hyperplanes, and the second identity follows from the first by way of [Alu, Theorem 1.1]. □

To make sense of the statement in Theorem 4.6, we need to recall the definition of *schön*. This term was coined by Tevelev in his study of tropical compactifications [Tevelev]. Let U be an arbitrary closed subvariety of the algebraic torus $(\mathbb{C}^*)^{n+1}$. In our application, $U = X \backslash \mathcal{H}$. We consider the closures $\overline{U}$ of U in various (not necessarily complete) normal toric varieties Y with dense torus $(\mathbb{C}^*)^{n+1}$. The closure $\overline{U}$ is complete if and only if the support of the fan of Y contains the tropicalization of U [Tevelev, Proposition 2.3]. We say that $\overline{U}$ is a *tropical compactification* of U if it is complete and the multiplication map

$$m : (\mathbb{C}^*)^{n+1} \times \overline{U} \longrightarrow Y, \quad (t, x) \longmapsto t \cdot x$$

is flat and surjective. Tropical compactifications exist, and they are obtained from toric varieties Y defined by sufficiently fine fan structures on the tropicalization of U [Tevelev, §2]. The very affine variety U is called *schön* if the multiplication is smooth for some tropical compactification of U. Equivalently, U is schön if the multiplication is smooth for every tropical compactification of U, by Tevelev [Tevelev, Theorem 1.4].

Two classes of schön very affine varieties are of particular interest. The first is the class of complements of essential hyperplane arrangements. The second is the class of nondegenerate hypersurfaces. What we need from the schön hypothesis is the existence of a simple normal crossings compactification which admits sufficiently many differential one-forms which have logarithmic singularities along the boundary. For complements of hyperplane arrangements, such a compactification is provided by the wonderful compactification of De Concini and Procesi [DP]. For nondegenerate hypersurfaces, and more generally for nondegenerate complete intersections, the needed compactification has been constructed by Khovanskii [Hovanskii].

Likelihood Geometry

We illustrate this in the setting of likelihood geometry by a d-dimensional linear subspace of $X \subset \mathbb{P}^n$. The intersection of X with distinguished hyperplanes $\mathcal{H}$ of $\mathbb{P}^n$ is an arrangement of $n+2$ hyperplanes in $X \simeq \mathbb{P}^d$, defining a matroid M of rank $d+1$ on $n+2$ elements.

Proposition 4.7. *If X is a linear space of dimension d then the CSM class of $X \backslash \mathcal{H}$ in $\mathbb{P}^n$ is*

$$c_{SM}(\mathbf{1}_{X \backslash \mathcal{H}}) = \sum_{i=0}^{d} (-1)^i h_i u^{d-i} p^{n-d+i}.$$

where the h_i are the signed coefficients of the shifted characteristic polynomial in (15).

Proof. This holds because the recursive formula for a triple of arrangement complements

$$c_{SM}(\mathbf{1}_{U_1}) = c_{SM}(\mathbf{1}_U - \mathbf{1}_{U_0}) = c_{SM}(\mathbf{1}_U) - c_{SM}(\mathbf{1}_{U_0}),$$

agrees with the usual deletion-restriction formula [OTBook, Theorem 2.56]:

$$\chi_{M_1}(q+1) = \chi_M(q+1) - \chi_{M_0}(q+1).$$

Here our notation is as in [Huh1, §3]. We now use induction on the number of hyperplanes. □

Proof of Theorem 1.20. The very affine variety $X \backslash \mathcal{H}$ is schön when X is linear. Hence the asserted formula for the ML bidegree of X follows from Theorem 4.6 and Proposition 4.7. □

Rank constraints on matrices are important both in statistics and in algebraic geometry, and they provide a rich source of test cases for the theory developed here. We close our discussion with the enumerative invariants of three hypersurfaces defined by 3×3-determinants. It would be very interesting to compute these formulas for larger determinantal varieties.

Example 4.8. We record the *ML bidegree*, the *CSM class*, the *sectional ML degree*, and the *sectional Euler characteristic* for three singular hypersurfaces seen earlier in this paper. These examples were studied already in [HKS]. The classes we present are elements of $H^*(\mathbb{P}_p^n \times \mathbb{P}_u^n)$ and of $H^*(\mathbb{P}_p^n; \mathbb{Z})$ respectively, and they are written as binary forms in (p, u) as before.

- The 3×3 determinantal hypersurface in $\mathbb{P}^8$ (Example 2.1) has

$$\begin{aligned} B_X(p,u) = {} & 10p^8 + 24p^7u + 33p^6u^2 + 38p^5u^3 + 39p^4u^4 \\ & + 33p^3u^5 + 12p^2u^6 + 3pu^7, \end{aligned}$$

$$c_{SM}(\mathbf{1}_{X\setminus\mathcal{H}}) = -11p^8 + 26p^7u - 37p^6u^2 + 44p^5u^3 - 45p^4u^4$$
$$+ 33p^3u^5 - 12p^2u^6 + 3pu^7,$$
$$S_X(p,u) = 11p^8 + 182p^7u + 436p^6u^2 + 518p^5u^3 + 351p^4u^4$$
$$+ 138p^3u^5 + 30p^2u^6 + 3pu^7,$$
$$\chi_{sec}(\mathbf{1}_{X\setminus\mathcal{H}}) = -11p^8 + 200p^7u - 470p^6u^2 + 542p^5u^3 - 357p^4u^4$$
$$+ 138p^3u^5 - 30p^2u^6 + 3pu^7.$$

- The 3×3 symmetric determinantal hypersurface in $\mathbb{P}^5$ (Example 2.7) has

$$B_X(p,u) = 6p^5 + 12p^4u + 15p^3u^2 + 12p^2u^3 + 3pu^4,$$
$$c_{SM}(\mathbf{1}_{X\setminus\mathcal{H}}) = 7p^5 - 14p^4u + 19p^3u^2 - 12p^2u^3 + 3pu^4,$$
$$S_X(p,u) = 6p^5 + 42p^4u + 48p^3u^2 + 21p^2u^3 + 3pu^4,$$
$$\chi_{sec}(\mathbf{1}_{X\setminus\mathcal{H}}) = 7p^5 - 48p^4u + 52p^3u^2 - 21p^2u^3 + 3pu^4.$$

- The secant variety of the rational normal curve in $\mathbb{P}^4$ (Example 3.18) has

$$B_X(p,u) = 12p^4 + 15p^3u + 12p^2u^2 + 3pu^3,$$
$$c_{SM}(\mathbf{1}_{X\setminus\mathcal{H}}) = -13p^4 + 19p^3u - 12p^2u^2 + 3pu^3,$$
$$S_X(p,u) = 12p^4 + 30p^3u + 18p^2u^2 + 3pu^3,$$
$$\chi_{sec}(\mathbf{1}_{X\setminus\mathcal{H}}) = -13p^4 + 34p^3u - 18p^2u^2 + 3pu^3.$$

In all known examples, the coefficients of $B_X(p,u)$ are less than or equal to the absolute value of the corresponding coefficients of $c_{SM}(\mathbf{1}_{X\setminus\mathcal{H}})$, and similarly for $S_X(p,u)$ and $\chi_{sec}(\mathbf{1}_{X\setminus\mathcal{H}})$. That this inequality holds for the first coefficient is Conjecture 1.8 which relates the ML degree of a singular X to the signed Euler characteristic of the very affine variety $X\setminus\mathcal{H}$. ◇

Acknowledgements We thank Paolo Aluffi and Sam Payne for helpful communications, and the Mathematics Department at KAIST, Daejeon, for hosting both authors in May 2013. Bernd Sturmfels was supported by NSF (DMS-0968882) and DARPA (HR0011-12-1-0011).

References

[ARSZ] E. Allmann, J. Rhodes, B. Sturmfels, P. Zwiernik, Tensors of nonnegative rank two, in *Linear Algebra and its Applications*. Special Issue on Statistics. http://www.sciencedirect.com/science/article/pii/S0024379513006812

[AluJSC]	P. Aluffi, Computing characteristic classes of projective schemes. J. Symb. Comput. **35**, 3–19 (2003)
[AluLectures]	P. Aluffi, Characteristic classes of singular varieties, in *Topics in Cohomological Studies of Algebraic Varieties*. Trends in Mathematics (Birkhäuser, Basel, 2005), pp. 1–32
[Alu]	P. Aluffi, Euler characteristics of general linear sections and polynomial Chern classes. Rend. Circ. Mat. Palermo. **62**, 3–26 (2013)
[AN]	S. Amari, H. Nagaoka, in *Methods of Information Geometry*. Translations of Mathematical Monographs, vol. 191 (American Mathematical Society, Providence, 2000)
[Bertini]	D.J. Bates, J.D. Hauenstein, A.J. Sommese, C.W. Wampler, *Bertini: Software for Numerical Algebraic Geometry* (2006). www.nd.edu/~sommese/bertini
[BHSW]	D.J. Bates, J.D. Hauenstein, A.J. Sommese, C.W. Wampler, *Numerically Solving Polynomial Systems with Bertini* (Society for Industrial and Applied Mathematics, Philelphia, 2013). http://www.ec-securehost.com/SIAM/SE25.html
[BFH]	Y. Bishop, S. Fienberg, P. Holland, *Discrete Multivariate Analysis: Theory and Practice* (Springer, New York, 1975)
[BoydVan]	S. Boyd, L. Vandenberghe, *Convex Optimization* (Cambridge University Press, Cambridge, 2004)
[CHKS]	F. Catanese, S. Hoşten, A. Khetan, B. Sturmfels, The maximum likelihood degree. Am. J. Math. **128**, 671–697 (2006)
[CDFV]	D. Cohen, G. Denham, M. Falk, A. Varchenko, Critical points and resonance of hyperplane arrangements. Can. J. Math. **63**, 1038–1057 (2011)
[DP]	C. De Concini, C. Procesi, Wonderful models of subspace arrangements. Selecta Math. New Ser. **1**, 459–494 (1995)
[Denham-Garrousian-Schulze]	G. Denham, M. Garrousian, M. Schulze, A geometric deletion-restriction formula. Adv. Math. **230**, 1979–1994 (2012)
[DR]	J. Draisma, J. Rodriguez, Maximum likelihood duality for determinantal varieties. Int. Math. Res. Not. [arXiv:1211.3196]. http://imrn.oxfordjournals.org/content/early/2013/07/03/imrn.rnt128.full.pdf
[LiAS]	M. Drton, B. Sturmfels, S. Sullivant, in *Lectures on Algebraic Statistics*. Oberwolfach Seminars, vol. 39 (Birkhäuser, Basel, 2009)
[Franecki-Kapranov]	J. Franecki, M. Kapranov, The Gauss map and a noncompact Riemann-Roch formula for constructible sheaves on semiabelian varieties. Duke Math. J. **104**, 171–180 (2000)
[Fulton]	W. Fulton, in *Intersection Theory*, 2nd edn. Ergebnisse der Mathematik und ihrer Grenzgebiete. A Series of Modern Surveys in Mathematics, vol. 2 (Springer, Berlin, 1998)
[Gabber-Loeser]	O. Gabber, F. Loeser, Faisceaux pervers l-adiques sur un tore. Duke Math. J. **83**, 501–606 (1996)
[GKZ]	I.M. Gel'fand, M. Kapranov, A. Zelevinsky, *Discriminants, Resultants, and Multidimensional Determinants* (Birkhäuser, Boston, 1994)
[GR]	E. Gross, J. Rodriguez, *Maximum likelihood geometry in the presence of data zeros*, http://front.math.ucdavis.edu/1310.4197
[HRS]	J. Hauenstein, J. Rodriguez, B. Sturmfels, Maximum likelihood for matrices with rank constraints. J. Algebr. Stat. (to appear) [arXiv:1210.0198]

[HKS] S. Hoşten, A. Khetan, B. Sturmfels, Solving the likelihood equations. Found. Comput. Math. **5**, 389–407 (2005)

[Hovanskii] A. Hovanskiĭ, Newton polyhedra and toroidal varieties. Akademija Nauk SSSR. Funkcional'nyi Analiz i ego Priloženija **11**, 56–64 (1977)

[Huh1] J. Huh, The maximum likelihood degree of a very affine variety. Compositio Math. **149**, 1245–1266 (2013)

[Huh0] J. Huh, *h-vectors of matroids and logarithmic concavity*. Preprint. http://arxiv.org/abs/1201.2915. [arXiv:1201.2915]

[Huh2] J. Huh, *Varieties with maximum likelihood degree one*. J. Algeb. Stat. (to appear). [arXiv:1301.2732]

[Kapranov] M. Kapranov, A characterization of A-discriminantal hypersurfaces in terms of the logarithmic Gauss map. Math. Ann. **290**, 277–285 (1991)

[KRS] K. Kubjas, E. Robeva, B. Sturmfels, Fixed points of the EM algorithm and nonnegative rank boundaries, http://arxiv.org/abs/1312.5634 (in preparation)

[Land] J.M. Landsberg, *Tensors: Geometry and Applications*. Graduate Studies in Mathematics, vol. 128 (American Mathematical Society, Providence, 2012)

[LM] J.M. Landsberg, L. Manivel, On ideals of secant varieties of Segre varieties. Found. Comput. Math. **4**, 397–422 (2004)

[Lau] S. Lauritzen, *Graphical Models* (Oxford University Press, Oxford, 1996)

[ch3:MS] E. Miller, B. Sturmfels, in *Combinatorial Commutative Algebra*. Graduate Texts in Mathematics, vol. 227 (Springer, New York, 2004)

[MSS] D. Mond, J. Smith, D. van Straten, Stochastic factorizations, sandwiched simplices and the topology of the space of explanations. R. Soc. Lond. Proc. Ser. A Math. Phys. Eng. Sci. **459**, 2821–2845 (2003)

[OTBook] P. Orlik, H. Terao, in *Arrangements of Hyperplanes*. Grundlehren der Mathematischen Wissenschaften, vol. 300 (Springer, Berlin, 1992)

[Orlik-Terao] P. Orlik, H. Terao, The number of critical points of a product of powers of linear functions. Inventiones Math. **120**, 1–14 (1995)

[PS] L. Pachter, B. Sturmfels, *Algebraic Statistics for Computational Biology* (Cambridge University Press, Cambridge, 2005)

[Rai] C. Raicu, Secant varieties of Segre–Veronese varieties. Algebra Number Theory **6**, 1817–1868 (2012)

[Rapallo] F. Rapallo, Markov bases and structural zeros. J. Symb. Comput. **41**, 164–172 (2006)

[SSV] R. Sanyal, B. Sturmfels, C. Vinzant, The entropic discriminant. Adv. Math. **244**, 678–707 (2013)

[Terao] H. Terao, Generalized exponents of a free arrangement of hyperplanes and the Shepherd-Todd-Brieskorn formula. Invent. Math. **63**, 159–179 (1981)

[Tevelev] J. Tevelev, Compactifications of subvarieties of tori. Am. J. Math. **129**, 1087–1104 (2007)

[Uhl] C. Uhler, Geometry of maximum likelihood estimation in Gaussian graphical models. Ann. Stat. **40**, 238–261 (2012)

[Varchenko] A. Varchenko, Critical points of the product of powers of linear functions and families of bases of singular vectors. Compositio Math. **97**, 385–401 (1995)

[Wat] S. Watanabe, in *Algebraic Geometry and Statistical Learning Theory*. Monographs on Applied and Computational Mathematics, vol. 25 (Cambridge University Press, Cambridge, 2009)

Linear Toric Fibrations

Sandra Di Rocco

1 Introduction

These notes are based on three lectures given at the 2013 CIME/CIRM summer school *Combinatorial Algebraic Geometry*.

The purpose of this series of lectures is to introduce the notion of a *toric fibration* and to give its geometrical and combinatorial characterizations.

Toric fibrations $f : X \to Y$, together with a choice of an ample line bundle L on X are associated to convex polytopes called *Cayley sums*. Such a polytope is a convex polytope $P \subset \mathbb{R}^n$ obtained by assembling a number of lower dimensional polytopes R_i, whose normal fan defines the same toric variety Y. Let $\mathbb{R}^n = M \otimes \mathbb{R}$, for a lattice M. The building-blocks R_i are glued together following their image via a surjective map of lattices $\pi : M \to \Lambda$, see Definition 3.7. In particular the normal fan of the polytope $\pi(P)$ defines the generic fiber of the map f. We will denote Cayley sums by $\mathrm{Cayley}(R_0, \ldots, R_t)_{\pi,Y}$. Our aim is to illustrate how classical notions in projective geometry are captured by certain properties of the associated Cayley sum.

When the image polytope $\pi(P)$ is a unimodular simplex Δ_k the generic fiber of the fibration f is a projective space $\mathbb{P}^k$ embedded linearly, i.e. $L|_F = \mathcal{O}_{\mathbb{P}^k}(1)$. For this reason the fibration is called a *linear toric fibration*. The following picture illustrates a linear toric fibration and the representation of the associated polytope as a Cayley sum.

S. Di Rocco (✉)
Department of Mathematics, Royal Institute of Technology (KTH), 10044 Stockholm, Sweden
e-mail: dirocco@kth.se
www.math.kth.se/~dirocco

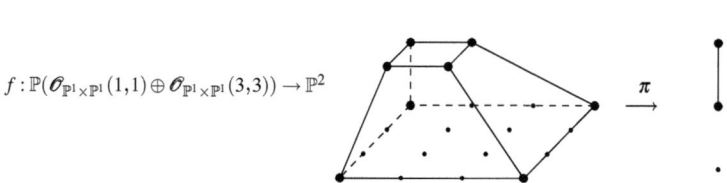

$$f : \mathbb{P}(\mathcal{O}_{\mathbb{P}^1 \times \mathbb{P}^1}(1,1) \oplus \mathcal{O}_{\mathbb{P}^1 \times \mathbb{P}^1}(3,3)) \to \mathbb{P}^2 \qquad \xrightarrow{\pi}$$

Section 3 will be devoted to define these concepts and to give the most relevant examples. In the following two sections we will present two characterizations of Cayley sums corresponding to linear toric fibrations. In both cases there are rich and interesting connections with classical projective geometry.

Section 4 discusses discriminants of polynomials. A polynomial supported on a subset $\mathcal{A} \subset \mathbb{Z}^n$ is a polynomial in n variables $x = (x_1, \ldots, x_n)$ of the form $p_{\mathcal{A}} = \sum_{a \in \mathcal{A}} c_a x^a$. The $\mathcal{A}$-discriminant is again a polynomial in $|\mathcal{A}|$ variables, $\Delta_{\mathcal{A}}(c_a)$, vanishing whenever the corresponding polynomial has at least one singularity in the torus $(\mathbb{C}^*)^n$. Understanding the existence and in that case the degree of the discriminant polynomial, for given classes of point-configurations $\mathcal{A}$, is highly desirable. Finite subsets $\mathcal{A} \subset \mathbb{Z}^n$ define toric projective varieties, $X_{\mathcal{A}} \subset \mathbb{P}^{|\mathcal{A}|-1}$. It is classical in Algebraic Geometry to associate to a given embedding, $X \subset \mathbb{P}^m$, the variety parametrizing hyperplanes singular along X. This variety is called the dual variety and it is denoted by $X^{\vee}$. Understanding when the codimension of the dual variety is higher that one and giving efficient formulas for its degree is a long standing problem. We will see that projective duality is a useful tool for describing the discriminants $\Delta_{\mathcal{A}}$ when the associated polytope $\mathrm{Conv}(\mathcal{A})$ is smooth or simple. In fact the case when $\Delta_{\mathcal{A}} = 1$ is completely characterized by Cayley sums and thus by toric fibrations.

In the non singular case the following holds.

Characterization 1. If $P_{\mathcal{A}} = \mathrm{Conv}(\mathcal{A})$ is a smooth polytope then the following assertions are equivalent:

(a) $P_{\mathcal{A}} = \mathrm{Cayley}_{\pi,Y}(R_0, \ldots, R_t)$ with $t \geq max(2, \frac{n+1}{2})$.
(b) $\mathrm{codim}(X_{\mathcal{A}}^{\vee}) > 1$.
(c) $\Delta_{\mathcal{A}} = 1$.

When the codimension of $X_{\mathcal{A}}^{\vee}$ is one then its degree is given by an alternating sum of volumes of the faces of the polytope $P_{\mathcal{A}}$. We will see that this formula corresponds to the top Chern class of the so called first jet bundle. This interpretation has a useful consequence. When the codimension of $X_{\mathcal{A}}^{\vee}$ is higher than one this Chern class has to vanish. This leads to another characterization of Cayley sums.

Characterization 2. If $P_{\mathcal{A}} = \mathrm{Conv}(\mathcal{A})$ is a smooth polytope then the following assertions are equivalent:

(a) $P_{\mathcal{A}} = \mathrm{Cayley}_{\pi,Y}(R_0, \ldots, R_t)$ with $t \geq max(2, \frac{n+1}{2})$.
(b) $\sum_{\emptyset \neq F \prec P_{\mathcal{A}}} (-1)^{\mathrm{codim}(F)} (\dim(F) + 1)! \, \mathrm{Vol}(F) = 0$.

In Sect. 5 we discuss the problem of classifying convex polytopes and algebraic varieties. A classification is typically done via invariants. In recent years much attention has been concentrated on the notion of codegree of a convex polytope.

$$\mathrm{codeg}(P) = \mathrm{min}_{\mathbb{Z}}\{t\,|\,tP \text{ has interior lattice points}\}$$

The unimodular simplex for example has $\mathrm{codeg}(\Delta_n) = n+1$. Batyrev and Nill conjectured that imposing this invariant to be large should force the polytope to be a Cayley sum.

It turned out that a $\mathbb{Q}$-version of this invariant, what we denote by $\mu(P)$, corresponds to a classical invariant in classification theory of algebraic varieties, called the log-canonical threshold. Let $(X_P, \mathscr{L}_P)$ be the toric variety and ample line bundle associated to the polytope P. The canonical threshold $\mu(\mathscr{L}_P)$ and the nef-value $\tau(\mathscr{L}_P)$ are the invariants used heavily in the classification theory of Gorenstein algebraic varieties. In particular Beltrametti-Sommese-Wisniewski conjectured that imposing $\mu(\mathscr{L}_P)$ to be large should force the variety to have the structure of a fibration.

Again in the toric setting we will see that these two stories intersect making it possible to prove the above conjectures, at least in the smooth case, and leading to yet another characterization of Cayley sums.

Characterization 3. Let P be a smooth polytope. The following assertions are equivalent:

(a) $\mathrm{codeg}(P) \geq (n+3)/2$.
(b) P is isomorphic to a Cayley sum $\mathrm{Cayley}(R_0, \ldots, R_t)_{\pi,Y}$ where $t + 1 = \mathrm{codeg}(P)$ with $k > \frac{n}{2}$.
(c) $\mu(\mathscr{L}_P) = \tau(\mathscr{L}_P) \geq (n+3)/2$.

In fact the characterizations above extend to more general classes of polytopes, not necessarily smooth, as we explain in Sects. 4 and 5. Section 6 is devoted to give a complete proof of these characterizations.

2 Conventions and Notation

We assume basic knowledge of toric geometry and refer to [EW, FU, ODA] for the necessary background on toric varieties. We will moreover assume some knowledge of projective algebraic geometry. We refer the reader to [HA, FUb] for further details. Throughout this paper, we work over the field of complex numbers $\mathbb{C}$. By a polarized variety we mean a pair (X, L) where X is an algebraic variety and L is an ample line bundle on X.

2.1 Toric Geometry

In this note a toric variety, X, is always assumed to be normal and thus defined by a fan $\Sigma_X \subset N \otimes \mathbb{R}$ for a lattice N. By $\Sigma_X(t)$ we will denote the collection of t-dimensional cones of Σ_X. The invariant sub-variety of codimension t associated to a cone $\sigma \in \Sigma(t)$ will be denoted by $V(\sigma)$.

For a lattice Δ we set $\Delta_\mathbb{R} = \Delta \otimes_\mathbb{Z} \mathbb{R}$. We denote by $\Delta^\vee = Hom(\Delta, \mathbb{Z})$ the dual lattice. If $\pi : \Delta \to \Gamma$ is a morphism of lattices we denote by $\pi_\mathbb{R} : \Delta_\mathbb{R} \to \Gamma_\mathbb{R}$ the induced $\mathbb{R}$-homomorphism. By a lattice polytope $P \subset \Delta_\mathbb{R}$ we mean a polytope with vertices in Δ.

Let $P \subset \mathbb{R}^n$ be a lattice polytope of dimension n. Consider the graded semigroup Π_P generated by $(\{1\} \times P) \cap (\mathbb{N} \times \mathbb{Z}^n)$. The polarized variety $(Proj(\mathbb{C}[\Pi_P]), \mathcal{O}(1))$ is a toric variety associated to the polytope P. It will be sometimes denoted by (X_P, L_P). Notice that the toric variety X_P is defined by the (inner) normal fan of P. Vice versa the symbol $P_{(X,L)}$ will denote the lattice polytope associated to a polarized toric variety (X, L).

Two polytopes are said to be *normally equivalent* if their normal fans are isomorphic.

The symbol Δ_n denotes the smooth (unimodular) simplex of dimension n. Recall that an n-dimensional polytope is *simple* if through every vertex pass exactly n edges. A lattice polytope is *smooth* if it is simple and the primitive vectors of the edges through every vertex form a lattice basis. Smooth polytopes are associated to smooth projective toric varieties. Simple polytopes are associated to $\mathbb{Q}$-factorial projective toric varieties.

When the toric variety is defined via a point configuration $\mathscr{A} \subset \mathbb{Z}^n$ we will use the symbol $(X_\mathscr{A}, \mathscr{L}_\mathscr{A})$ for the associated polarized toric variety and $P_\mathscr{A} = Conv(\mathscr{A})$ for the associated polytope. The corresponding fan is denoted by $\Sigma_\mathscr{A}$.

2.2 Vector Bundles

The notion of Chern classes of a vector bundle is an essential tool in some of the proofs. Let E be a vector bundle of rank k over an n-dimensional algebraic variety X. Recall that the i-th Chern class of E, $c_i(E)$, is the class of a codimension i cycle on X modulo rational equivalence. The top Chern class of a rank $k \geq n$ vector bundle is $c_n(E)$. The same symbol $c_n(E)$ will be used to denote the degree of the associated zero-dimensional subvariety.

The projectivization of a vector bundle plays a fundamental role throughout these notes. Let $S^l(E)$ denote the l-th symmetric power of a rank $r+1$ vector bundle E. The projectivization of E is $\mathbb{P}(E) = Proj(\oplus_{l=0}^\infty S^l(E))$. It is a projective bundle with fiber $F = \mathbb{P}(E)_x = \mathbb{P}(E_x) = \mathbb{P}^r$. Let $\pi : \mathbb{P}(E) \to Y$ be the bundle map. There is a line bundle ξ on $\mathbb{P}(E)$, called the tautological line bundle, defined by the property that $\xi_F \cong \mathcal{O}_{\mathbb{P}^r}(1)$. When E is a vector bundle on a toric variety Y then the projective bundle $\mathbb{P}(E)$ has the structure of a toric variety if and only if $E = L_1 \oplus \ldots \oplus L_k$, [DRS04, Lemma 1.1.]. When the line bundles L_i are ample then the tautological line bundle ξ is also ample.

We refer to [FU] for the necessary background on vector bundles and their characteristic classes.

3 Toric Fibrations

Definition 3.1. A *toric fibration* is a surjective flat map $f : X \to Y$ with connected fibers where

(a) X is a toric variety
(b) Y is a normal algebraic variety
(c) $\dim(Y) < \dim(X)$.

Remark 3.2. A surjective morphism $f : X \to Y$, with connected fibers between normal projective varieties, induces a homomorphism from the connected component of the identity of the automorphism group of X to the connected component of the identity of the automorphism group of Y, with respect to which f is equivariant. It follows that if $f : X \to Y$ is a toric fibration then Y and a general fiber F admit a toric structure with respect to which f becomes an equivariant morphism. Moreover if X is smooth, respectively $\mathbb{Q}$-factorial, then Y and F are also smooth, respectively $\mathbb{Q}$-factorial.

Example 3.3. Let $L_0, \ldots, L_k$ be line bundles over a toric variety Y. The total space $\mathbb{P}(L_0 \oplus \ldots \oplus L_k)$ is a toric variety, Lemma 1.1, and the projective bundle $\pi : \mathbb{P}(L_0 \oplus \ldots \oplus L_k) \to Y$ is a toric fibration.

3.1 Combinatorial Characterization

A toric fibration has the following combinatorial characterization, see [EW, Chapter VI] for further details. Let $N \cong \mathbb{Z}^n$ be a lattice, $\Sigma \subset N \otimes \mathbb{R}$ be a fan and $X = X_\Sigma$, the associated toric variety. Let $i : \Delta \hookrightarrow N$ be a sub-lattice.

Proposition 3.4 ([EW]). *The inclusion i induces a toric fibration, $f : X \to Y$ if and only if:*

(a) Δ *is a primitive lattice, i.e.* $(\Delta \otimes \mathbb{R}) \cap N = \Delta$.
(b) *For every* $\sigma \in \Sigma(n)$, $\sigma = \tau + \eta$, *where* $\tau \in \Delta$ *and* $\eta \cap \Delta = \{0\}$ *(i.e.* Σ *is a split fan).*

We briefly outline the construction. The projection $\pi : N \to N/\Delta$ induces a map of fans $\Sigma \to \pi(\Sigma)$ and thus a map of toric varieties $f : X \to Y$. The general fiber

F is a toric variety defined by the fan $\Sigma_F = \{\sigma \in \Sigma \cap \Delta\}$. The invariant varieties $V(\tau)$ in X, where $\tau \in \Sigma$ is a maximal-dimensional cone in Σ_F, are called invariant sections of the fibration. The subvariety $V(\tau)$ is the invariant section passing through the fixed point of F corresponding to the cone $\tau \in \Sigma_F$. Observe that they are all isomorphic, as toric varieties, to Y.

Example 3.5. In Example 3.3 let $\Gamma \subset \mathbb{R}^n$ be the fan defining Y, and let $D_1, \ldots, D_s$ be the generators of $Pic(Y)$ associated to the rays $\eta_i, \ldots, \eta_s \subset \Gamma$. The line bundle L_i can be written as $L_i = \sum \phi_i(\eta_j) D_j$ where $\phi_i : \Gamma_{\mathbb{R}} \to \mathbb{R}$ are piecewise linear functions. Let $e_1, \ldots, e_k \in \mathbb{Z}^k$ be a lattice basis and let $e_0 = -e_1 - \ldots - e_k$. One can define a map:

$$\psi : \mathbb{R}^n \to \mathbb{R}^{n+k} \text{ as } \psi(v) = (v, \sum \phi_i(v) e_i).$$

Consider now the fan $\Sigma' \subset \mathbb{R}^{n+k}$ given by the image of Γ under ψ, $\Sigma' = \{\psi(\sigma), \sigma \subset \Gamma\}$. Let $\Pi \subset \mathbb{Z}^k$ be the fan defining $\mathbb{P}^k$. The fan $\Sigma = \{\sigma' + \tau | \sigma' \in \Sigma', \tau \in \Pi\}$ is a split fan, defining the toric fibration $\pi : \mathbb{P}(L_0 \oplus \ldots \oplus L_k) \to Y$. See also [ODAb, Proposition 1.33].

Definition 3.6. A *polarized toric fibration* is a pair $(f : X \to Y, L)$, where f is a toric fibration and L is an ample line bundle on X.

Observe that for a general fiber F, the pair $(F, L|_F)$ is also a polarized toric variety. It follow that both pairs (X, L) and $(F, L|_F)$ define lattice polytopes $P_{(X,L)}, P_{(F,L|_F)}$. The polytope $P_{(X,L)}$ is in fact a "twisted sum" of a finite number of lattice polytopes fibering over $P_{(F,L|_F)}$.

Definition 3.7. Let $R_0, \ldots, R_k \subset M_{\mathbb{R}}$ be lattice polytopes and let $k \geq 1$. Let $\pi : M \to \Lambda$ be a surjective map of lattices such that $\pi_{\mathbb{R}}(R_i) = v_i$ and such that $v_0, \cdots, v_k$ are distinct vertices of $\text{Conv}(v_0, \ldots, v_k)$. We will call a *Cayley π-twisted sum (or simply a Cayley sum) of $R_0, \ldots, R_k$* a polytope which is affinely isomorphic to $\text{Conv}(R_0, \ldots, R_k)$.

We will denote it by: $[R_0 \star \ldots \star R_k]_\pi$.

If the polytopes R_i are additionally normally equivalent, i.e. they define the same normal fan Σ_Y, we will denote the Cayley sum by:

$$\text{Cayley}(R_0, \ldots, R_k)_{(\pi, Y)}.$$

We will see that these are the polytopes that are associated to polarized toric fibrations.

Proposition 3.8 ([CDR08]). *Let $X = X_\Sigma$ be a toric variety of dimension n, where $\Sigma \subset N_{\mathbb{R}} \cong \mathbb{R}^n$, and let $i : \Delta \hookrightarrow N$ be a sublattice. Let L be an ample line bundle* on X. Then the inclusion i induces a polarized toric fibration $(f : X \to Y, L)$ if and only if $P_{(X,L)} = \text{Cayley}(R_0, \ldots, R_k)_{(\pi, Y)}$, where $R_0, \ldots, R_k$ are normally equivalent polytopes on Y and $\pi : M \to \Lambda$ is the lattice map dual to i.

Proof. We first prove the implication "$\Rightarrow$".

Assume that $i : \Delta \hookrightarrow N$ induces a toric fibration $f : X \to Y$ and consider the polarization L on X. We will prove that $P_{(X,L)} = \text{Cayley}(R_0, \ldots, R_k)_{(\pi, Y)}$ for some normally equivalent polytopes $R_0, \ldots, R_k$.

Notice first that the fact that Δ is a primitive sub-lattice of N implies that the dual map $\pi : M \to \Lambda$ is a surjection. Let F be a general fiber of f, and let $S := P_{(F,L|F)} \subset \Lambda_\mathbb{R}$. Denote by $v_0, \ldots, v_k$ the vertices of S. Every v_i corresponds to a fixed point of F; call $Y_i = V(\tau_i)$ the invariant section of f passing through that point. Note that $\tau_i \in \Sigma_X$, $\dim \tau_i = \dim F$ and $\tau_i \subset \Delta_\mathbb{R}$. Let R_i be the face of $P_{(X,L)}$ corresponding to Y_i.

Observe that $\text{Aff}(\tau_i) = \Delta_\mathbb{R}$, so that $\text{Aff}(\tau_i^\perp) = \Delta_\mathbb{R}^\perp = \ker \pi_\mathbb{R}$. Then there exists $u_i \in M$ such that:

- $\text{Aff}(R_i) + u_i = \ker(\pi_\mathbb{R})$;
- $R_i + u_i = P_{(Y_i, L|Y_i)}$.

This says that $R_0, \ldots, R_k$ are normally equivalent (because every Y_i is isomorphic to Y), and that $\pi_\mathbb{R}(R_i)$ is a point. Since the Y_i's are pairwise disjoint, the same holds for the R_i's. If s is the number of fixed points of Y, then each R_i has s vertices. On the other hand, we know that F has $(k+1)$ fixed points, and therefore X must have $s(k+1)$ fixed points. So $P_{(X,L)}$ has $s(k+1)$ vertices, namely the union of all vertices of the R_i's. We can conclude that

$$P_{(X,L)} = \text{Conv}(R_0, \ldots, R_k)$$

Let $D = \sum_{x \in \Sigma(1)} a_x D_x$ be an invariant Cartier divisor on X such that $L = \mathcal{O}_X(D)$. Since F is a general fiber, we have $D_x \cap F \neq \emptyset$ if and only if $x \in \Delta$, and $D_{|F} = \sum_{x \in \Delta} a_x D_{x|F}$. This implies that $\pi_\mathbb{R}(R_i) = v_i$ and $\pi_\mathbb{R}(P_{(X,L)}) = S$. We conclude that $P_{(X,L)} = \text{Cayley}(R_0, \ldots, R_k)_{(\pi, Y)}$.

We now show the other direction: "$\Leftarrow$".

Assume that $P_{(X,L)} = \text{Cayley}(R_0, \ldots, R_k)_{(\pi, Y)}$. We will prove that the associated polarized toric variety is a polarized toric fibration. First observe that the fact that the dual map π is a surjection implies that the sublattice Δ is primitive. Since v_i is a vertex of $\pi_\mathbb{R}(P_{(X,L)})$, R_i is a face of $P_{(X,L)}$ for every $i = 0, \ldots, k$. Let Y be the projective toric variety defined by the polytopes R_i. Observe that $\text{Aff}(R_i)$ is a translate of $\ker \pi_\mathbb{R}$, and $(\ker \pi)^\vee = N/\Delta$. So the fan Σ_Y is contained in $(N/\Delta)_\mathbb{R}$.

Let $\gamma \in \Sigma_Y(\dim(Y))$ and for every $i = 0, \ldots, k$ let w_i be the vertex of R_i corresponding to γ. We will show that $Q := \text{Conv}(w_0, \ldots, w_k)$ is a face of $P_{(X,L)}$.

Observe first that $(\pi_\mathbb{R})_{\text{Aff}(Q)} : \text{Aff}(Q) \to \Lambda_\mathbb{R}$ is bijective. Let H be the linear subspace of $M_\mathbb{R}$ which is a translate of $\text{Aff}(Q)$. Then we have $M_\mathbb{R} = H \oplus \ker \pi_\mathbb{R}$. Dually $N_\mathbb{R} = \Delta_\mathbb{R} \oplus H^\perp$, where $H^\perp$ projects isomorphically onto $(N/\Delta)_\mathbb{R}$.

Let $u \in H^\perp$ be such that its image in $(N/\Delta)_\mathbb{Q}$ is contained in the interior of γ. Then for every $i = 0, \ldots, k$ we have that (see [FU, §1.5]):

$$(u, x) \geqslant (u, w_i) \text{ for every } x \in R_i,$$
$$(u, x) = (u, w_i) \text{ if and only if } x = w_i.$$

Moreover u is constant on $Aff(Q)$, namely there exists $m_0 \in \mathbb{Q}$ such that $(u, z) = m_0$ for every $z \in Q$.

Any $z \in P$ can be written as $z = \sum_{i=1}^{l} \lambda_i z_i$, with $z_i \in R_i$, $\lambda_i \geqslant 0$ and $\sum_{i=0}^{l} \lambda_i = 1$. Then

$$(u, z) = \sum_{i=1}^{l} \lambda_i (u, z_i) \geqslant \sum_{i=1}^{l} \lambda_i (u, w_i) = \sum_{i=1}^{l} \lambda_i m_0 = m_0.$$

Moreover, $(u, z) = m_0$ if and only if $\lambda_i > 0$ for every i such that $(u, z_i) = (u, w_i)$, and $\lambda_i = 0$ otherwise. This happens if and only if $z \in Q$, implying that Q is a face of $P_{(X,L)}$.

Let $\sigma \in \Sigma_X$ be a cone of maximal dimension, and let w be the corresponding vertex of $P_{(X,L)}$. Then $\pi(w)$ is a vertex, say v_1, of $\pi_\mathbb{Q}(P_{(X,L)})$ and hence w lies in R_1. Since R_1 is also a face of $P_{(X,L)}$, w is a vertex of R_1 and hence it corresponds to a maximal dimensional cone in Σ_Y. In each R_i, consider the vertex w_i corresponding to the same cone of Σ_Y. We set $w_1 = w$. We have shown that $Q := \text{Conv}(w_0, \ldots, w_k)$ is a face of $P_{(X,L)}$, and $w = Q \cap R_1$. Now call τ and η the cones of Σ_X corresponding respectively to R_1 and Q. It follows that $\sigma = \tau + \eta$, $\tau \subset \Delta_\mathbb{Q}$, and $\eta \cap \Delta_\mathbb{Q} = \{0\}$. This concludes the proof. □

The previous proof shows the following corollary.

Corollary 3.9. *Let $(f : X \to Y, L)$ be a polarized toric fibration and let $P_{(X,L)} = \text{Cayley}(R_0, \ldots, R_k)_{(\pi, Y)}$ be the associated polytope. Let F be a general fiber of the fibration, $Y_0, \ldots, Y_k$ be the invariant sections and $\pi(R_i) = v_i$. The following holds.*

(a) *The polarized toric variety $(F, L|_F)$ corresponds to the polytope $P_{(F,L|_F)} = \text{Conv}(v_0, \ldots, v_k)$.*
(b) *The polarized toric varieties (Y_i, L_{Y_i}) correspond to the polytopes*

$$R_0 - u_0, \cdots, R_k - u_k,$$

where $u_i \in M$ is a point such that $\pi(u_i) = \pi(R_i)$.

Example 3.10. The toric surface obtained by blowing up $\mathbb{P}^2$ at a fixed point has the structure of a toric fibration, $\mathbb{P}(\mathcal{O}_{\mathbb{P}^1} \oplus \mathcal{O}_{\mathbb{P}^1}(1)) \to \mathbb{P}^1$. It is often referred to as the Hirzebruch surface $\mathbb{F}_1$. Consider the polarization given by the tautological line bundle $\xi = 2\phi^*(\mathcal{O}_{\mathbb{P}^2}(1)) - E$ where ϕ is the blow-up map and E is the exceptional divisor. The associated polytope is $P = \text{Cayley}(\Delta_1, 2\Delta_1)$, see the figure below.

Linear Toric Fibrations

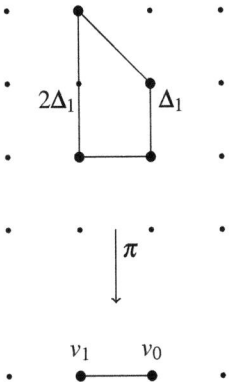

Remark 3.11. The following are important classes of polarized toric fibrations, relevant both in Combinatorics and Algebraic Geometry.

Projective Bundles. When $\pi(P) = \Delta_t$ the polytope Cayley$(R_0, \ldots, R_t)_{(\pi,Y)}$ defines the polarized toric fibration $(\mathbb{P}(L_0 \oplus \ldots \oplus L_t) \to Y, \xi)$, where the L_i are ample line bundles on the toric variety Y and ξ is the tautological line bundle. In particular $L|_F = \mathcal{O}_{\mathbb{P}^t}(1)$. These fibrations play an important role in the theory of discriminants and resultants of polynomial systems. See Sect. 4 for more details.

Mori Fibrations. When $\pi(P)$ is a simplex (not necessarily smooth) the Cayley polytope Cayley$(R_0, \ldots, R_k)_{(\pi,Y)}$ defines a Mori fibration, i.e. a surjective flat map onto a $\mathbb{Q}$-factorial toric variety whose generic fiber is reduced and has Picard number one. This type of fibrations are important blocks in the Minimal Model Program for toric varieties. See [CDR08] and [Re83] for more details.

$\mathbb{P}^k$-Bundles. When $\pi(P) = k\Delta_t$ then again the variety has the structure of a $\mathbb{P}^t$-fibration whose general fiber F is embedded as an k-Veronese variety: $(F, L|_F) = (\mathbb{P}^t, \mathcal{O}_{\mathbb{P}^t}(k))$. These fibrations arise in the study of k-th toric duality, see [DDRP12].

In the polarized toric fibration $(\mathbb{P}(L_0 \oplus \ldots \oplus L_t), \xi)$ the fibers are embedded as linear spaces. For this reason the associated Cayley polytopes Cayley$(R_0, \ldots, R_t)_{(\pi,Y)}$ can be referred to as *linear toric fibrations*.

Remark 3.12. For general Cayley sums, $[R_0 \star \ldots \star R_k]_\pi$, one has the following geometrical interpretation. Let (X, L) be the associated polarized toric variety and let Y be the toric variety defined by the Minkowski sum $R_0 + \ldots + R_k$. The fan defining Y is a refinement of the normal fans of the R_i for $i = 0, \ldots, k$. Consider the associated birational maps $\phi_i : Y \to Y_i$, where (Y_i, L_i) is the polarized toric variety defined by the polytope R_i. The line bundles $H_i = \phi_i^*(L_i)$ are nef line bundles on Y and the polytopes $P_{(Y,H_i)}$ are affinely isomorphic to R_i. In particular $[R_0 \star \ldots \star R_k]_\pi$ is the polytope defined by the tautological line bundle on the toric fibration $\mathbb{P}(H_0 \oplus \ldots \oplus H_k) \to Y$. Notice that in this case the line bundle ξ may not be ample.

If we want to relate $[R_0 \star \ldots \star R_k]_\pi$ to a polarized toric fibration we need to enlarge the polytopes R_i is order to get an ample tautological line bundle. Consider the polytopes $P_i = P_{(Y,H_i)} + \sum_0^k R_j$. The normal fan of P_i is isomorphic to the fan defining the common resolution, Y, for $i = 0, \ldots, k$. Hence the polytopes P_i are normally equivalent. Let (Y, M_i) be the polarized toric variety associated to the polytope P_i. One can then define the Cayley sum $\mathrm{Cayley}(P_0, \ldots, P_k)_{(\pi,Y)}$, whose normal fan is in fact a refinement of the one defining $[R_0 \star \ldots \star R_k]_\pi$. Let $(\mathbb{P}(M_0 \oplus \ldots \oplus M_k) \to Y, \xi)$ be the polarized toric fibration associated to $\mathrm{Cayley}(P_0, \ldots, P_k)_{(\pi,Y)}$. There is a birational morphism $\phi: \mathbb{P}(M_0 \oplus \ldots \oplus M_k) \to X$.

Example 3.13. Consider the polytopes $R_0 = \Delta_2$, $R_1 = \Delta_1 \times \Delta_1$ in $\mathbb{Q}^2$. Consider the projection onto the first component $\pi: \mathbb{Z}^3 \to \mathbb{Z}$ and $P = \mathrm{Conv}(R_0 \times \{0\}, R_1 \times \{1\})$. The polytope P is then isomorphic to $[R_0 \star R_1]_\pi$, and $\pi_\mathbb{Q}(P) = \Delta_1$. The common refinement defined by $R_0 + R_1$ is the fan of the blow up of $\mathbb{P}^2$ at two fixed points, $\phi: Y \to \mathbb{P}^2$. The polytopes P_0 defines the polarized toric variety $(Y, \phi^*(\mathcal{O}_{\mathbb{P}^2}(4)) - E_1 - E_2)$ and the polytope P_1 the pair $(Y, \phi^*(\mathcal{O}_{\mathbb{P}^2}(5)) - 2E_1 - 2E_2)$, where E_i are the exceptional divisors. The polarized toric fibration $(\mathbb{P}(M_0 \oplus M_1) \to Y, \xi)$ is then

$$(\mathbb{P}([\phi^*(\mathcal{O}_{\mathbb{P}^2}(4)) - E_1 - E_2] \oplus [\phi^*(\mathcal{O}_{\mathbb{P}^2}(5)) - 2E_1 - 2E_2]) \to Y, \xi).$$

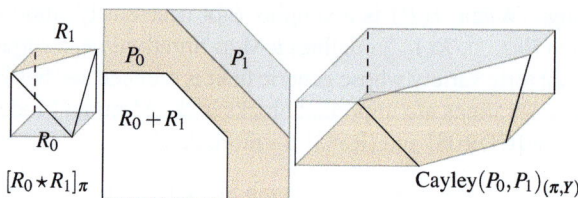

3.2 Historical Remark

The definition of a Cayley polytope originated from what is "classically" referred to as the *Cayley trick*, in connection with the Resultant and Discriminant of a system of polynomials. A system of n polynomials in n variables $x = (x_1, \ldots, x_n)$, $f_1(x), \ldots, f_n(x)$, is supported on $(\mathscr{A}_1, \mathscr{A}_2, \ldots, \mathscr{A}_n)$, where $\mathscr{A}_i \subset \mathbb{Z}^n$ if $f_i = \prod_{a_j \in \mathscr{A}_i} c_j x^{a_j}$.

The $(\mathscr{A}_1, \mathscr{A}_2, \ldots, \mathscr{A}_n)$-*resultant* is a polynomial, $R(\ldots, c_j, \ldots)$, in the coefficients c_j, which vanishes whenever the corresponding polynomials have a common zero.

The discriminant of a finite subset $\mathscr{A} \subset \mathbb{Z}^n$, $\Delta_\mathscr{A}$, is also a polynomial $\Delta_\mathscr{A}(\ldots, c_j, \ldots)$ in the variables c_j, which vanishes whenever the corresponding polynomial supported on $\mathscr{A}$, $f = \Pi_{a_j \in \mathscr{A}} c_j x^{a_j}$, has a singularity in the torus $(\mathbb{C}^*)^n$.

Theorem 3.14 ([GKZ] Cayley Trick). *The $(\mathscr{A}_1, \mathscr{A}_2, \ldots, \mathscr{A}_n)$-resultant equals the $\mathscr{A}$-discriminant where*

$$\mathscr{A} = (\mathscr{A}_1 \times \{0\}) \cup (\mathscr{A}_2 \times \{e_1\}) \cup \ldots \cup (\mathscr{A}_n \times \{e_{n-1}\}) \subset \mathbb{Z}^{2n-1}$$

where $(e_1, \ldots, e_{n-1})$ is a lattice basis for $\mathbb{Z}^{n-1}$.

Let $R_i = N(f_i) \subset \mathbb{R}^n$ be the Newton polytopes of the polynomials f_i supported on $\mathscr{A}_i$. The Newton polytope of the polynomial f supported on $\mathscr{A}$ is the Cayley sum

$$N(f) = [R_1 \star \ldots \star R_n]_\pi,$$

where $\pi : \mathbb{Z}^{2n-1} \to \mathbb{Z}^{n-1}$ is the natural projection such that $\pi_\mathbb{R}([R_1 \star \ldots \star R_n]_\pi) = \Delta_{n-1}$.

4 Toric Discriminants and Toric Fibrations

The term "discriminant" is well known in relation with low degree equations or ordinary differential equations. We will study discriminants of polynomials in n variables with prescribed monomials, i.e. polynomials whose exponents are given by lattice points in $\mathbb{Z}^n$.

Polynomials in n-variables describe locally the hyperplane sections of a projective n-dimensional algebraic variety, $\phi : X \hookrightarrow \mathbb{P}^m$. The monomials are prescribed by the local representation of a basis of the vector space of global sections $H^0(X, \phi^*(\mathscr{O}_{\mathbb{P}^m}(1)))$. For this reason the term discriminant has also been classically used in Algebraic Geometry.

In what follows we will describe discriminants from a combinatorial and an algebraic geometric prospective. The two points of view coincide when the projective embedding is toric.

4.1 The $\mathscr{A}$ Discriminant

Let $\mathscr{A} = \{a_0, \ldots, a_m\}$ be a subset of $\mathbb{Z}^n$. The discriminant of $\mathscr{A}$ (when it exists) is an irreducible homogeneous polynomial $\Delta_\mathscr{A}(c_0, \ldots, c_m)$ vanishing when the corresponding Laurent polynomial supported on $\mathscr{A}$, $f(x) = \sum_{a_i \in \mathscr{A}} c_i x^{a_i}$, has at least one singularity in the torus $(\mathbb{C}^*)^n$. Geometrically, the zero-locus of the discriminant is an irreducible algebraic variety of codimension one in the dual projective space $\mathbb{P}^{m\vee}$, called the *dual variety* of the embedding $X_\mathscr{A} \hookrightarrow \mathbb{P}^m$.

Example 4.1. Consider the point configuration

$$\mathscr{A} = \{(0,0), (1,0), (0,1), (1,1)\} \subset \mathbb{Z}^2.$$

The discriminant is given by an homogeneous polynomial $\Delta_{\mathscr{A}}(a_0, a_1, a_2, a_3)$ vanishing whenever the quadric $a_0 + a_1 x + a_2 y + a_3 xy$ has a singular point in $(\mathbb{C}^*)^2$. It is well known that this locus correspond to singular 2×2 matrices and it is thus described by the vanishing of the determinant: $\Delta_{\mathscr{A}}(a_0, a_1, a_2, a_3) = a_0 a_3 - a_1 a_2$. Similarly, one can associate the polynomials supported on $\mathscr{A}$ with local expansions of global sections in $H^0(\mathbb{P}^1 \times \mathbb{P}^1, \mathscr{O}(1,1))$ defining the Segre embedding of $\mathbb{P}^1 \times \mathbb{P}^1$ in $\mathbb{P}^3$.

Example 4.2. The 2-Segre embedding $\nu_2 : \mathbb{P}^2 \hookrightarrow \mathbb{P}^5$ defined by the global sections of the line bundle $\mathscr{O}_{\mathbb{P}^2}(2)$ can be associated to the point configuration $\mathscr{A} = \{a_0, a_1, a_2, a_3, a_4, a_5\} = \{(0,0), (0,1), (1,0), (0,2), (1,1), (2,0)\}$

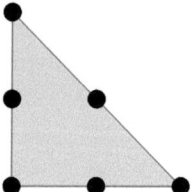

A simple computation shows that $\Delta_{\mathscr{A}} = \det \begin{bmatrix} c_0 & c_1 & c_2 \\ c_1 & c_3 & c_4 \\ c_2 & c_4 & c_5 \end{bmatrix}$

Projective duality is a classical subject in algebraic geometry. Given en embedding $i : X \hookrightarrow \mathbb{P}^m$ of an n-dimensional algebraic variety, the dual variety, $X^\vee \subset (\mathbb{P}^m)^\vee$ is defined as the Zariski-closure of all the hyperplanes $H \subset \mathbb{P}^m$ tangent to X at some non singular point. We can speak of a defining homogeneous polynomial $\Delta(c_0, \ldots, c_m)$, and thus of a discriminant, only when the dual variety has codimension one. Embeddings whose dual variety has higher codimension are called *dually defective* and the discriminant is set to be 1. Finding formulas for the discriminant Δ and giving a classification of the embeddings with discriminant 1 is a long standing problem in algebraic geometry. In the case of a toric embedding defined by a point-configuration, $X_{\mathscr{A}} \hookrightarrow \mathbb{P}^{|\mathscr{A}|-1}$, the problem is equivalent to finding formulas for the discriminant $\Delta_{\mathscr{A}}$ and giving a classification of the *dually defective point-configurations,* i.e. the point-configurations with discriminant $\Delta_{\mathscr{A}} = 1$.

Dickenstein-Sturmfels [DS02] characterized the case when $m = n+2$, Cattani-Curran [CC07] extended the classification to $m = n+3, n+4$. In these cases the corresponding embedding is possibly very singular and the methods used are purely combinatorial. In [DiR06] and [CDR08] we completely characterize the case when $P_{\mathscr{A}} = \text{Conv}(\mathscr{A})$ is smooth or simple. The latter characterisation relies on tools from Algebraic Geometry which will be explained in the next paragraph.

4.2 The Dual Variety of a Projective Variety

The dual variety corresponds to the locus of singular hyperplane sections of a given embedding. By requiring the singularity to be of a given order, one can define more general dual varieties. Singularities of fixed multiplicity k correspond to hyperplanes tangent "to the order k." Consider an embedding $i : X \hookrightarrow \mathbb{P}^m$ of an n-dimensional variety, defined by the global sections of the line bundle $\mathscr{L} = i^*(\mathscr{O}_{\mathbb{P}^m}(1))$. For any smooth point x of the embedded variety let:

$$jet_x^k : H^0(X, \mathscr{L}) \to H^0(X, \mathscr{L} \otimes \mathscr{O}_X/\mathfrak{m}_x^{k+1})$$

be the map assigning to a global section s in $H^0(X, \mathscr{L})$ the tuple

$$jet_x^k(s) = (s(x), \ldots, (\partial^t s/\partial \underline{x}^t)(x), \ldots)_{t \leq k}$$

where $\underline{x} = (x_1, \ldots, x_n)$ is a local system of coordinates around x. The k-th *osculating space* at x is defined as $\mathbb{O}sc_x^k = \mathbb{P}(Im(jet_x^k))$. As the map jet_x^1 is surjective, the first osculating space is always isomorphic to $\mathbb{P}^n$ and it is classically called the *projective tangent space*. The jet maps of higher order do not necessarily have maximal rank and thus the dimension of the osculating spaces of order bigger than 1 can vary. The embeddings admitting osculating space of maximal dimension at every point are called k-jet spanned.

Definition 4.3. A line bundle $\mathscr{L}$ on X is called *k-jet spanned* at x if the map jet_x^k is surjective. It is called k-jet spanned if it is k-jet spanned at every smooth point $x \in X$.

Example 4.4. A line bundle $\mathscr{L} = \mathscr{O}_{\mathbb{P}^n}(a)$ on $\mathbb{P}^n$ is k-jet spanned for all $a \geq k$. In fact the map

$$jet_x^k : H^0(\mathbb{P}^n, \mathscr{O}_{\mathbb{P}^n}(a)) \to J_k(\mathscr{O}_{\mathbb{P}^n}(a))_x$$

is surjective for all $x \in \mathbb{P}^2$, as a local basis of the global sections of $\mathscr{O}_{\mathbb{P}^n}(a)$ consists of all the monomials in n variables of degree up to a and we are assuming $a \geq k$.

Example 4.5. Let $\mathscr{L}$ be a line bundle on a non singular toric variety X. Then the following statements are equivalent, see [DiR01]:

(a) $\mathscr{L}$ is k-jet spanned.
(b) all the edges of $P_{\mathscr{L}}$ have length at least k.
(c) $\mathscr{L} \cdot C \geq k$ for every invariant curve C on X.

As an example consider the polytope P in figure below. The associated toric embedding is the embedding of the blow up of $\mathbb{P}^2$ at the three fixed points, $\phi : X \to \mathbb{P}^2$, defined by the anticanonical line bundle $\phi^*(\mathcal{O}_{\mathbb{P}^2}(3)) - E_1 - E_2 - E_3$. Here E_i denote the exceptional divisors. The embedded variety is a Del Pezzo surface of degree 6. Let F be the set of the 6 fixed points on X and $E = \{\phi^*(\mathcal{O}_{\mathbb{P}^2}(3)) - E_i - E_j, i \neq i\} \cup \{E_1, E_2, E_3\}$ be the set of invariant curves. The osculating spaces can easily seen to be:

$$\mathbb{O}sc_p^2 = \begin{cases} \mathbb{P}^3 = & <jet_p^2(1), jet_p^2(x), jet_p^2(y), jet_p^2(xy)> \\ & \text{if } x \in F. \\ \mathbb{P}^4 = & <jet_p^2(1), jet_p^2(x), jet_p^2(y), jet_p^2(xy), jet_p^2(x^2y)> \\ & \text{if } x \in E \setminus F. \\ \mathbb{P}^5 = & <jet_p^2(1), jet_p^2(x), jet_p^2(y), jet_p^2(xy), jet_p^2(x^2y), jet_p^2(xy^2)> \\ & \text{at a general point } p \in X \setminus E. \end{cases}$$

The embedding defined by P is not 2-jet spanned on the whole X. It is 2-jet spanned at every point in $X \setminus E$.

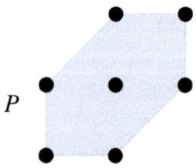

P

Definition 4.6. A hyperplane $H \subset \mathbb{P}^m$ is tangent at x to the order k if it contains the k-th osculating space at x: $\mathbb{O}sc_x^k \subset H$.

Definition 4.7. The k-th order dual variety X^k is:

$$X^k = \overline{\{H \in \mathbb{P}^{m*} \text{ tangent to the order } k \text{ to } X \text{ at some non singular point}\}}.$$

Notice that $X^1 = X^\vee$ and that X^2 is contained in the singular locus of $X^\vee$. General properties of the higher order dual variety have been studied by S. Kleiman and R. Piene. Because the definition is related to local osculating properties and generation of jets, it is useful to introduce the sheaf of jets, $J_k(\mathscr{L})$, associated to a polarized variety $(X, \mathscr{L})$. In the classical literature it is sometime referred to as the *sheaf of principal parts*.

Consider the projections $\pi_i : X \times X \to X$ and let $\mathscr{I}_{\Delta_X}$ be the ideal sheaf of the diagonal in $X \times X$. The sheaf of k-th order jets of the line bundle $\mathscr{L}$ is defined as

$$J_k(\mathscr{L}) = \pi_{2*}(\pi_1^*(\mathscr{L}) \otimes (\mathscr{O}_{X \times X}/\mathscr{I}_{\Delta_X}^{k+1})).$$

When the variety X is smooth $J_k(\mathscr{L})$ is a vector bundle of rank $\binom{n+k}{n}$, called the *k-jet bundle*.

Example 4.8. If $\mathscr{L} \neq \mathscr{O}_X$ is a globally generated line bundle then $J_k(\mathscr{L})$ splits as a sum of line bundles only if $X = \mathbb{P}^n$ and $\mathscr{L} = \mathscr{O}_{\mathbb{P}^n}(a)$. In fact:

$$J_k(\mathscr{O}_{\mathbb{P}^n}(a)) = \bigoplus_1^{\binom{n+k}{k}} \mathscr{O}_{\mathbb{P}^n}(a-k)$$

See [DRS01] for more details.

It is important to note that when the map jet_x^k is surjective for all smooth points x, properties of the higher dual variety X^k can be related to vanishing of Chern classes of the associated k-th jet bundle, $J_k(\mathscr{L})$. We start by identifying the k-th dual variety with a projection of the conormal bundle. Let X be a smooth algebraic variety and let $\mathscr{L}$ be a k-jet spanned line bundle on X. Consider the following commutative diagram.

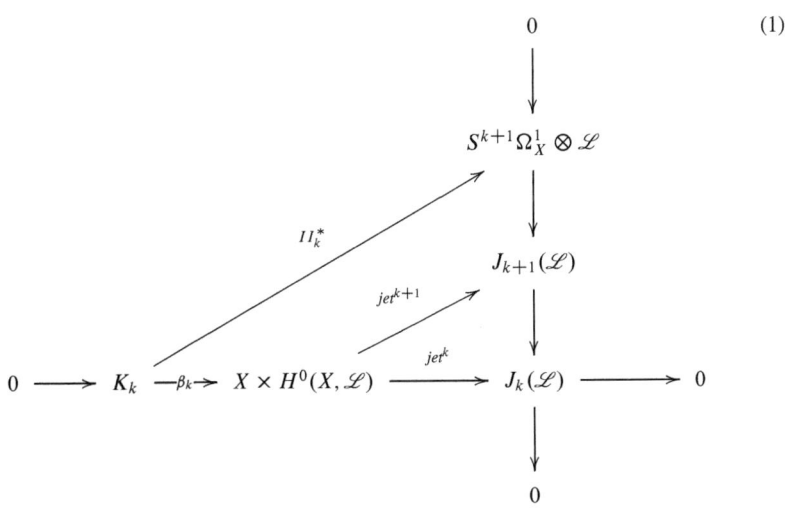
(1)

The vertical exact sequence is often called the k-jet sequence. The map jet^k is defined as $jet^k(s,x) = jet_x^k(s)$. The vector bundle K_k is the kernel of the map jet^k (which has maximal rank!). The induced map $II_k^\vee$ can be identified with the dual of the k-th fundamental form. See [L94, GH79] for more details. By dualizing the map β_k and projectivizing the corresponding vector bundles one gets the following maps:

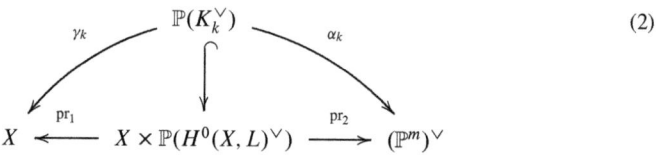
(2)

It is straightforward to see that $X^k = Im(\alpha_k)$. A simple dimension count shows that when the map jet^k has maximal rank one expects the codimension of the k-th dual variety to be $\text{codim}(X^k) = \binom{n+k}{k} - n$. Notice that this is equivalent to requiring that the map α_k is generically finite. When the codimension is higher than the expected one the embedding is said to be k-th dually defective.

The commutativity in diagram (1) has the following useful consequence.

Lemma 4.9. *Let $(X, \mathscr{L})$ be a polarized variety, where X is smooth and the line bundle $\mathscr{L}$ is $(k+1)$-jet spanned. Then the dual variety X^k has the expected dimension.*

Proof. We follow diagram (1). Because the line bundle $\mathscr{L}$ is $(k+1)$-jet ample the map Π_k^* is surjective. This means that for every $x \in X$ and for every monomial $\Pi_{\sum t_i = k+1} x_1^{t_1} \cdots x_n^{t_n}$ there is an hyperplane section that locally around x is defined as

$$C \cdot \Pi_{\sum t_i = k+1} x_1^{t_1} \cdots x_n^{t_n} + \text{higher order terms} = 0, \text{ where } C \neq 0$$

In other words, hyperplanes tangent at a point x to the order k are in one-to-one correspondence with elements of the linear system $|\mathscr{O}_{\mathbb{P}^{n-1}}(k+1)|$. The map α_k having positive dimensional fibers is equivalent to saying that hyperplanes tangent at a point x to the order k are also tangent to nearby points $y \neq x$, which in turn implies that the linear system $|\mathscr{O}_{\mathbb{P}^{n-1}}(k+1)|$ has base points. This is a contradiction as the linear system is $k+1$-jet spanned and thus base-point free. $\square$

When $k = 1$ the contact locus of a general singular hyperplane H, $\gamma_k(\alpha_k^{-1}(H))$ is always a linear subspace. This implies that if finite then $\deg(\alpha_1) = 1$. For higher order tangencies, $k > 1$, the degree can be higher. When the map α_k is finite we set $n_k = \deg(\alpha_k)$.

Lemma 4.10 ([LM00, DDRP12]). *Let X be a smooth variety and let $\mathscr{L}$ be a k-jet spanned line bundle. Then $\text{codim}(X^k) > \binom{n+k}{k} - n$ if and only if $c_n(J_k(\mathscr{L})) = 0$. Moreover when $\text{codim}(X^k) = \binom{n+k}{k} - n$ the degree of the k-dual variety is given by:*

$$n_k \deg(X^k) = c_n((J_k(\mathscr{L})).$$

Proof. Observe first that because the map jet_k is of maximal rank the vector bundle $J_k(\mathscr{L})$ is spanned by the global sections of the line bundle $\mathscr{L}$. This implies that, after fixing a basis $\{s_1, \ldots, s_{m+1}\}$ of $H^0(X, \mathscr{L}) \cong \mathbb{C}^{m+1}$, the Chern class $c_n(J_k(\mathscr{L}))$ is represented by the set:

$$\{x \in X \mid \dim(\mathrm{Span}(jet_x^k(s_1), jet_x^k(s_2))) \leq 1\}$$

Notice that an hyperplane in the linear span $\mathbb{P}^t = \langle s_1,\ldots,s_{t+1}\rangle$ is tangent at a point x to the order k exactly when $\dim(\mathrm{Span}(jet_x^k(s_1),\ldots,jet_x^k(s_{t+1}))) = t+1$. The map γ_k in diagram (2) defines a projective bundle of rank $m - \binom{n+k}{k}$. The statement $c_n(J_k(\mathscr{L})) = 0$ is then equivalent to $\alpha_k(\gamma^{-1}(x)) \cap \mathbb{P}^1 = \emptyset$ for every $x \in X$ and for a general $\mathbb{P}^1 = \langle s_1, s_2\rangle$. By Bertini this is equivalent to $\mathrm{codim}(X^k) > \binom{n+k}{k} - n$. Assume now that $c_n(J_k(\mathscr{L})) \neq 0$ and thus that the generic fiber of the map α_k is finite. The degree of $X^k = im(\alpha_k)$ times the degree of the map α_k is given by the degree of the line bundle $\alpha_k^*(\mathscr{O}_{\mathbb{P}^{m\vee}}(1))$ which corresponds to the tautological line bundle $\mathscr{O}_{\mathbb{P}(K_k^\vee)}(1)$.

$$n_k \deg(X^k) = c_1(\alpha_k^*(\mathscr{O}_{(\mathbb{P}^m)^\vee} 1))^{m+n-\binom{n+k}{k}} = c_1(\mathscr{O}_{\mathbb{P}(K_k^\vee)}(1))^{m+n-\binom{n+k}{k}}.$$

From diagram (1) we see that

$$c_n(J_k(\mathscr{L})) = c_n(K_k^\vee)^{-1} = s_n(K_k^\vee)$$

Finally let $\pi : \mathbb{P}(K_k^\vee) \to X$ be the bundle map. By relating the Segre class $s_n(K_k^\vee)$ to the tautological bundle [FU, 3.1] $s_n(K_k^\vee) = \pi_*(c_1(\mathscr{O}_{\mathbb{P}(K_k^\vee)}(1))^{m+n-\binom{n+k}{k}}) = c_1(\mathscr{O}_{\mathbb{P}(K_k^\vee)}(1))^{m+n-\binom{n+k}{k}}$ we conclude that: $n_k \deg(X^k) = c_n(J_k(\mathscr{L}))$. □

The case of $k = 1$ is referred to as classical projective duality. When the codimension of the dual variety is one, the homogeneous polynomial in $m + 1$ variables defining it is called the *discriminant of the embedding*. For a polarized variety the discriminant, when it exists, parametrizes the singular hyperplane sections.

4.3 The Toric Discriminant

In the case of singular varieties the sheaves of k-jets are not necessarily locally free and thus it is not possible to use Chern-classes techniques.

For toric varieties however estimates of the degree of the dual varieties are possible, even in the singular case, and rely on properties of the associated polytope. In the classical case $k = 1$ there is a precise characterization in any dimension. For higher order duality, results in dimension 3 and for $k = 2$ can be found in [DDRP12]. A generalization to higher dimension and higher order is an open problem.

Proposition 4.11 ([GKZ, DiR06, MT11]). *Let $(X_\mathscr{A}, L_\mathscr{A})$ be a polarized toric variety associated to the polytope $P_\mathscr{A}$. Set*

$$\delta_i = \sum_{\emptyset \neq F \prec P} (-1)^{\mathrm{codim}(F)} \left\{ \binom{\dim(F)+1}{i} + ((-1)^{i-1}(i-1)) \right\} \mathrm{Vol}(F)\,\mathrm{Eu}(V(F)).$$

Then $\mathrm{codim}(X_{\mathscr{A}}^{\vee}) = r = \min\{i, \delta_i \neq 0\}$ and $\deg(X_{\mathscr{A}}^{\vee}) = \delta_r$.

The function Eu : {invariant subvarieties of X_A} $\to \mathbb{Z}$ in the above proposition assigns an integer to all invariant subvarieties. Its value is different from 1 only when the variety is singular. In particular, when $X_{\mathscr{A}}$ is smooth we have:

$$\mathrm{codim}(X_A^{\vee}) > 1 \Leftrightarrow \sum_{\emptyset \neq F \prec P} (-1)^{\mathrm{codim}(F)} (\dim(F)+1)!\,\mathrm{Vol}(F) = 0$$

In fact in the smooth case one can prove this characterization using the vector bundle of 1-jets.

Proposition 4.12. *Let $(X_{\mathscr{A}}, L_{\mathscr{A}})$ be an n-dimensional non singular polarized toric variety associated to the polytope $P_{\mathscr{A}}$. Assume $\mathscr{A} = P_{\mathscr{A}} \cap \mathbb{Z}^n$. Then*

$$c_n(J_1(\mathscr{L}_{\mathscr{A}})) = \sum_{\emptyset \neq F \prec P} (-1)^{\mathrm{codim}(F)} (\dim(F)+1)!\,\mathrm{Vol}(F)$$

Proof. Chasing the diagram (1) one sees:

$$c_n(J_1(\mathscr{L}_{\mathscr{A}})) = \sum_{i=0}^{n} (n+1-i) c_i(\Omega_{X_{\mathscr{A}}}^1) \cdot \mathscr{L}_{\mathscr{A}}^i$$

Consider now the generalized Euler sequence for smooth toric varieties [BC94, 12.1]:

$$0 \to \Omega_{X_{\mathscr{A}}}^1 \to \bigoplus_{\xi \in \Sigma_{\mathscr{A}}(1)} \mathscr{O}_{X_{\mathscr{A}}}(V(\xi)) \to \mathscr{O}_{X_{\mathscr{A}}}^{|\Sigma_{\mathscr{A}}(1)|-n} \to 0$$

Where $V(\xi)$ is the invariant divisor associated to the ray $\xi \in \Sigma_{\mathscr{A}}(1)$. It follows that: $c_i(\Omega_{X_{\mathscr{A}}}^1) = (-1)^i \sum_{\xi_1 \neq \xi_2 \neq \ldots \neq \xi_i} [V(\xi_1)] \cdot \ldots \cdot [V(\xi_i)]$. Recall that the intersection products $[V(\xi_1)] \cdot \ldots \cdot [V(\xi_i)]$ correspond to codimension i invariant subvarieties and thus faces of $P_{\mathscr{A}}$ of dimension $n-i$. Moreover the degree of the embedded subvariety $[V(\xi_1)] \cdot \ldots \cdot [V(\xi_i)]$ is equal to $\mathscr{L}^{n-i} \cdot ([V(\xi_1)] \cdot \ldots \cdot [V(\xi_i)]) = (n-i)!\,\mathrm{Vol}(F)$, where F is the corresponding face. We can then conclude:

$$c_n(J_1(\mathscr{L}_{\mathscr{A}})) = \sum_{\emptyset \neq F \prec P_{\mathscr{A}}} (n+1-i)(n-i)!(-1)^i \mathrm{Vol}(F) =$$
$$= \sum_{\emptyset \neq F \prec P_{\mathscr{A}}} (-1)^{\mathrm{codim}(F)} (\dim(F)+1)!\,\mathrm{Vol}(F)$$

□

Example 4.13. Consider the simplex $2\Delta_2$ in Example 4.2. All the edges have length equal to two and therefore the toric embedding is 2-jet spanned. The dual variety is then an hypersurface and the degree of the discriminant is given by $c_2(J_1(\mathcal{O}_{\mathbb{P}^2}(2))) = c_2(\mathcal{O}_{\mathbb{P}^2}(1) \oplus \mathcal{O}_{\mathbb{P}^2}(1) \oplus \mathcal{O}_{\mathbb{P}^2}(1)) = 3$. The volume formula gives in fact:

$$c_2(J_1(\mathcal{O}_{\mathbb{P}^2}(2))) = 6\text{Vol}(2\delta_2) - 2\sum_1^3 \text{Vol}(2\Delta_1) + 3 = 12 - 12 + 3 = 3.$$

Example 4.14. Consider the Segre embedding $\mathbb{P}^1 \times \mathbb{P}^2 \hookrightarrow \mathbb{P}^5$, associated to the polytope Q. Then $c_3(J_1(\mathcal{L})) = 4!\frac{1}{2} - 3!(1 + 1 + 1 + \frac{1}{2} + \frac{1}{2}) + 2(9) - 6 = 0$. This embedding is therefore dually defective.

The following is an amusing observation, which is a simple consequence of the previous characterization.

Corollary 4.15. *Let $P_{\mathscr{A}}$ be a smooth polytope such that $\mathscr{A} = P_{\mathscr{A}} \cap \mathbb{Z}^n$. Then*

$$\sum_{\emptyset \neq F \prec P_{\mathscr{A}}} (-1)^{\text{codim}(F)}(\dim(F) + 1)!\,\text{Vol}(F) \geq 0$$

Proof. Because the associated line bundle $\mathcal{L}_{\mathscr{A}}$ defines an embedding of the variety $X_{\mathscr{A}}$, the map jet^1 has maximal rank and thus the vector bundle $J_1(\mathcal{L}_{\mathscr{A}})$ is spanned (by the global sections of $\mathcal{L}_{\mathscr{A}}$). It follows that the degree of its Chern classes must be non negative which implies the assertion. □

Now we can state the characterization of $\mathbb{Q}$-factorial and non singular toric embeddings admitting discriminant $\Delta_{\mathscr{A}} = 1$. The theorem will include the combinatorial characterization and the equivalent algebraic geometry description. The proof in the non singular case will be given in Sect. 6.

Theorem 4.16 ([DiR06, CDR08]). *Let $\mathscr{A} = P_{\mathscr{A}} \cap \mathbb{Z}^n$ and assume that $X_{\mathscr{A}}$ is $\mathbb{Q}$-factorial. Then the following equivalent statements hold.*

(a) *The point-configuration $\mathscr{A}$ is dually defective if and only if $P_{\mathscr{A}}$ is a Cayley sum of the form $P_{\mathscr{A}} \cong \text{Cayley}(R_0, \ldots, R_t)_{(\pi, Y)}$, where $\pi(P)$ is a simplex (not necessarily unimodular) in $\mathbb{R}^t$ and $R_0, \ldots, R_t$ are normally equivalent polytopes with $t > \frac{n}{2}$. If moreover $P_{\mathscr{A}}$ is smooth then $\pi(P)$ is a unimodular simplex.*

(b) *The projective dual variety of the toric embedding $X_\mathscr{A} \hookrightarrow \mathbb{P}^{|\mathscr{A} \cap \mathbb{Z}^n|-1}$ has codimension $s \geq 2$ if and only if $X_\mathscr{A}$ is a Mori-fibration, $X_\mathscr{A} \to Y$ and $\dim(Y) < \dim(X)/2$. If moreover $X_\mathscr{A}$ is non singular then $(X_\mathscr{A}, L_\mathscr{A}) = (\mathbb{P}(L_0 \oplus \cdots \oplus L_t), \xi)$, where L_i are line bundles on a toric variety Y of dimension $m < t$.*

Proposition 4.16 provides a characterization of the class of smooth polytopes achieving the minimal value 0.

Corollary 4.17. *Let P be a convex smooth lattice polytope. Then*

$$\sum_{\emptyset \neq F \prec P_\mathscr{A}} (\dim(F) + 1)!(-1)^{\operatorname{codim}(F)} \operatorname{Vol}(F) = 0$$

If and only if $P = \operatorname{Cayley}(R_0, \ldots, R_t)$ for normally equivalent smooth lattice polytopes R_i with $\dim(R_i) < t$.

5 Toric Fibrations and Adjunction Theory

The classification of projective algebraic varieties is a central problem in Algebraic Geometry dating back to early nineteenth century. The way one can realistically carry out a classification theory is through invariants, such as the degree, genus, Hilbert polynomial. Modern adjunction theory and Mori theory are the basis for major advances in this area.

Let $(X, \mathscr{L})$ be a polarized n-dimensional variety. Assume that X is Gorenstein (i.e. the canonical class K_X is a Cartier divisor). The two key invariants occurring in classification theory, see [Fuj90], are the *effective log threshold* $\mu(\mathscr{L})$ and the *nef value* $\tau(\mathscr{L})$:

$$\mu(\mathscr{L}) := \sup_{\mathbb{R}}\{s \in \mathbb{Q} : \dim(H^0(K_X + s\mathscr{L})) = 0\}$$
$$\tau(\mathscr{L}) := \min_{\mathbb{R}}\{s \in \mathbb{R} : K_X + s\mathscr{L} \text{ is nef}\}.$$

Both invariants are at most equal to $n + 1$. Kawamata proved that $\mu(\mathscr{L})$ is indeed a rational number and recent advances in the minimal model program establish the same for $\mu(\mathscr{L})$. They can be visualized as follows.

Traveling from $\mathscr{L}$ in the direction of the vector K_X in the Neron-Severi space $\operatorname{NS}(X) \otimes \mathbb{R}$ of divisors, $\mathscr{L} + (1/\mu(\mathscr{L}))K_X$ is the meeting point with the cone of effective divisors $\operatorname{Eff}(X)$ and $\mathscr{L} + (1/\tau(\mathscr{L}))K_X$ is the meeting point with the cone of nef-divisors $\operatorname{Nef}(X)$, see Fig. 1.

A multiple of the nef line bundle $K_X + \tau \mathscr{L}$ defines a morphism $X \to \mathbb{P}^M$ which can be decomposed (Remmert-Stein factorization) as a composition of a morphism $\phi_\tau : X \to Y$ with connected fibers onto a normal variety Y and finite-to-one morphism $Y \to \mathbb{P}^M$. The map ϕ_τ is called the *nef-value morphism*. Kawamata showed that if one writes $r\tau = u/v$ for coprime integers u, v, then:

Fig. 1 Illustrating $\mu(\mathscr{L})$ and $\tau(\mathscr{L})$

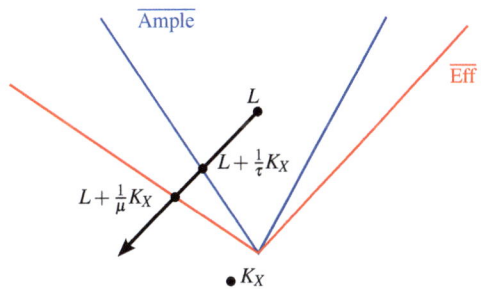

$$u \leq r(1 + \max_{y \in Y}(\dim(\phi_\tau^{-1}(y)))).$$

Corollary 5.1. *Let $(X, \mathscr{L})$ be a polarized variety. Then the nef-value achieves the maximum value $\tau(\mathscr{L}) = n + 1$ if and only if $(X, \mathscr{L}) = (\mathbb{P}^n, \mathscr{O}_{\mathbb{P}^n}(1))$.*

Proof. Consider the nef value morphism $\phi_\tau : X \to Y$ and observe that

$$(n + 1) \leq (1 + \max_{y \in Y}(\dim(\phi_\tau^{-1}(y)))).$$

This implies that the dimension of a fiber of ϕ_τ must be n and thus that the morphism contracts the whole space X to a point. By construction, the fact that ϕ_τ contracts the whole space implies that $K_X + (n+1)\mathscr{L} = \mathscr{O}_X$. A celebrated criterion in projective geometry, called the Kobayashi-Ochiai theorem, asserts that if L is an ample line bundle such that $K_X + (n+1)\mathscr{L} = \mathscr{O}_X$ then $(X, \mathscr{L}) = (\mathbb{P}^n, \mathscr{O}_{\mathbb{P}^n}(1))$. □

Remark 5.2. Recall that the interior of the closure of the effective cone is the cone of big divisors, $(\overline{\text{Eff}(X)})^\circ = \text{Big}(X)$, and that the closure of the ample cone is the nef cone, $\overline{\text{Ample}(X)} = \text{Nef}(X)$. In particular the equality $\tau(\mathscr{L}) = \mu(\mathscr{L})$ occurs if and only if the line bundle $K_X + \tau(\mathscr{L})\mathscr{L}$ is nef and *not* big, which implies that ϕ_τ defines a fibration structure on X.

A fibration structure on an algebraic variety is a powerful geometrical tool as many invariants are induced by corresponding invariants on the (lower dimensional) basis and generic fiber. Criteria for a space to be a fibration are therefore highly desirable. Beltrametti, Sommese and Wisniewski conjectured the if the effective log threshold is strictly bigger than half the dimension then the nef-value morphism should be a fibration.

Conjecture 5.3 ([BS94]). *If X is non singular and $\mu(\mathscr{L}) > (n + 1)/2$ then $\mu(\mathscr{L}) = \tau(\mathscr{L})$.*

Let us now assume that the algebraic variety is toric. In this case it is immediate to see that the defined invariants are rational numbers as the cones $\text{Eff}(X), \text{Big}(X), \text{Ample}(X), \text{Nef}(X)$ are all rational cones.

We have seen in Sect. 3 that toric fibrations are associated to certain Cayley polytopes. Analogously to the classification theory of projective algebraic varieties it is important to find invariants of polytopes that would characterize a Cayley structure. One invariant which has attracted increasing attention in recent years is the *codegree* of a lattice polytope:

$$\text{codeg}(P) = \min\{t \in \mathbb{Z}_{>0} \text{ such that } tP \text{ contains interior lattice points}\}.$$

Via Ehrhart theory one can conclude that $\text{codeg}(P) \leq n+1$ and that $\text{codeg}(P) = n+1$ if and only if $P = \Delta_n$. This is in fact a simple consequence of our previous observations.

Corollary 5.4. *Let P be a Gorenstein lattice polytope. Then $\text{codeg}(P) = n+1$ if and only if $P = \Delta_n$.*

Proof. Let $(X, \mathscr{L})$ be the Gorenstein toric variety associated to P. Notice that, because $K_X = -\sum D_i$ where the D_i are the invariant divisors, the polytope defined by the line bundle $K_X + t\mathscr{L}$ is the convex hull of the interior points of tP. The equality $\text{codeg}(P) = n+1$ is equivalent to $H^0(K_X + t\mathscr{L}) = 0$ for $t \leq n$. Because nef line bundles must have sections (in particular being nef is equivalent to being globally generated on toric varieties) we have $\tau(\mathscr{L}) \geq \text{codeg}(P) = n+1$. It follows from Corollary 5.1 that $(X, \mathscr{L}) = (\mathbb{P}^n, \mathscr{O}_{\mathbb{P}^n}(1))$ and thus $P = \Delta_n$. □

Let us now examine the class of Cayley polytopes we encountered in the characterization of dually defective toric embeddings. We will see that this is a class of polytopes satisfying the strong lower bound $\text{codeg}(P) \geq \frac{\dim(P)}{2} + 1$ and the equality $\text{codeg}(P) = \mu(\mathscr{L})$.

Lemma 5.5. *Let $P = \text{Cayley}_{h,Y}(R_0, \ldots, R_t)$ with $t > \frac{n}{2}$, then:*

$$\tau(\mathscr{L}_P) = \mu(\mathscr{L}_P) = \text{codeg}(P) = t + 1 \geq \frac{n+3}{2}.$$

Proof. Observe that $X_P = \mathbb{P}(L_0 \oplus \ldots \oplus L_t)$ for ample line bundles L_i on the toric variety Y and $\mathscr{L} = \xi$ is the tautological line bundle. Consider the projective bundle map $\pi : X_P \to Y$. The Picard group of X_P is generated by the pull back of generators of $\text{Pic}(Y)$ and by the tautological line bundle ξ. Moreover the canonical line bundle is given by the following expression:

$$K_{X_P} = \pi^*(K_Y + L_0 + \ldots + L_t) - (t+1)\xi.$$

The toric nefness criterion says that a line bundle on a toric variety is nef if and only if the intersection with all the invariant curves is non-negative, see for example [ODA]. On the toric variety $\mathbb{P}(L_0 \oplus \ldots \oplus L_t)$ there are two types of rational invariant curves. The ones contained in the fibers $F \cong \mathbb{P}^t$ and the pull back of rational invariant curves in Y which will be denoted by $\pi^*(C)_i$ when contained in the invariant section defined by the polytope R_i. For any rational invariant curve

$C \subset F$, it holds that $\xi|_C = \mathcal{O}_{\mathbb{P}^1}(1)$ and $\pi^*(D) \cdot C = 0$, for all divisors D on Y. For every curve of the form $\pi^*(C)_i$ it holds that $\pi^*(C)_i \cdot \pi^*(D) = C \cdot D$ and $\xi \cdot \pi^*(C)_i = L_i \cdot C$. See [DiR06, Remark 3] for more details. We conclude that $K_{X_P} + s\mathcal{L}$ is nef if the following is satisfied:

$$[\pi^*(K_Y + L_0 + \ldots + L_t) + (s-t-1)\xi]C = s-t-1 \geq 0 \quad \text{if } C \subset F$$
$$(K_Y + L_0 + \ldots + (s-t)L_i + \ldots + L_t) \cdot C \geq 0 \quad \text{if } C = \pi^*(C)_i$$

In [MU02] Mustata proved a toric-Fujita conjecture showing that if for a line bundle H on an n-dimensional toric variety, $H \cdot C \geq n$ for every invariant curve C, then the adjoint bundle $K + H$ is globally generated, unless $H = \mathcal{O}_{\mathbb{P}^n}(n)$. Because

$$[L_0 + \ldots + (s-t)L_i + \ldots + L_t] \cdot C \geq (s-t) + t$$

it follows that $(K_Y + L_0 + \ldots + (s-t)L_i + \ldots + L_t) \cdot C \geq 0$ for all invariant curves $C = \pi^*(C)_i$ if $s \geq t$. This implies that $K_{X_P} + s\mathcal{L}$ is nef if and only if $s \geq t + 1$ and thus $\tau(\xi) = t + 1$.

Consider now the projection $h : \mathbb{R}^n \to \mathbb{R}^t$ such that $h(P) = \Delta_t$. Under this projection interior points of a dilation sP are mapped to interior points of the corresponding dilation $s\Delta_t$. This implies that $\text{codeg}(P) = t + 1$. Notice that $\mu(\mathcal{L}) \leq \text{codeg}(P) = t + 1$ as interior points of sP correspond to global sections of $K_{X_P} + s\mathcal{L}$. On the other hand, see [HA, Ex. 8.4]:

$$H^0(u(\pi^*(K_Y + L_0 + \ldots + L_t)) + (v - u(t+1))\xi) =$$
$$= H^0(\pi_*(u(\pi^*(K_Y + L_0 + \ldots + L_t)) + (v - u(t+1))\xi)) =$$
$$= H^0(u(K_Y + L_0 + \ldots + L_t) + \pi_*((u - v(t+1))\xi) = 0 \text{ if } v - u(t+1) < 0.$$

This implies that $\mu(\mathcal{L}) \geq t + 1$, which proves the assertion. □

Recently Batyrev and Nill in [BN08] classified polytopes with $\text{codeg}(P) = n$ and conjectured the following.

Conjecture 5.6 ([BN08]). There is a function $f(n)$ such that any n-dimensional polytope P with $\text{codeg}(P) \geq f(n)$ decomposes as a Cayley sum of lattice polytopes.

The above conjecture was proven by Haase, Nill and Payne in [HNP09]. They showed that $f(n)$ is at most quadratic in n. It is important to observe that, as interior lattice points of tP correspond to global sections of $K_X + t\mathcal{L}$ for the associated toric embedding, $\text{codeg}(P)$ can be considered as the *integral variant* of $\mu(\mathcal{L})$. This observation, techniques from toric Mori theory and adjunction theory led to prove a stronger version of Conjectures 5.3 and 5.6 for smooth polytopes giving yet another characterization of Cayley sums.

Theorem 5.7 ([DDRP09, DN10]). *Let $P \subset \mathbb{R}^n$ be a smooth n-dimensional polytope. Then the following statements are equivalent.*

(a) $\mathrm{codeg}(P) \geq (n+3)/2$.
(b) P is affinely isomorphic to a Cayley sum $\mathrm{Cayley}(R_0,\ldots,R_t)_{\pi,Y}$ where $t+1 = \mathrm{codeg}(P)$ with $t > \frac{n}{2}$.
(c) $\mu(\mathscr{L}_P) = \tau(\mathscr{L}_P) = t+1$ and $t > \frac{n}{2}$.
(d) $(X_P, \mathscr{L}_P) = (\mathbb{P}(L_0 \oplus \cdots \oplus L_t), \xi)$ for ample line bundles L_i on a non singular toric variety Y.

Notice that Theorem 5.7 proves the reverse statement of Lemma 5.5.

Conjectures 5.3 and 5.6, made independently in two apparently unrelated fields, constitute a beautiful example of the interplay between classical projective (toric) geometry and convex geometry. In view of the results above one could hope that in the toric setting the conjectures should hold in more generally.

Conjecture 5.8. Let $(X, \mathscr{L})$ be an n-dimensional toric polarized variety (not necessarily smooth or even Gorenstein), then $\mu(\mathscr{L}) > (n+1)/2$ implies that $\mu(\mathscr{L}) = \tau(\mathscr{L})$.

The invariants $\mu(\mathscr{L}), \tau(\mathscr{L})$ in the non Gorenstein case can be defined using corresponding invariants, $\mu(P), \tau(P)$ of the associated polytope, see below for a definition.

Conjecture 5.9. If an n-dimensional lattice polytope P satisfies $\mathrm{codeg}(P) > (n+2)/2$, then it decomposes as a Cayley sum of lattice polytopes of dimension at most $2(n+1-\mathrm{codeg}(P))$.

Conjecture 5.8 is a toric version of Conjecture 5.3, extending the statement to possibly singular and non Gorenstein varieties. Conjecture 5.9 states that the function $f(n)$ in Conjecture 5.6 should be equal to $(n+2)/2$. An important step to prove these conjectures is to define the convex analog of $\mu(\mathscr{L}_P)$.

Let $P \subseteq \mathbb{R}^n$ be a rational polytope of dimension n. Any such polytope P can be described in a unique minimal way as

$$P = \{x \in \mathbb{R}^n : \langle a_i, x \rangle \geq b_i,\ i = 1,\ldots,m\}$$

where the a_i are the rows of an $m \times n$ integer matrix A, and $b \in \mathbb{Q}^m$.

For any $s \geq 0$ we define the *adjoint polytope* $P^{(s)}$ as

$$P^{(s)} := \{x \in \mathbb{R}^n : Ax \geq b + s\mathbf{1}\},$$

where $\mathbf{1} = (1,\ldots,1)^T$.

We call the study of such polytopes $P^{(s)}$ *polyhedral adjunction theory* (Fig. 2).

Definition 5.10. We define the $\mathbb{Q}$-*codegree* of P as

$$\mu(P) := (\sup\{s > 0 : P^{(s)} \neq \emptyset\})^{-1},$$

and the *core* of P to be $\mathrm{core}(P) := P^{(1/\mu(P))}$.

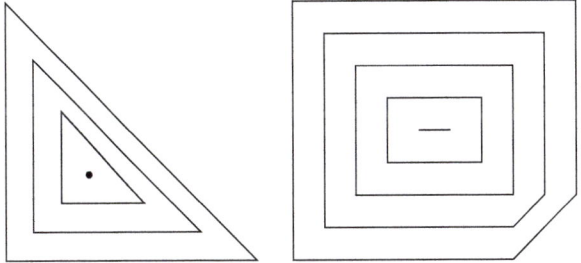

Fig. 2 Two examples of polyhedral adjunction

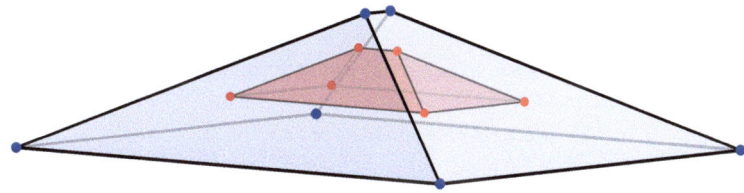

Fig. 3 $P^{(4)} \subseteq P$ for a three-dimensional lattice polytope P

Notice that in this case the supremum is actually a maximum. Moreover, since P is a rational polytope, $\mu(P)$ is a positive rational number.

One sees that for a *lattice* polytope P

$$\mu(P) \leq \operatorname{codeg}(P) \leq n+1$$

Definition 5.11. The *nef value* of P is given as

$$\tau(P) := (\sup\{s > 0 : \mathcal{N}(P^{(s)}) = \mathcal{N}(P)\})^{-1} \in \mathbb{R}_{>0} \cup \{\infty\}$$

where $\mathcal{N}(P)$ denotes the normal fan of the polytope P.

Note that in contrast to the definition of the $\mathbb{Q}$-codegree, here the supremum is never a maximum.

Figure 3 illustrates a polytope P with $\tau(P)^{-1} = 2$ and $\mu(P)^{-1} = 6$. In this case core(P) is an interval.

In [DRHNP13] the precise analogue of Conjecture 5.9 for the $\mathbb{Q}$-codegree is proven.

Theorem 5.12 ([DRHNP13]). *Let P be an n-dimensional lattice polytope. If n is odd and $\mu(P) > (n+1)/2$, or if n is even and $\mu(P) \geq (n+1)/2$, then P is a Cayley polytope.*

Results from [DRHNP13] show Conjecture 5.9 in two interesting cases: when $\lceil \mu(P) \rceil = \operatorname{codeg}(P)$ and when the normal fan of P is Gorenstein and $\mu(P) = \tau(P)$.

6 Connecting the Three Characterizations

In Sect. 4 we have seen that a certain class of Cayley polytopes characterizes dually defective configuration points. Moreover this class corresponds to the polytopes achieving the equality in Corollary 4.15. In Sect. 5 the same class of Cayley polytopes was characterized as corresponding to smooth configurations with codegree larger than slightly more that half the dimension. We will here assemble the three characterizations and provide proofs in the non singular case.

Theorem 6.1. *Let $\mathscr{A} \subset \mathbb{Z}^n$ be a point configuration such that $P_{\mathscr{A}} \cap \mathbb{Z}^n = \mathscr{A}$, $\dim(P_{\mathscr{A}}) = n$ and such that $P_{\mathscr{A}}$ is a smooth polytope. Then the following statements are equivalent.*

(a) *$P_{\mathscr{A}}$ is affinely isomorphic to a Cayley sum $\operatorname{Cayley}(R_0, \ldots, R_t)_{\pi,Y}$ where $t + 1 = \operatorname{codeg}(P_{\mathscr{A}})$ and $t > \frac{n}{2}$.*
(b) *$\operatorname{codeg}(P_{\mathscr{A}}) \geq \frac{n+1}{2} + 1$ and $\tau(P_{\mathscr{A}}) = \mu(P_{\mathscr{A}})$.*
(c) *The discriminant $\Delta_{\mathscr{A}} = 1$.*
(d) *$\sum_{\emptyset \neq F \prec P_{\mathscr{A}}} (\dim(F) + 1)!(-1)^{\operatorname{codim}(F)} \operatorname{Vol}(F) = 0$.*

Proof. [(d) ⇔ (c)]. The implication (d) ⤳ (c) follows from Lemma 4.10 and Proposition 4.12. The reverse implication follows from Corollary 4.17.

[(c) ⇒ (b).] Assume now (c), i.e. assume that the configuration is dually defective. Consider the associated polarized toric manifold $(X_{\mathscr{A}}, \mathscr{L}_{\mathscr{A}})$. It is a classical result that the generic tangent hyperplane is in fact tangent along a linear space in $X_{\mathscr{A}}$. Therefore if $\operatorname{codim}(X_{\mathscr{A}}^{\vee}) = k > 1$ then there is a linear $\mathbb{P}^k$ through a general point of $X_{\mathscr{A}}$. By linear $\mathbb{P}^k$ we mean a subspace $Y \cong \mathbb{P}^k$ such that $\mathscr{L}_{\mathscr{A}}|_Y = \mathscr{O}_{\mathbb{P}^k}(1)$. Moreover, by a result of Ein [E86] $N_{\mathbb{P}^k/X} = (\bigoplus_1^{\frac{n-k}{2}} \mathscr{O}_{\mathbb{P}^k}) \oplus (\bigoplus_1^{\frac{n-k}{2}} \mathscr{O}_{\mathbb{P}^k}(1))$. Observe that if we fix a point $x \in X_{\mathscr{A}}$, a sequence $\{F_j\}$ of general linear subspaces $F_J \cong \mathbb{P}^k$ can be chosen so that $x \in \lim(F_j)$. Since the F_i are all linear the limit space has to be also a linear $\mathbb{P}^k$. We can then assume that there is a linear $\mathbb{P}^k$ through every point of $X_{\mathscr{A}}$. Let L now be an invariant line in one of the $\mathbb{P}^k$ through a fixed point. Then:

$$[K_X + t\mathscr{L}]_L = \mathscr{O}_{\mathbb{P}^1}((-n-2-k)/2+t)$$

which implies $\tau(\mathscr{L}_{\mathscr{A}}) \geq \frac{n+k}{2} + 1$. Assume now that $\tau(\mathscr{L}_{\mathscr{A}}) > \frac{n+k}{2} + 1$ and let L be again a line in the family of linear spaces covering X. The quantity $-K_X \cdot L - 2 = \nu$ is called the normal degree of the family. In our case $\nu = (n+k)/2 - 1 > n/2$. By a result of Beltrametti-Sommese-Wisniewski [BSW92], this assumption implies $\nu = \tau - 2$, proving $\tau(\mathscr{L}_A) = \frac{n+k}{2} + 1$. Notice that the nef-morphism ϕ_τ contracts all the linear $\mathbb{P}^k$ of the covering family and thus it is a fibration. As a consequence the line bundle $K_{X_{\mathscr{A}}} + \tau \mathscr{L}_{\mathscr{A}}$ is not big and thus $\tau(\mathscr{L}_{\mathscr{A}}) = \mu(\mathscr{L}_{\mathscr{A}})$. The inequality

$$\operatorname{codeg}(P_{\mathscr{A}}) \geq \mu(\mathscr{L}_{\mathscr{A}}) = \tau(\mathscr{L}_{\mathscr{A}}) = \frac{n+k}{2} + 1$$

shows the implication (c) ⤳ (b).

Linear Toric Fibrations

[(b) ⇒ (a)]. Assume now (b). The nef-value morphism is then a fibration and

$$\tau(\mathcal{L}_\mathscr{A}) > \operatorname{codeg}(P_\mathscr{A}) - 1 > \frac{n}{2}.$$

Notice that the nef-morphism ϕ_τ contracts a face of the Mori-cone and thus faces of the lattice polytope $P_\mathscr{A}$, i.e. all the invariant curves with 0-intersection with the line bundle $K_{X_\mathscr{A}} + \tau \mathcal{L}_\mathscr{A}$. Let now C be a generator of an extremal ray contracted by the morphism ϕ_τ. If $\mathcal{L}_\mathscr{A} \cdot C \geq 2$, then $-K_X \cdot C > n + 1$ which is impossible. We can conclude that C is a line and $\tau(\mathcal{L}_\mathscr{A}) = -K_{X_\mathscr{A}} \cdot C$ is an integer. It follows that $\tau(\mathcal{L}_\mathscr{A}) \geq \frac{n+1}{2} + 1$. This inequality implies that ϕ_τ is the contraction of one extremal ray, by [BSW92, Cor. 2.5]. These morphisms are analyzed in detail in [Re83]. Because $X_\mathscr{A}$ is smooth and toric and this contraction has connected fibres, the general fiber F of the contraction is a smooth toric variety with Picard number one. It follows that F is a projective space and thus ϕ_τ is a $\mathbb{P}^t$ bundle. Let $L|_F = \mathcal{O}_{\mathbb{P}^t}(a)$. Observe that by construction $K_{X_\mathscr{A}}|_F = K_F$. Consider a line $l \subset F$. It follows that

$$0 = (K_{X_\mathscr{A}} + \tau \mathcal{L}_\mathscr{A}) \cdot l = K_F \cdot l + \tau \mathcal{L}_\mathscr{A} \cdot l = -t - 1 + a\tau$$

and thus $\tau = \frac{t+1}{a} > \frac{n+1}{2} + 1$ which implies $a = 1$ and $t > \frac{n+1}{2}$. Since $a = 1$ the fibers are embedded linearly and thus $(X_\mathscr{A}, \mathcal{L}_\mathscr{A}) = (\mathbb{P}(L_0 \oplus \ldots \oplus L_t), \xi)$, for ample line bundles L_i on a smooth toric variety Y. This proves the implication (b) ⤳ (a).

[(a) ⇒ (c)]. Assume now (a). Using notation as in (2), consider the commutative diagram:

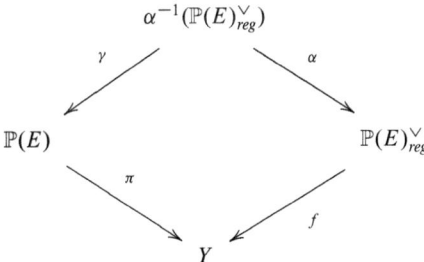

where $E = L_0 \oplus \ldots \oplus L_t$ and (Y, L_i) is the smooth polarized variety associated to the polytope R_i. The commutativity of the diagram and the existence of f follows from [DeB01, Lemma 1.15]. Let $y \in Y$ and let $F \cong \mathbb{P}^t \subset \mathbb{P}^{|\mathscr{A}|-1}$ be the fiber $\pi^{-1}(y)$. Commutativity of the diagram implies that the contact locus $\gamma(\alpha^{-1}(H))$ is included in F for all $H \in f^{-1}(y)$. Moreover $\operatorname{Osc}_{F,y} \subset \operatorname{Osc}_{\mathbb{P}(E),y} \subset H$ implies that H belongs to the dual variety $F^\vee$, with contact locus at least of the same dimension. Because the map f is dominant we can conclude that: $\dim(F^\vee) \geq \dim(\mathbb{P}(E)^\vee) - \dim(Y)$, which implies

$$\mathrm{codim}(\mathbb{P}(E)^{\vee}) \geq \mathrm{codim}(F^{\vee}) - \dim(Y).$$

Recall that the fibers F are embedded linearly and thus $\mathrm{codim}(F^{\vee}) = \dim(F) + 1$. It follows that $\mathrm{codim}(\mathbb{P}(E)^{\vee}) \geq \dim(F) + 1 - \dim(Y) > 1$ and thus $\Delta_{\mathscr{A}} = 1$. This proves $(a) \rightsquigarrow (c)$. □

Acknowledgements The author was supported by a grant from the Swedish Research Council (VR). Special thanks to A. Lundman, B. Nill and B. Sturmfels for reading a preliminary version of the notes.

References

[BN08] V. Batyrev, in *Combinatorial Aspects of Mirror Symmetry*, vol. 452 of *Contemp. Math.*, ed. by Matthias Beck and et. al., (AMS, 2008), pp. 35–66

[BC94] V.V. Batyrev, D. Cox, On the Hodge structures of projective hyper surfaces in toric varieties. Duke Math. **75**, 293–338 (1994)

[BSW92] M. Beltrametti, in *Complex Algebraic Varieties (Bayreuth, 1990), Lecture Notes in Math.*, vol. 1507 (Springer, Berlin, 1992), pp. 16–38. DOI 10.1007/BFb0094508. URL http://dx.doi.org/10.1007/BFb0094508

[BS94] M. Beltrametti, *Classification of Algebraic Varieties (L'Aquila, 1992)*, vol. 162 of *Contemp. Math.* (Amer. Math. Soc., Providence, RI, 1994), pp. 31–48

[CDR08] C. Casagrande, Commun. Contemp. Math. **10**(3), 363 (2008)

[CC07] E. Cattani, J. Symbolic Comput. **42**(1-2), 115 (2007). DOI 10.1016/j.jsc.2006.02.006. URL http://dx.doi.org/10.1016/j.jsc.2006.02.006

[DeB01] O. Debarre, *Higher-Dimensional Algebraic Geometry*. Universitext (Springer, New York, 2001)

[DS02] A. Dickenstein, J. Symbolic Comput. **34**(2), 119 (2002). DOI 10.1006/jsco.2002.0545. URL http://dx.doi.org/10.1006/jsco.2002.0545

[DDRP09] A. Dickenstein, S. Di Rocco, R. Piene, Classifying smooth lattice polytopes via toric fibrations. Adv. Math. **222**(1), 240–254 (2009)

[DN10] A. Dickenstein, B. Nill, A simple combinatorial criterion for projective toric manifolds with dual defect. Math. Res. Lett. **17**, 435–448 (2010)

[DDRP12] A. Dickenstein, S. Di Rocco, R. Piene, Higher order duality and toric embeddings. Ann. l'Institut Fourier (to appear)

[DiR01] S. Di Rocco, Generation of k-jets on toric varieties. Math. Z. **231**, 169–188 (1999)

[DiR06] S. Di Rocco, Projective duality of toric manifolds and defect polytopes. Proc. Lond. Math. Soc. (3) **93**(1), 85–104 (2006)

[DRS01] S. Di Rocco, A.J. Sommese, Line bundles for which a projectivized jet bundle is a product. Proc. A.M.S. **129**(6), 1659–1663 (2001)

[DRS04] S. Di Rocco, A.J. Sommese, Chern numbers of ample vector bundles on toric surfaces. Trans. Am. Math. Soc. **356**(2), 587–598 (2004)

[DRHNP13] S. Di Rocco, B. Nill, C. Haase, A. Paffenholz, Polyhedral adjunction theory. Algebra Number Theory (to appear)

[E86] L. Ein, Varieties with small dual varieties. Inv. Math. **96**, 63–74 (1986)

[EW] G. Ewald, *Combinatorial Convexity and Algebraic Geometry, Graduate Texts in Mathematics*, vol. 168 (Springer, New York, 1996)

[FU] W. Fulton, *Introduction to Toric Varieties, Annals of Mathematics Studies*, vol. 131 (Princeton University Press, Princeton, NJ, 1993). The William H. Roever Lectures in Geometry

[FUb]	W. Fulton, *Intersection Theory, Ergebnisse der Mathematik und ihrer Grenzgebiete. 3. Folge. A Series of Modern Surveys in Mathematics [Results in Mathematics and Related Areas. 3rd Series. A Series of Modern Surveys in Mathematics]*, vol. 2, 2nd edn. (Springer, Berlin, 1998). DOI 10.1007/978-1-4612-1700-8. URL http://dx.doi.org/10.1007/978-1-4612-1700-8
[Fuj90]	T. Fujita, in *Classification Theories of Polarized Varieties*. Lecture Note Series, vol. 155 (London Mathematical Society/Cambridge University Press, London/Cambridge, 1990)
[GKZ]	I. Gel'fand, *Discriminants, Resultants, and Multidimensional Determinants. Mathematics: Theory & Applications* (Birkhäuser Boston, Boston, MA, 1994). DOI 10.1007/978-0-8176-4771-1. URL http://dx.doi.org/10.1007/978-0-8176-4771-1
[GH79]	P. Griffiths, Ann. Sci. École Norm. Sup. (4) **12**(3), 355 (1979). URL http://www.numdam.org/item?id=ASENS_1979_4_12_3_355_0
[HNP09]	C. Haase, J. Reine Angew. Math. **637**, 207 (2009)
[HA]	R. Hartshorne, *Algebraic Geometry* (Springer, New York, 1977). Graduate Texts in Mathematics, No. 52
[L94]	J. Landsberg, Invent. Math. **117**(2), 303 (1994). DOI 10.1007/BF01232243. URL http://dx.doi.org/10.1007/BF01232243
[LM00]	A.A. Lanteri, R. Mallavibarrena, Higher order dual varieties of generically k-regular surfaces. Arch. Math. (Basel) **75**(1), 75–80 (2000). doi:10.1007/s000130050476
[MT11]	Y. Matsui, Adv. Math. **226**(2), 2040 (2011). DOI 10.1016/j.aim.2010.08.020. URL http://dx.doi.org/10.1016/j.aim.2010.08.020
[MU02]	M. Mustata, Tohoku Math. J. **54**(3), 4451 (2002)
[ODA]	T. Oda, *Algebraic Geometry Seminar (Singapore, 1987)* (World Scitific, Singapore, 1988), pp. 89–94
[ODAb]	T. Oda, *Convex Bodies and Algebraic Geometry, Ergebnisse der Mathematik und ihrer Grenzgebiete (3) [Results in Mathematics and Related Areas (3)]*, vol. 15 (Springer, Berlin, 1988). An introduction to the theory of toric varieties, Translated from the Japanese
[Re83]	M. Reid, *Arithmetic and Geometry* (Progress in Math. 36, Birkhäuser, Boston, 1983), pp. 395–418

A Tour on Hermitian Symmetric Manifolds

Filippo Viviani

1 Introduction

A Hermitian symmetric manifold (or **HSM** for short) is a Hermitian manifold which is homogeneous and such that every point has a symmetry preserving the Hermitian structure. First studied by Élie Cartan [Car35], they are the specialization of the notion of Riemannian symmetric manifolds (introduced by Élie Cartan himself in [Car26-27]) to complex manifolds.

HSMs (or more generally Riemannian symmetric manifolds) arise in a wide variety of mathematical contexts: representation theory, harmonic analysis, automorphic forms, complex analysis, differential geometry, algebra (Lie theory and Jordan theory), number theory and algebraic geometry. For example, in algebraic geometry, HSMs arise often as (orbifold) fundamental covers of moduli spaces, such as the moduli space of polarized abelian varieties (possibly with level structures or with fixed endomorphism algebras), the moduli space of polarized K3 surfaces, the moduli space of polarized irreducible symplectic manifolds, etc.

Due to their frequent occurrence in different areas of mathematics, there is a vast literature on HSMs (e.g. [AMRT10, Bor52, BJ06, FKKLR00, Hel78, Koe69, Loo69a, Loo69b, Loo77, Mok89, PS69, Sat80, Wol67]) dealing with the various aspects of the theory. This vast literature, however, makes it difficult for a non-expert to have a global overview on the subject. The aim of these notes is to survey the different points of view on HSMs, so that a beginner can orient himself inside the vast literature. For this reason, we have chosen to give very few proofs of the results presented, referring the reader to the relevant literature for complete proofs.

F. Viviani (✉)
Dipartimento di Matematica e Fisica, Università degli Studi Roma Tre, Largo S. Leonardo Murialdo 1, 00146 Roma, Italy
e-mail: filippo.viviani@gmail.com

Let us now examine more in detail the contents of the paper. In studying HSMs, the reader should keep in mind the (well known) classification of HSMs of (complex) dimension one. Namely, a HSM of complex dimension one is isomorphic to one of the following HSMs:

1. The complex manifold $\mathbb{C}/\Lambda$, where $\Lambda \subset \mathbb{C}$ is a discrete additive subgroup, endowed with the Hermitian structure induced by the standard Euclidean metric $g = dx\,dx + dy\,dy$ on $\mathbb{C} = \mathbb{R}^2_{(x,y)}$ (which has constant zero curvature). The translations $z \mapsto z + a$ (with $a \in \mathbb{C}$) act transitively via holomorphic isometries and the inversion symmetry at $[0] \in \mathbb{C}/\Lambda$ is given by $s_{[0]} : z \mapsto -z$.
2. The upper half space $\mathcal{H} := \{z = x + iy \in \mathbb{C} \;:\; \operatorname{Im} z = y > 0\}$ endowed with a Hermitian structure induced by the hyperbolic metric $g = \frac{dx\,dy}{y^2}$ (which has constant negative curvature). The group $\operatorname{SL}_2(\mathbb{R})$ acts transitively via Möbius transformations (which are holomorphic isometries)

$$\begin{pmatrix} a & b \\ c & d \end{pmatrix} \cdot z := \frac{az+b}{cz+d}$$

and the inversion symmetry at $i \in \mathcal{H}$ is given by $s_i : z \mapsto -\frac{1}{z}$.
3. The complex projective line $\mathbb{P}^1_{\mathbb{C}}$ with the Fubini–Studi Hermitian metric (with constant positive curvature), which is induced by pulling back the Euclidean metric on the two dimensional sphere $\mathbb{S}^2 \subset \mathbb{R}^3$ via the diffeomorphism $\mathbb{P}^1_{\mathbb{C}} \cong \mathbb{S}^2$ induced via stereographic projection from the north pole $N = (1, 0, 0) \in \mathbb{S}^2$. The group $\operatorname{SO}_3(\mathbb{R})$ acts transitively on $\mathbb{S}^2$ via rotations (which are holomorphic isometries of $\mathbb{P}^1_{\mathbb{C}} \cong \mathbb{S}^2$) and the inversion symmetry at the north pole N is given by the rotation $s_N : (x, y, z) \mapsto (x, -y, -z)$.

Note that, according to the Riemann's uniformization theorem, the unique simply connected complex manifolds of dimension one are $\mathbb{C}$, $\mathcal{H}$ and $\mathbb{P}^1_{\mathbb{C}}$:

The above trichotomy in dimension one extends to arbitrary dimensions (see the *Decomposition Theorem* 2.9): any HSM can be written uniquely as the product of a HSM of the form $\mathbb{C}^n/\Lambda$ for some discrete additive subgroup $\Lambda \subset \mathbb{C}^n$ (which is called the Euclidean factor of the HSM), of a HSM of non-compact type (i.e. a product of irreducible non-compact HSMs) and a HSM of compact type (i.e. a product of irreducible compact HSMs).

The rest of Sect. 2 is devoted to the study of non-Euclidean HSM, i.e. those for which the Euclidean factor in the above decomposition is trivial.

Hermitian symmetric manifolds of compact or of non-compact type admit another natural incarnation. Namely, HSMs of compact type are exactly the *cominuscule rational homogeneous projective varieties*, i.e. those varieties isomorphic to a quotient of the form G/P, where G is a semisimple complex Lie group and $P \subset G$ is a parabolic subgroup whose unipotent radical is abelian. We review this description in Sect. 2.6.

HSMs of non-compact type admit a canonical embedding (the so called Harish–Chandra embedding) inside a complex vector space in such a way that they become *bounded symmetric domains*. And, conversely, any bounded symmetric domain becomes a HSM of non-compact type when it is endowed with the Bergman metric. We review this correspondence between HSMs of non-compact type and bounded symmetric domains in Sect. 2.5.

There is a natural correspondence between HSMs of non-compact type and HSMs of compact type, which we review in Sect. 2.3. Moreover, this correspondence satisfies the property that each HSM of non-compact type is canonically realized (via the so called *Borel embedding*) as an open subset inside the associated HSM of compact type (which is called its compact dual). We review the Borel embedding in Sect. 2.4.

Irreducible HSMs of compact or non-compact type can be classified using *Lie theory*. Indeed, they are diffeomorphic to a quotient of the form G/K where G is a simple Lie group (compact in the compact type case and non-compact in the non-compact type case) and $K \subset G$ is a maximal compact proper subgroup whose center is equal to $\mathbb{S}^1$. We review this description in Sect. 2.1.

By passing to the associated Lie algebras, we get a correspondence between non-Euclidean HSMs and irreducible *Hermitian symmetric Lie algebras*, which is the datum of a simple real Lie algebra $\mathfrak{g}$ together with an involution $\theta : \mathfrak{g} \to \mathfrak{g}$ such that its $+1$ eigenvalue has a one-dimensional center. See Sect. 2.2 for more details.

There is an alternative approach to the study of HSMs of non-compact type based on Jordan theory rather than Lie theory. Indeed, there is a natural bijection between HSMs of non-compact type and *Hermitian Jordan triple systems*; see Sect. 2.7 for more details.

Using either Lie theory or Jordan theory, it is possible to give a classification of irreducible HSMs of non-compact type (and of their compact duals). They are divided into four infinite families, called (following Siegel's notation) $I_{p,q}, II_n, III_n$ and IV_n, and two exceptional cases, called V and VI. Section 3 is devoted to a detailed analysis of each of the above mentioned irreducible HSMs. In particular, we make explicit, in each of the cases, the general properties of HSMs presented in Sect. 2.

Section 4 is devoted to the study of the boundary components of HSMs of non-compact type. More precisely, fix a HSM of non-compact type and realize it as a bounded symmetric domain $D \subset \mathbb{C}^N$ via its Harish–Chandra embedding. The closure $\overline{D}$ of D inside $\mathbb{C}^N$ can be partitioned into several equivalence classes for the equivalence relation of being connected through a chain of holomorphic disks. Each of these equivalence classes, called *boundary components* of D, is indeed again a HSM of non-compact type which is realized as a bounded symmetric domain inside its linear span in $\mathbb{C}^N$.

Boundary components can be classified via their normalizer subgroups, which turn out to be all the maximal parabolic subgroups of the group G of automorphisms of D (see Theorem 4.5). The structure of the normalizer subgroups of the boundary components is analyzed in detail in Sect. 4.1.

In Sect. 4.2, we show that, for every boundary component F of D, the domain D can be decomposed into the product of F, a real vector space $W(F)$ and a *symmetric cone* $C(F)$ associated to F, i.e. an open homogeneous cone inside a real vector space which is self-dual with respect to a suitable scalar product.

In Sect. 4.3, we show how symmetric cones correspond bijectively to *Euclidean Jordan algebras* and we present the classification of irreducible symmetric cones via the classification of simple Euclidean Jordan algebras.

In Sect. 4.4, we show how bounded symmetric domains can be realized in a unique way as *Siegel domains* (of the second type) associated to a suitable symmetric cone and to a suitable representation of the associated Euclidean Jordan algebra.

In Sect. 4.5, we describe explicitly the boundary components of each of the irreducible bounded symmetric domains by computing their normalizer subgroups and their associated symmetric cones.

These notes were written for a Ph.D. course (held at the University of Roma Tre in Spring 2013) entitled "Toroidal compactifications of locally symmetric varieties" and a course in the Summer School "Combinatorial Algebraic Geometry" (held in Levico Terme in June 2013) with the same title. Thus, our original motivation was to write a survey on the construction of toroidal compactifications of locally symmetric varieties (i.e. quotients of Hermitian symmetric manifolds of non-compact type by arithmetic subgroups), by revisiting the original work of Ash–Mumford–Rapoport–Tai [AMRT10] (see also the work of Namikawa [Nam80], where the case of Siegel spaces is worked out explicitly). Due to limitations in space and time, we were unable to complete this project and we ended up with an attempt to write a survey on the beautiful and rich theory of Hermitian symmetric manifolds. We plan to write a sequel to these notes on the construction of toroidal compactifications of locally symmetric varieties.

Notations

1.1. Given a Lie group G, we denote by G^o the connected component of G containing the identity and by Z_G its center. A semisimple Lie group G is said to be adjoint if it has trivial center, or in symbols if $Z_G = \{e\}$.

1.2. Given a Lie algebra $\mathfrak{g}$, we denote its center by $Z(\mathfrak{g})$. For a real Lie algebra $\mathfrak{g}$, we denote by $\mathfrak{g}_\mathbb{C} := \mathfrak{g} \otimes_\mathbb{R} \mathbb{C} = \mathfrak{g} \oplus i\mathfrak{g}$ its complexification.

1.3. Given a real (resp. complex) finite-dimensional vector space V and two symmetric (resp. Hermitian) linear operators $F, G \in \text{End}(V)$, we write:

(a) $F > G$ (or $G < F$) if and only if $F - G$ is positive definite;
(b) $F \geq G$ (or $G \leq F$) if and only if $F - G$ is positive semidefinite.

1.4. Given a matrix $M \in M_{n,n}(F)$ with entries in $F = \mathbb{R}, \mathbb{C}, \mathbb{H}$, we will denote by M^t its transpose and by $\overline{M}$ its conjugate with respect to:

- the trivial conjugation if $F = \mathbb{R}$;
- the conjugation $x_0 + ix_1 \mapsto x_0 - ix_1$ if $F = \mathbb{C}$;
- the conjugation $x_0 + ix_1 + jx_2 + kx_3 \mapsto x_0 - ix_1 - jx_2 - kx_3$ if $F = \mathbb{H}$.

Moreover, we set $M^* = \overline{M}^t$.

1.5. We will denote by 0 the zero matrix of any size, by I_n the $n \times n$ identity matrix, by J_n the $2n \times 2n$ standard symplectic matrix, i.e. $J_n := \begin{pmatrix} 0 & I_n \\ -I_n & 0 \end{pmatrix}$, and we set $S_n := \begin{pmatrix} 0 & I_n \\ I_n & 0 \end{pmatrix}$.

1.6. For the notation on simple real Lie groups, we will follow [Hel78, Chap. X, §2] (see also [Kna96, Chap. I, §17]).

2 Hermitian Symmetric Manifolds

The aim of this section is to introduce Hermitian symmetric manifolds and to establish their basic properties.

Let us begin by recalling the definition of a complex structure and of an almost complex structure on a differentiable manifold M.

Definition 2.1.

(i) A *complex manifold* is a pair $(M, \mathcal{O}_M)$ consisting of a (connected) differentiable manifold M and a sheaf $\mathcal{O}_M$ of $\mathbb{C}$-valued smooth functions on M such that $(M, \mathcal{O}_M)$ is locally isomorphic to $(\mathbb{C}^N, \mathcal{O}_{\mathbb{C}^N})$, where $\mathcal{O}_{\mathbb{C}^N}$ is the sheaf of holomorphic functions on $\mathbb{C}^N$. The sheaf $\mathcal{O}_M$ is said to be a complex structure on the manifold M.

(ii) A *quasi-complex manifold* is a pair (M, J) consisting of a (connected) differentiable manifold M and a smooth tensor field J of type $(1,1)$ such that for every $p \in M$ the induced linear map $J_p : T_pM \to T_pM$ satisfies $J_p^2 = -\mathrm{id}$, i.e. J_p is a complex structure on the vector space T_pM. The tensor field J is said to be a quasi-complex structure on the manifold M.

Given a complex manifold $(M, \mathcal{O}_M)$, the local isomorphism of $(M, \mathcal{O}_M)$ with $(\mathbb{C}^n, \mathcal{O}_{\mathbb{C}^n})$ together with the natural complex structure on each tangent space $T_q\mathbb{C}^n \cong \mathbb{C}^n$ given by multiplication by i, induces a quasi-complex structure J on M. A quasi-complex structure J on M induced by a complex structure is said to be *integrable*. Integrable quasi-complex structures are characterized by the following well-known theorem of Newlander–Nirenberg (see [Hel78, Chap. VIII, Thm. 1.2] and the references therein).

Theorem 2.2 (Newlander–Nirenberg). *A quasi-complex structure J on M is induced by a complex structure on M (i.e. it is integrable) if and only if*

$$[JX, JY] = J[JX, Y] + J[X, JY] + [X, Y],$$

for any two vector fields X and Y on M. In this case, the complex structure on M is uniquely determined by the almost complex structure J on M.

Therefore giving a complex manifold is equivalent to giving an almost complex manifold (M, J) such that J is integrable.

There are three equivalent ways of giving an Hermitian structure on a complex manifold (M, J), which we now recall.

Lemma - Definition 2.3. Let (M, J) be a complex manifold. A Hermitian structure on (M, J) is the assignment of one of the following equivalent structures:

(i) A smooth tensor field h of type $(0, 2)$ such that $h_p : T_pM \times T_pM \to \mathbb{C}$ is a positive definite Hermitian form with respect to J_p for any $p \in M$ (called a *Hermitian metric*), i.e.

- $h_p(x, y) = \overline{h_p(y, x)}$ for any $x, y \in T_pM$;
- $h_p(J_px, y) = i h_p(x, y)$ for any $x, y \in T_pM$;
- $h_p(x, x) > 0$ for any $0 \neq x \in T_pM$.

(ii) A Riemannian metric g such that $g_p : T_pM \times T_pM \to \mathbb{R}$ is compatible with J_p for any $p \in M$, i.e.

- $g_p(x, y) = g_p(y, x)$ for any $x, y \in T_pM$;
- $g_p(J_px, J_py) = g_p(x, y)$ for any $x, y \in T_pM$;
- $g_p(x, x) > 0$ for any $0 \neq x \in T_pM$.

(iii) A 2-form ω such that $\omega_p : T_pM \times T_pM \to \mathbb{R}$ is compatible with J_p and positive definite with respect to J_p for any $p \in M$, i.e.

- $\omega_p(x, y) = -\omega_p(y, x)$ for any $x, y \in T_pM$;
- $\omega_p(J_px, J_py) = \omega_p(x, y)$ for any $x, y \in T_pM$;
- $\omega_p(x, J_px) > 0$ for any $0 \neq x \in T_pM$.

One can pass from one assignment to the other two by means of the following formulas:

$$g(X, Y) = \operatorname{Re} h(X, Y) = \omega(X, JY),$$
$$\omega(X, Y) = -\operatorname{Im} h(X, Y) = g(JX, Y),$$
$$h(X, Y) = g(X, Y) - i g(JX, Y) = \omega(X, JY) - i\omega(X, Y).$$

for X, Y any smooth vector fields on M.

We say that (M, J, h) (resp. (M, J, g), resp. (M, J, ω)) is a Hermitian manifold if (M, J) is a complex structure and h (resp. g, resp. ω) defines a Hermitian structure on (M, J). Sometimes, we will say that M is a Hermitian manifold if there is no need to specify the complex structure and the Hermitian structure.

Using partitions of unity, it is easy to show that every complex manifold can be endowed with a Hermitian structure. In what follows, we will be interested in complex manifolds that admits a special Hermitian structure in the following sense.

Definition 2.4. Let (M, J, h) be a Hermitian manifold. Denote by $\mathrm{Aut}(M, J, h)$ the group of holomorphic isometries, i.e. the group of self-diffeomorphisms $\phi : M \to M$ such that $\phi^* J = J$ and $\phi^* h = h$. We say that

(i) (M, J, h) is homogeneous if $\mathrm{Aut}(M, J, h)$ acts transitively on M;
(ii) (M, J, h) is symmetric (or **HSM** for short) if it is homogeneous and for some $p \in M$ (or, equivalently, for any $p \in M$) there exists $s_p \in \mathrm{Aut}(M, J, h)$ (called a *symmetry* at p) such that $s_p^2 = \mathrm{id}$ and p is an isolated fixed point of s_p.

Remark 2.5. (i) If every point $p \in M$ admits a symmetry s_p as above then (M, J, h) is automatically homogeneous (see [Mil05, Prop. 1.6]).

(ii) The symmetry s_p at p can be characterized as the unique $s_p \in \mathrm{Aut}(M, J, h)$ such that $s_p(p) = p$ and $ds_p = -\mathrm{id}_{T_p M}$. It follows that s_p is a geodesic symmetry at p, i.e. if $\gamma : (-a, a) \to M$ is any geodesic such that $\gamma(0) = p$ then $s_p(\gamma(t)) = \gamma(-t)$ for any $-a < t < a$ (see [Mil05, Prop. 1.11]).

(iii) If $(M, J, h = g - i\omega)$ is a Hermitian symmetric manifold then:

- (M, g) is a (geodesically) complete, i.e. M is a complete metric space or, equivalently, every geodesic of the Riemannian manifold (M, g) can be defined on the entire real line (see [Mil05, Prop. 1.11]);
- (M, J, ω) is Kähler, i.e. ω is a closed 2-form (see [Hel78, Chap. VIII, Thm. 4.1]).

(iv) A Riemannian manifold (M, g) such that the group $\mathrm{Aut}(M, g)$ of isometries acts transitively on M and for some $p \in M$ (or, equivalently, for any $p \in M$) there exists $s_p \in \mathrm{Aut}(M, g)$ which is a geodesic symmetry at p is called a *Riemannian symmetric manifold*; see [Hel78] for an extensive study of Riemannian symmetric manifold.

Note that Hermitian symmetric manifolds are in particular Riemannian symmetric manifolds.

In dimension one, every Hermitian symmetric manifold is isomorphic to one of the following examples.

Example 2.6.

(1) Let $\Lambda \subset \mathbb{C}$ be a discrete additive subgroup (note that Λ is isomorphic to (0), $\mathbb{Z}$ or $\mathbb{Z}^2$). The quotient $\mathbb{C}/\Lambda$ is a complex manifold which we endow with the Hermitian structure induced by the standard Euclidean metric $g = dx\,dx + dy\,dy$ on $\mathbb{C} = \mathbb{R}^2_{(x,y)}$ (which has constant zero curvature). The translations $z \mapsto z + a$ (with $a \in \mathbb{C}$) act transitively via holomorphic isometries, so that $\mathbb{C}/\Lambda$ is a

homogeneous Hermitian manifold. If o denotes the class of 0 in the quotient $\mathbb{C}/\Lambda$, then the map $s_o : z \mapsto -z$ is an isometry at o, which shows that $\mathbb{C}/\Lambda$ is a Hermitian symmetric manifold.

(2) Let $\mathcal{H} := \{z = x + iy \in \mathbb{C} : \operatorname{Im} z = y > 0\}$ be the upper half space. Then $\mathcal{H}$ inherits from $\mathbb{C}$ a complex structure and we endow it with a Hermitian structure induced by the hyperbolic metric $g = \frac{dx\,dy}{y^2}$ (which has constant negative curvature). The group $\operatorname{SL}_2(\mathbb{R})$ acts transitively via Möbius transformations (which are holomorphic isometries)

$$\begin{pmatrix} a & b \\ c & d \end{pmatrix} \cdot z := \frac{az+b}{cz+d},$$

which shows that $\mathcal{H}$ is a homogeneous Hermitian manifold. The Möbius transformation $s_i : z \mapsto -\frac{1}{z}$ is a symmetry at $i \in \mathcal{H}$, so that $\mathcal{H}$ is a Hermitian symmetric manifold.

(3) Let $\mathbb{P}^1_\mathbb{C}$ be the complex projective line. Via stereographic projection from the north pole $N = (1, 0, 0)$, the two dimensional sphere $\mathbb{S}^2 \subset \mathbb{R}^3_{(x,y,z)}$ is diffeomorphic to $\mathbb{P}^1_\mathbb{C}$. Via this diffeomorphism, the restriction of the Euclidean metric to $\mathbb{S}^2$ induces a metric g on $\mathbb{P}^1_\mathbb{C}$ (with constant positive curvature) which is compatible with its complex structure, i.e. it induces a Hermitian structure on $\mathbb{P}^1_\mathbb{C}$, which is called the Fubini–Studi metric. The group $\operatorname{SO}_3(\mathbb{R})$ acts transitively on $\mathbb{S}^2$ via rotations, which are holomorphic isometries of $\mathbb{P}^1_\mathbb{C} \cong \mathbb{S}^2$; hence $\mathbb{P}^1_\mathbb{C}$ is a homogeneous Hermitian manifold. The rotation $s_N : (x, y, z) \mapsto (x, -y, -z)$ is a symmetry at the north pole $N \in \mathbb{S}^2 \cong \mathbb{P}^1_\mathbb{C}$, which show that $\mathbb{P}^1_\mathbb{C}$ is a Hermitian symmetric manifold.

The trichotomy of the previous Example 2.6 extends to arbitrary dimension.

Definition 2.7. Let M be a Hermitian symmetric manifold (HSM).

(i) M is said to be of *Euclidean type* if M is isomorphic to $\mathbb{C}^n/\Lambda$ for some discrete additive subgroup $\Lambda \subset \mathbb{C}^n$, where $\mathbb{C}^n/\Lambda$ is endowed with the complex structure and the Hermitian metric induced by $\mathbb{C}^n$.

(ii) M is said to be *irreducible* if it is not Euclidean and it cannot be written as the product of two non-trivial HSMs.

(iii) M is said to be *non-Euclidean* if it is the product of irreducible HSMs.

(iv) M is said to be of *compact type* (resp. *non-compact type*) if it is the product of compact (resp. non-compact) irreducible HSMs.

Hermitian symmetric manifolds of non-compact type are also called *Hermitian symmetric domains*, due to the fact that they are biholomorphic to bounded symmetric domains (see Sect. 2.5).

Remark 2.8. Clearly, Euclidean HSMs have identically zero Riemannian sectional curvature. On the other hand, HSMs of compact type (resp. of non-compact type) have semipositive (resp. seminegative) Riemannian sectional curvature (see

[Hel78, Chap. V, Thm. 3.1]) and therefore also semipositive (resp. seminegative) holomorphic bisectional curvature (see [Mok89, Chap. 2, (3.3), Prop. 1]). Moreover, irreducible HSMs of compact type (resp. of non-compact type) have positive (resp. negative) Ricci curvature (see [Mok89, Chap. 3, (1.3), Prop. 2]).

Every Hermitian symmetric manifold can be decomposed uniquely in the following way (see [Hel78, Chap. VIII, Prop. 4.4, Thm. 4.6, Prop. 5.5]).

Theorem 2.9 (Decomposition theorem). *Every Hermitian symmetric manifold M decomposes uniquely as*

$$M = M_0 \times M_- \times M_+,$$

where M_0 is a Euclidean HSM, M_- is a HSM of compact type and M_+ is a HSM of non-compact type. Moreover, M_- (resp. M_+) is simply connected and it decomposes uniquely as a product of compact (resp. non-compact) irreducible HSMs.

In particular, any HSM is the product of a Euclidean HSM and of a non-Euclidean HSM. Since Euclidean HSMs are easy to understand (being isomorphic to $\mathbb{C}^n/\Lambda$, for some discrete additive subgroup $\Lambda \subset \mathbb{C}^n$), from now on we will focus on non-Euclidean HSMs.

An important invariant of a non-Euclidean HSM is its rank, which we are now going to define following [Hel78, Chap. V, §6]. Recall that a submanifold S of a Riemannian manifold (M, g) is called *totally geodesic* if for every $p \in S$ it holds that all the geodesics of M through p that are tangent to S are contained in S. Moreover, in the case where (M, g) is a Riemannian symmetric manifold [in the sense of Remark 2.5(iv)], N is totally geodesic if and only if for every $p \in N$ we have that $s_p(N) = N$ (see [Mok89, Chap. 5, (1.1), Lemma 1.1]). Furthermore, N is said to be *flat* if the restriction of the Riemannian metric g to N has identically zero curvature tensor.

Definition 2.10. Let M be a non-Euclidean HSM. The **rank** of M is the maximal dimension of a flat totally geodesic submanifold of M.

2.1 Classifying Non-Euclidean HSM Via Lie Groups

The aim of this subsection is to classify non-Euclidean Hermitian symmetric manifolds in terms of Lie groups.

Let $M = (M, J, h)$ be a non-Euclidean Hermitian symmetric manifold and fix a point $o \in M$. The group $\mathrm{Aut}(M) = \mathrm{Aut}(M, J, h)$ of holomorphic isometries of M, endowed with the compact-open topology, becomes a (real) Lie group (see [Hel78, Chap. VIII, §4]). We denote by $\mathrm{Aut}(M)^o$ the connected component of $\mathrm{Aut}(M)$ containing the identity and by $\mathrm{Stab}(o)$ the Lie subgroup of $\mathrm{Aut}(M)^o$ consisting of all the elements that fix $o \in M$. The symmetry s_o at o induces the following involutive automorphism of $\mathrm{Aut}(M)^o$:

$$\sigma : \mathrm{Aut}(M)^o \longrightarrow \mathrm{Aut}(M)^o$$

$$g \mapsto s_o g s_o.$$

Denote by $\mathrm{Fix}(\sigma)$ the closed Lie subgroup of $\mathrm{Aut}(M)^o$ consisting of all the elements that are fixed by σ and let $\mathrm{Fix}(\sigma)^o$ be the connected component of $\mathrm{Fix}(\sigma)$ containing the origin.

Theorem 2.11. *Notations as above.*

(i) $\mathrm{Aut}(M)^o$ *is a semisimple adjoint (i.e. with trivial center) Lie group and* $\mathrm{Stab}(o)$ *is a compact Lie subgroup of* $\mathrm{Aut}(M)^o$ *such that*

$$\mathrm{Fix}(\sigma)^o \subseteq \mathrm{Stab}(o) \subseteq \mathrm{Fix}(\sigma).$$

(ii) *The map*

$$\mathrm{Aut}(M)^o / \mathrm{Stab}(o) \longrightarrow M$$

$$[g] \mapsto g \cdot o$$

is a $\mathrm{Aut}(M)^o$*-equivariant diffeomorphism.*

(iii) *The symmetry s_o is contained in the identity component of the center of* $\mathrm{Stab}(o)$.

Proof. See [Hel78, Chap. IV, Thm. 3.3; Chap. VIII, Thm. 4.5] □

A pair (G, K) consisting of a connected semisimple Lie group G and a compact subgroup K for which there exists an involutive automorphism σ of G such that $\mathrm{Fix}(\sigma)^o \subseteq K \subseteq \mathrm{Fix}(\sigma)$ is a particular case of a *Riemann symmetric pair* (see [Hel78, Chap. IV, §3]). In particular, Theorem 2.11(i) is saying that for any non-Euclidean Hermitian symmetric manifold M, the pair $(\mathrm{Aut}(M)^o, \mathrm{Stab}(o))$ is a Riemannian symmetric pair.

Conversely, starting with a Riemannian symmetric pair (G, K), the quotient manifold G/K can always be endowed with a Riemannian metric g such that $(G/K, g)$ is a Riemannian symmetric space (see [Hel78, Chap. IV, Prop. 3.4]). However, in order to endow G/K with a complex structure J and a Hermitian metric h such that $(G/K, J, h)$ becomes a Hermitian symmetric manifold, the pair (G, K) must satisfy some extra conditions. Theorem 2.11(iii) says that a necessary condition for a Riemannian symmetric pair (G, K) to come from a Hermitian symmetric manifold is that the center of K is not finite. Indeed, in the irreducible case, this last condition is also sufficient.

Theorem 2.12. (i) *Every irreducible HSM of non-compact type is diffeomorphic to G/K for a unique pair (G, K) such that G is a connected non-compact*

Table 1 Irreducible HSMs: G/K is of non-compact type and G^c/K is its compact dual

Type	Group G	Group G^c	Compact subgroup K	$\dim_\mathbb{R} M$	Rank M
$I_{p,q}$ ($p \geq q \geq 1$)	$\mathrm{PSU}(p,q)$	$\mathrm{PSU}(p+q)$	$\mathrm{S}(U_p \times U_q)$	$2pq$	q
II_n ($n \geq 2$)	$\cong \mathrm{PSO}^*(2n)$	$\cong \mathrm{PSO}(2n)$	$\overline{U(n)}$	$n(n-1)$	$\lfloor \frac{n}{2} \rfloor$
III_n ($n \geq 1$)	$\cong \mathrm{PSp}(n,\mathbb{R})$	$\mathrm{PSp}(n)$	$\overline{U(n)}$	$n(n+1)$	n
IV_n ($2 \neq n \geq 1$)	$\cong \mathrm{PSO}(2,n)$	$\mathrm{PSO}(2+n)$	$\mathrm{SO}(2) \times \mathrm{SO}(n)$	$2n$	$\min\{2,n\}$
V	$E_{6(-14)}$	E_6^c	$\mathrm{SO}(10) \times \mathrm{SO}(2)$	32	2
VI	$E_{7(-25)}$	E_7^c	$E_6^c \times \mathrm{SO}(2)$	54	3

simple adjoint Lie group and K is a maximal connected and compact Lie subgroup of G with non discrete center Z_K (or, equivalently, with $Z_K = \mathbb{S}^1$).

(ii) Every irreducible HSM of compact type is diffeomorphic to G/K for a unique pair (G, K) such that G is a connected compact simple adjoint Lie group and K is a maximal connected and compact proper Lie subgroup of G with non discrete center Z_K (or, equivalently, with $Z_K = \mathbb{S}^1$).

Proof. See [Hel78, Chap. VIII, §6]. □

Using the Lie-theoretic representation given by Theorem 2.12, È. Cartan in [Car35] (based upon his previous work [Car26-27] in which he classifies Riemannian symmetric manifolds) was able to classify the irreducible HSMs of non-compact type (rep. of compact type) into six types (see also [Bor52] for a nice exposition of the work of Cartan). We list the six types (together with their real dimensions and their rank) in Table 1, referring to Sect. 3 for more details on each type and to Sect. 2.3 for an explanation of the duality between HSMs of non-compact type and HSMs of compact type.

Using the above presentation of HSMs, it is possible to give a Lie theoretic description of the rank of a HSM of non-compact type (as in Definition 2.10).

Proposition 2.13. *Let M be a HSM of non-compact type and write $M \cong \mathrm{Aut}(M)^o/\mathrm{Stab}(o) = G/K$ as in Theorem 2.11. Then the rank of M is equal to the dimension of any maximal $\mathbb{R}$-split torus T contained in G.*

Proof. See [Mor, Sec. 8C]. □

2.2 Classifying Non-Euclidean HSM Via Lie Algebras

The aim of this subsection is to classify non-Euclidean HSM in terms of Lie algebras data.

Let $M = (M, J, h)$ be a non-Euclidean HSM with a fixed base point $o \in M$. We shorten the notation used in Sect. 2.1 by setting $G := \mathrm{Aut}(M)^o$ and $K := \mathrm{Stab}(o)$. Let $\mathfrak{g} = \mathrm{Lie}(G)$ be the real Lie algebra of G [which is semisimple by

Theorem 2.11(i)] and let $\theta = d\sigma : \mathfrak{g} \to \mathfrak{g}$ be the involution of $\mathfrak{g}$ given by the differential of the involution σ of G. Denote by $\mathfrak{k}$ (resp. $\mathfrak{p}$) the eigenspace for θ relative to the eigenvalue $+1$ (resp. -1). Since $\theta^2 = \mathrm{id}$, we have a direct sum decomposition

$$\mathfrak{g} = \mathfrak{k} \oplus \mathfrak{p} \tag{1}$$

in such a way that the Lie bracket $[,]$ of $\mathfrak{g}$ satisfies

$$[\mathfrak{k}, \mathfrak{k}] \subseteq \mathfrak{k}, \quad [\mathfrak{k}, \mathfrak{p}] \subseteq \mathfrak{p}, \quad [\mathfrak{p}, \mathfrak{p}] \subseteq \mathfrak{k}. \tag{2}$$

From Theorem 2.11(i), it follows that the Lie subalgebra $\mathfrak{k} \subset \mathfrak{g}$ is equal to the Lie subalgebra $\mathrm{Lie}(K) \subset \mathrm{Lie}(G)$ of the Lie subgroup $K \subset G$. In particular, since K is a compact Lie group, it follows that $\mathfrak{k}$ is a compactly embedded subalgebra of $\mathfrak{g}$ (in the sense of [Hel78, p. 130]). Therefore, the pair $(\mathfrak{g}, \theta)$ above defined is a special case of an *effective orthogonal symmetric Lie algebra* in the sense of [Hel78, Chap. V, §1].

Moreover, the G-equivariant diffeomorphism $G/K \stackrel{\cong}{\to} M$ of Theorem 2.11(i) induces a canonical identification $\mathfrak{p} \cong T_o M$, where $T_o M$ is the tangent space of M at $o \in M$. The above identification, together with the complex structure J on M, induces a complex structure J_o on $\mathfrak{p}$. We extend J_o to a linear operator on $\mathfrak{g}$ by setting:

$$J = \begin{cases} 0 & \text{on } \mathfrak{k}, \\ J_o & \text{on } \mathfrak{p}. \end{cases} \tag{3}$$

Using the fact that the complex structure J on M is compatible with the Hermitian metric h on M, one can prove the following (see [Sat80, Chap. II, §3])

Lemma 2.14. *J is a derivation of $\mathfrak{g}$.*

Since $\mathfrak{g}$ is semisimple, each derivation on $\mathfrak{g}$ is inner (see [Kna96, Prop. 1.98]) and the center of $\mathfrak{g}$ is trivial; therefore, $J = \mathrm{ad}\, H$ for a unique element $H \in \mathfrak{g}$. Using (3), it follows that H belongs to the center $Z(\mathfrak{k})$ of $\mathfrak{k}$.

The properties of the triple $(\mathfrak{g}, \theta, H)$ are summarized in the following definition, which is a slight adaptation of [Sat80, p. 54].

Definition 2.15. A *Hermitian symmetric Lie algebra* (or, for short, a **Hermitian SLA**) is a triple $(\mathfrak{g}, \theta, H)$ consisting a semisimple real Lie algebra $\mathfrak{g}$, an involution $\theta : \mathfrak{g} \to \mathfrak{g}$ whose associated decomposition $\mathfrak{g} = \mathfrak{k} \oplus \mathfrak{p}$ into eigenvalues for $+1$ and -1 is such that $\mathfrak{k}$ is a compactly embedded subalgebra of $\mathfrak{g}$ and an element $H \in Z(\mathfrak{k})$ such that $\mathrm{ad}(H)^2_{|\mathfrak{p}} = -\mathrm{id}_{\mathfrak{p}}$.

A Hermitian SLA $(\mathfrak{g}, \theta, H)$ is said to be *irreducible* if the complexification $\mathfrak{g}_\mathbb{C}$ of $\mathfrak{g}$ is a simple complex Lie algebra.

A Hermitian SLA $(\mathfrak{g}, \theta, H)$ is said to be

(i) *of compact type* if $\mathfrak{g}$ is a compact Lie algebra, i.e. its Killing form B is negative definite;
(ii) *of non-compact type* if $\mathfrak{g}$ does not contain compact simple factors and θ is a Cartan involution of $\mathfrak{g}$, i.e. B is negative definite on $\mathfrak{k}$ and positive definite on $\mathfrak{p}$.

We have seen before how to associate to a non-Euclidean HSM M a Hermitian SLA $(\mathfrak{g}, \theta, H)$. Indeed, this construction is bijective and it is compatible with the decomposition of HSMs as in Theorem 2.9.

Theorem 2.16. *There is a bijection*

$$\{\text{non-Euclidean HSMs}\} \xrightarrow{\cong} \{\text{Hermitian SLAs}\} \qquad (4)$$

obtained by sending a non-Euclidean Hermitian symmetric manifold M to its associated Hermitian SLA $(\mathfrak{g}, \theta, H)$.

(i) *If $M = M_1 \times \cdots \times M_r$ is the decomposition of M into irreducible HSMs and $(\mathfrak{g}_i, \theta, H_i)$ is the Hermitian SLA associated to M_i, then*

$$(\mathfrak{g}, \theta, H) = (\mathfrak{g}_1, \theta_1, H_1) \oplus \cdots \oplus (\mathfrak{g}_r, \theta_r, H_r), \qquad (5)$$

meaning that $\mathfrak{g}$ decomposes a direct sum of ideals $\mathfrak{g} = \mathfrak{g}_1 \oplus \cdots \oplus \mathfrak{g}_r$, θ is the unique involution on $\mathfrak{g}$ that preserves the above decomposition and such that $\theta_{|\mathfrak{g}_i} = \theta_i$ and $H = H_1 + \cdots + H_r$.
(ii) *M is an irreducible HSM if and only if $(\mathfrak{g}, \theta, H)$ is irreducible.*
(iii) *M is of non-compact type if and only if $(\mathfrak{g}, \theta, H)$ is of non-compact type.*
(iv) *M is of compact type if and only if $(\mathfrak{g}, \theta, H)$ is of compact type.*

Proof. Part (i) follows from [Hel78, Prop. 4.4, Prop. 5.5].

According to [Hel78, Prop. 5.5, Thm. 5.3, Thm. 5.4], M is irreducible if and only if $(\mathfrak{g}, \theta)$ belongs to one of the following four types:

(1) *Type I*: $\mathfrak{g}$ is a simple compact Lie algebra;
(2) *Type II*: $\mathfrak{g}$ is a compact Lie algebra which is the sum of two simple ideals $\mathfrak{g} = \mathfrak{g}_1 \oplus \mathfrak{g}_2$ which are interchanged by θ;
(3) *Type III*: $\mathfrak{g}$ is a simple non-compact Lie algebras such that $\mathfrak{g}_\mathbb{C}$ is simple;
(4) *Type IV*: $\mathfrak{g}$ is a simple complex Lie algebra (regarded as a real Lie algebra).

Moreover, the existence of a non-trivial element $H \in Z(\mathfrak{k})$ rules out Type II and Type IV (see [Hel78, p. 518]). The remaining cases (Type I and Type III) are exactly those cases for which $\mathfrak{g}_\mathbb{C}$ is a simple complex Lie algebra. Part (ii) follows.

Parts (iii) and (iv) follow easily from (ii) and the Definitions 2.7 and 2.15.

It remains to prove that the map from non-Euclidean HSM to Hermitian SLAs is bijective. In order to prove that, we will construct the inverse map of (4).

First of all, we have the following

Claim: Any Hermitian SLA admits a decomposition (as in (5)) into the direct sum of irreducible Hermitian SLAs.

Indeed, given a Hermitian SLA $(\mathfrak{g}, \theta, H)$, using [Hel78, Prop. 5.2, Thm. 5.3, Thm. 5.4], we can write $\mathfrak{g}$ as the direct sum of ideals

$$\mathfrak{g} = \mathfrak{g}_1 \oplus \cdots \oplus \mathfrak{g}_r,$$

in such a way that θ preserves this decomposition and each factor $(\mathfrak{g}_i, \theta_i := \theta_{|\mathfrak{g}_i})$ belongs to one of the four Types mentioned before. The element H can be written as a sum $H = H_1 + \cdots + H_r$ in such a way that $(\mathfrak{g}_i, \theta_i, H_i)$ is a Hermitian SLA. As observed before, the existence of such an element $H_i \in Z(\mathfrak{k}_i)$ forces $(\mathfrak{g}_i, \theta_i)$ to be of Type I or Type III, so that each $(\mathfrak{g}_i, \theta_i, H_i)$ is irreducible, q.e.d.

Using the Claim and part (ii), it is now enough to construct an inverse of (4) for irreducible Hermitian SLAs.

Let $(\mathfrak{g}, \theta, H)$ be an irreducible Hermitian SLA and assume first that it is of non-compact type. Let G the unique connected adjoint Lie group with $\text{Lie}(G) = \mathfrak{g}$ and let K be the unique Lie subgroup of G corresponding to the Lie subalgebra $\mathfrak{k} \subset \mathfrak{g}$. Since θ is a Cartan involution of $\mathfrak{g}$ and $\mathfrak{g}$ is simple, we deduce that G is a simple non-compact Lie group and K is a maximal compact Lie subgroup of G. On the quotient manifold G/K (with base point $o = [e]$), consider the unique G-invariant almost complex structure J such that J_o is equal to $\text{ad}(H)_{|\mathfrak{p}}$ via the identification $T_o M \cong \mathfrak{p}$ and the unique G-invariant Riemannian metric g such that g_o is equal to the Killing form of $\mathfrak{g}$ restricted to $\mathfrak{p}$. It follows from [Hel78, Chap. VIII, Prop. 4.2] that $(G/K, J, g)$ is a HSM, which is irreducible of non-compact type by Theorem 2.12. Moreover, it is easy to check that the Hermitian SLA associated to $(G/K, J, g)$ is the Hermitian SLA $(\mathfrak{g}, \theta, H)$ we started with, and we are done.

The case where $(\mathfrak{g}, \theta, H)$ is irreducible of compact type can be dealt with similarly and therefore it is left to the reader. □

Remark 2.17. The bijection (4) becomes an equivalence of categories if the two sets are endowed with the following morphisms:

(i) A *symmetric* (or equivariant) morphism between two pointed non-Euclidean HSMs (M, o) and (M', o') is a pointed holomorphic map $\phi : (M, o) \to (M', o')$ such that

$$\phi \circ s_p = s'_{\phi(p)} \circ \phi \text{ for any } p \in M,$$

where s_p is the symmetry of M at $p \in M$ and $s'_{\phi(p)}$ is the symmetry of M' at $\phi(p) \in M'$.

(ii) A morphism of Hermitian SLAs (also called a H_1-morphism) between two Hermitian SLAs $(\mathfrak{g}, \theta, H)$ and $(\mathfrak{g}', \theta', H')$ is a morphism of Lie algebras $\rho : \mathfrak{g} \to \mathfrak{g}'$ such that

Table 2 Irreducible classical Hermitian SLAs of non-compact type

Type	Lie algebra $\mathfrak{g}$	Involution θ	Central element $H \in Z(\mathfrak{k})$
$I_{p,q}$	$\mathfrak{su}(p,q)$	$\theta \begin{pmatrix} Z_1 & Z_2 \\ \overline{Z}_2^t & Z_3 \end{pmatrix} = \begin{pmatrix} Z_1 & -Z_2 \\ -\overline{Z}_2^t & Z_3 \end{pmatrix}$	$i\begin{pmatrix} \frac{q}{p+q}I_p & 0 \\ 0 & \frac{-p}{p+q}I_q \end{pmatrix}$
II_n	$\cong \mathfrak{so}^*(2n)$	$\theta \begin{pmatrix} Z_1 & Z_2 \\ \overline{Z}_2^t & -Z_1^t \end{pmatrix} = \begin{pmatrix} Z_1 & -Z_2 \\ -\overline{Z}_2^t & -Z_1^t \end{pmatrix}$	$\frac{i}{2}\begin{pmatrix} I_n & 0 \\ 0 & -I_n \end{pmatrix}$
III_n	$\cong \mathfrak{sp}(n,\mathbb{R})$	$\theta \begin{pmatrix} Z_1 & Z_2 \\ \overline{Z}_2^t & -Z_1^t \end{pmatrix} = \begin{pmatrix} Z_1 & -Z_2 \\ -\overline{Z}_2^t & -Z_1^t \end{pmatrix}$	$\frac{i}{2}\begin{pmatrix} I_n & 0 \\ 0 & -I_n \end{pmatrix}$
IV_n	$\cong \mathfrak{so}(2,n)$	$\theta \begin{pmatrix} X_1 & iX_2 \\ -iX_2^t & X_3 \end{pmatrix} = \begin{pmatrix} X_1 & -iX_2 \\ iX_2^t & X_3 \end{pmatrix}$	$\begin{pmatrix} 0 & 0 \\ 0 & J_1 \end{pmatrix}$

Table 3 Irreducible classical Hermitian SLAs of compact type

Type	Lie algebra $\mathfrak{g}^*$	Involution θ^*	Central element $H^* \in Z(\mathfrak{k})$
$I_{p,q}$	$\mathfrak{su}(p+q)$	$\theta^* \begin{pmatrix} Z_1 & Z_2 \\ -\overline{Z}_2^t & Z_3 \end{pmatrix} = \begin{pmatrix} Z_1 & -Z_2 \\ \overline{Z}_2^t & Z_3 \end{pmatrix}$	$i\begin{pmatrix} \frac{q}{p+q}I_p & 0 \\ 0 & \frac{-p}{p+q}I_q \end{pmatrix}$
II_n	$\cong \mathfrak{so}(2n)$	$\theta^* \begin{pmatrix} Z_1 & Z_2 \\ -\overline{Z}_2^t & -Z_1^t \end{pmatrix} = \begin{pmatrix} Z_1 & -Z_2 \\ \overline{Z}_2^t & -Z_1^t \end{pmatrix}$	$\frac{i}{2}\begin{pmatrix} I_n & 0 \\ 0 & -I_n \end{pmatrix}$
III_n	$\mathfrak{sp}(n)$	$\theta^* \begin{pmatrix} Z_1 & Z_2 \\ -\overline{Z}_2^t & -Z_1^t \end{pmatrix} = \begin{pmatrix} Z_1 & -Z_2 \\ \overline{Z}_2^t & -Z_1^t \end{pmatrix}$	$\frac{i}{2}\begin{pmatrix} I_n & 0 \\ 0 & -I_n \end{pmatrix}$
IV_n	$\mathfrak{so}(2+n)$	$\theta^* \begin{pmatrix} X_1 & X_2 \\ -X_2^t & X_3 \end{pmatrix} = \begin{pmatrix} X_1 & -X_2 \\ X_2^t & X_3 \end{pmatrix}$	$\begin{pmatrix} 0 & 0 \\ 0 & J_1 \end{pmatrix}$

$$\rho \circ \theta = \theta' \circ \rho,$$
$$\rho \circ \mathrm{ad}(H) = \mathrm{ad}(H') \circ \rho.$$

See [Sat80, Chap. II, §8] and [AMRT10, Chap. III, §2.2].

The correspondence in Theorem 2.16 together with the classification of irreducible HSMs recalled in Sect. 2.1 gives a classification of irreducible Hermitian SLAs. We record the list of irreducible classical Hermitian SLAs of non-compact type (resp. of compact type) into Table 2 (resp. Table 3). Moreover, to each irreducible Hermitian SLA $(\mathfrak{g}, \theta, H)$ of non-compact type in Table 2, its dual Hermitian SLA of compact type (as defined in Sect. 2.3) is denoted by $(\mathfrak{g}^*, \theta^*, H^*)$ in Table 3. We refer to Sect. 3 for more details.

The rank of a non-Euclidean HSM as defined in Definition 2.10 can be read off from the associated Hermitian SLA as it follows.

Proposition 2.18. *Let M be a non-Euclidean HSM with base point $o \in M$ and let $(\mathfrak{g}, \theta, H)$ its associated Hermitian SLA. Consider the decomposition $\mathfrak{g} = \mathfrak{k} \oplus \mathfrak{p}$ as in (1). Then any two maximal abelian Lie subalgebras of $\mathfrak{g}$ contained in $\mathfrak{p}$ are conjugate by an element of $\mathrm{Stab}(o) \subset \mathrm{Aut}(M)^o$, acting on $\mathfrak{g}$ via the adjoint representation, and their common dimension is equal to the rank of M.*

Proof. See [Hel78, Chap. V, Prop. 6.1 and Lemma 6.3]. □

2.3 Duality Between Compact and Non-compact HSMs

We are now going to define an involution on non-Euclidean HSMs that exchanges compact HSMs with non-compact HSMs. The involution is defined most easily in terms of Hermitian SLAs, using the bijection of Theorem 2.16.

Given a Hermitian SLA $(\mathfrak{g}, \theta, H)$, define a new Hermitian SLA $(\mathfrak{g}, \theta, H)^* = (\mathfrak{g}^*, \theta^*, H^*)$ as it follows. The Lie algebra $\mathfrak{g}^*$ is the subalgebra of the complexification $\mathfrak{g}_\mathbb{C}$ given by

$$\mathfrak{g}^* := \mathfrak{k} \oplus i\mathfrak{p} \subseteq \mathfrak{g}_\mathbb{C} = (\mathfrak{k} \oplus \mathfrak{p}) \oplus i(\mathfrak{k} \oplus \mathfrak{p}),$$

where as usual $\mathfrak{k}$ and $\mathfrak{p}$ are the eigenspaces for θ relative to the eigenvalues $+1$ and -1. Since $\mathfrak{g}_\mathbb{C} = (\mathfrak{g}^*)_\mathbb{C}$ and the property of being semisimple is preserved by the complexification functor, it follows that $\mathfrak{g}^*$ is a semisimple real Lie algebra. The involution θ^* on $\mathfrak{g}^*$ is defined by

$$\theta^* = \begin{cases} +1 & \text{on } \mathfrak{k}, \\ -1 & \text{on } i\mathfrak{p}. \end{cases}$$

In other words, $\mathfrak{g}^* = \mathfrak{k} \oplus i\mathfrak{p}$ is the decomposition of $\mathfrak{g}^*$ into eigenspaces for θ^* relative to the eigenvalues $+1$ and -1. Finally, we set

$$H^* := H.$$

Note that $H^* = H \in Z(\mathfrak{k})$ and that $\operatorname{ad}(H^*)_{|i\mathfrak{p}} = -\operatorname{id}_{i\mathfrak{p}}$. Therefore, $(\mathfrak{g}, \theta, H)^*$ is a Hermitian SLA, which is called the *dual Hermitian SLA* of $(\mathfrak{g}, \theta, H)$.

Theorem 2.19. *The map (called the* duality map*)*

$$\{\text{Hermitian SLAs}\} \longrightarrow \{\text{Hermitian SLAs}\}$$
$$(\mathfrak{g}, \theta, H) \mapsto (\mathfrak{g}, \theta, H)^* := (\mathfrak{g}^*, \theta^*, H^*)$$

is an involution which satisfies the following properties:

(i) If

$$(\mathfrak{g}, \theta, H) = (\mathfrak{g}_1, \theta_1, H_1) \oplus \cdots \oplus (\mathfrak{g}_r, \theta_r, H_r)$$

is the decomposition of $(\mathfrak{g}, \theta, H)$ into irreducible Hermitian SLAs, then the dual Hermitian SLA $(\mathfrak{g}, \theta, H)^*$ admits the following decomposition into irreducible Hermitian SLAs

$$(\mathfrak{g}, \theta, H)^* = (\mathfrak{g}_1, \theta_1, H_1)^* \oplus \cdots \oplus (\mathfrak{g}_r, \theta_r, H_r)^*.$$

(ii) $(\mathfrak{g}, \theta, H)$ *is of compact type (resp. of non-compact type) if and only if* $(\mathfrak{g}, \theta, H)^*$ *is of non-compact type (resp. of compact type).*

Proof. The fact that the duality map is an involution follows immediately from the definition.

Part (i) follows from easily from the definitions of the dual and of the direct sum of Hermitian SLAs, together with the observation that $(\mathfrak{g}, \theta, H)$ is irreducible if and only if $(\mathfrak{g}, \theta, H)^*$ is irreducible since $\mathfrak{g}_\mathbb{C} = (\mathfrak{g}^*)_\mathbb{C}$.

Part (ii) follows from the well-know fact that an involution θ on a semisimple real Lie algebra $\mathfrak{g}$, with associated decomposition $\mathfrak{g} = \mathfrak{k} \oplus \mathfrak{p}$ into $+1$ and -1 eigenvalues, is a Cartan involution of $\mathfrak{g}$ if and only if $\mathfrak{g}^* = \mathfrak{k} \oplus i\mathfrak{p}$ is a compact (semisimple) Lie algebra (see e.g. [Hel78, Chap. III, Prop. 7.4]). □

We can now define the dual of a non-Euclidean HSM, using the bijection of Theorem 2.16.

Definition 2.20. Let M be a non-Euclidean HSM whose associated Hermitian SLA is $(\mathfrak{g}, \theta, s)$. The *dual HSM* of M, denoted by M^*, is the unique non-Euclidean HSM whose associated Hermitian SLA is $(\mathfrak{g}, \theta, H)^*$.

From Theorems 2.19 and 2.16, we can deduce the following

Corollary 2.21. *The map (called the* duality map*)*

$$\{\text{non-Euclidean HSMs}\} \longrightarrow \{\text{non-Euclidean HSMs}\}$$

$$M \mapsto M^*$$

is an involution which satisfies the following properties:

(i) *If*

$$M = M_1 \times \cdots \times M_r$$

is the decomposition of M into irreducible HSMs, then M^* admits the following decomposition into irreducible HSMs

$$M^* = M_1^* \times \cdots \times M_r^*.$$

(ii) *M is of compact type (resp. of non-compact type) if and only if M^* is of non-compact type (resp. of compact type).*

2.4 Harish–Chandra and Borel Embeddings

The aim of this subsection is to realize a given HSM of non-compact type as an open subset of a complex vector space (Harish–Chandra embedding) and as an open subset of its dual HSM (Borel embedding), which is also called its *compact dual*.

Fix a HSM of non-compact type $M = (M, J, h)$ together with a base point $o \in M$. By Theorem 2.11(ii), we have a diffeomorphism $M \cong G/K$ where $G = \text{Aut}(M)^o$ and $K = \text{Stab}(o)$. Since G is an adjoint semisimple Lie group [by Theorem 2.11(i)], the adjoint representation $\text{Ad} : G \to \text{GL}(\mathfrak{g})$ is faithful. Therefore G admits a natural complexification $G_{\mathbb{C}}$ (in the sense of [Kna96, p. 437]), namely the complex connected (semisimple) Lie subgroup of $\text{GL}(\mathfrak{g}, \mathbb{C})$ whose Lie algebra is the Lie subalgebra $\mathfrak{g}_{\mathbb{C}} \xrightarrow{\text{ad}} \mathfrak{gl}(\mathfrak{g}, \mathbb{C})$. Denote by $K_{\mathbb{C}}$ the unique connected Lie subgroup of $G_{\mathbb{C}}$ whose Lie algebra is $\mathfrak{k}_{\mathbb{C}} \subset \mathfrak{g}_{\mathbb{C}}$. Note that $K_{\mathbb{C}}$ is a complex reductive Lie group which is a complexification of K.

Denote by $(\mathfrak{g}, \theta, H)$ the Hermitian SLA associated to M, as in Sect. 2.2. Since $\text{ad}(H)_{|\mathfrak{p}} = -\text{id}_{\mathfrak{p}}$, the complexification $\mathfrak{p} \otimes_{\mathbb{R}} \mathbb{C}$ admits a decomposition

$$\mathfrak{p} \otimes_{\mathbb{R}} \mathbb{C} = \mathfrak{p}_+ \oplus \mathfrak{p}_-,$$

where $\mathfrak{p}_+$ (resp. $\mathfrak{p}_-$) is the eigenspace for $\text{ad}(H)_{|\mathfrak{p}}$ relative to the eigenvalue $+i$ (resp. $-i$). Therefore, $\mathfrak{g}_{\mathbb{C}}$ admits the following decomposition

$$\mathfrak{g}_{\mathbb{C}} = \mathfrak{k}_{\mathbb{C}} \oplus \mathfrak{p}_+ \oplus \mathfrak{p}_- \qquad (6)$$

which satisfies the relations (see [Sat80, p. 53]):

$$[\mathfrak{k}_{\mathbb{C}}, \mathfrak{p}_{\pm}] \subset \mathfrak{p}_{\pm} \quad [\mathfrak{p}_+, \mathfrak{p}_-] \subset \mathfrak{k}_{\mathbb{C}} \quad [\mathfrak{p}_+, \mathfrak{p}_+] = 0 \quad [\mathfrak{p}_-, \mathfrak{p}_-] = 0.$$

In particular, $\mathfrak{p}_+$ and $\mathfrak{p}_-$ are abelian subalgebras of $\mathfrak{g}_{\mathbb{C}}$ which are normalized by $\mathfrak{k}_{\mathbb{C}}$. Denote by P_+ (resp. P_-) the connected Lie subgroups of $G_{\mathbb{C}}$ whose Lie algebra is $\mathfrak{p}_+ \subset \mathfrak{g}_{\mathbb{C}}$ (resp. $\mathfrak{p}_- \subset \mathfrak{g}_{\mathbb{C}}$). Then P_+ and P_- are abelian unipotent Lie groups that are stabilized by $K_{\mathbb{C}}$. It turns out that P_+ and P_- are simply connected so that the exponential map

$$\exp : \mathfrak{p}_{\pm} \to P_{\pm}$$

is a diffeomorphism (see [Hel78, Chap. VIII, Lemma 7.8]). Moreover the multiplication map $(P_+ \times P_-) \rtimes K_{\mathbb{C}} \to G_{\mathbb{C}}$ is injective and the image contains G.

Finally, let G_c be the Lie subgroup of $G_{\mathbb{C}}$ corresponding to the Lie subalgebra $\mathfrak{g}^* \subset (\mathfrak{g}^*)_{\mathbb{C}} = \mathfrak{g}_{\mathbb{C}}$. By Theorem 2.19 and Corollary 2.21, G_c is a compact Lie group containing K such that

$$M^* \cong G_c/K.$$

We can summarize the above discussion into the following commutative diagram

$$G \hookrightarrow (P_+ \times P_-) \rtimes K_{\mathbb{C}} \hookrightarrow G_{\mathbb{C}} \longleftarrow G_c \qquad (7)$$

$$\uparrow \qquad \uparrow \qquad \uparrow \qquad \uparrow$$

$$K \hookrightarrow P_- \rtimes K_{\mathbb{C}} = P_- \rtimes K_{\mathbb{C}} \longleftarrow K$$

Theorem 2.22. *By taking quotients in (7), we get a diagram of complex manifolds*

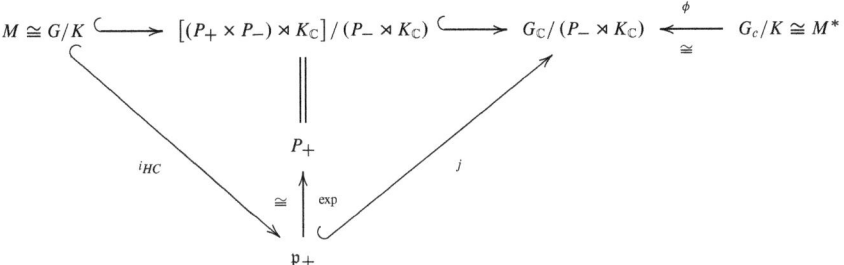

in which ϕ is a biholomorphism, i_{HC} is a holomorphic open embeddings and j is a Zariski open embedding onto the homogeneous projective variety $G_{\mathbb{C}}/(P_- \rtimes K_{\mathbb{C}})$

Proof. See [Hel78, Chap. VIII, §7] or [Sat80, Chap. II, §4]. □

The embedding i_{HC} is called the *Harish–Chandra embedding* while the composition $\phi^{-1} \circ j \circ i_{HC}$ is called the *Borel embedding*. Harish–Chandra embeddings and Borel embeddings for each irreducible HSM will be studied in detail in Sect. 3.

2.5 HSMs of Non-compact Type as Bounded Symmetric Domains

The Harish–Chandra embedding $i_{HC} : M \hookrightarrow \mathfrak{p}_+$ defined in Sect. 2.4 allows to realize canonically a given HSM of non-compact type M as a bounded symmetric domain.

Definition 2.23. A **domain** $D \subseteq \mathbb{C}^N$ (i.e. an open connected subset of $\mathbb{C}^N$) is said to be

(i) *bounded* it is bounded as a subset of $\mathbb{C}^N$;
(ii) *homogeneous* if the group $\mathrm{Hol}(D)$ of biholomorphisms of D acts transitively on D;
(iii) *symmetric* if it is homogeneous and for some $p \in D$ (or, equivalently, for any $p \in D$) there exists $s_p \in \mathrm{Hol}(D)$ (called a symmetry at p) such that $s_p^2 = \mathrm{id}$ and p is an isolated fixed point of s_p.

(iv) *irreducible* if there does not exist a non-trivial decomposition $\mathbb{C}^N = \mathbb{C}^{N_1} \times \mathbb{C}^{N_2}$ and two domains $D_1 \subset \mathbb{C}^{N_1}$ and $D_2 \subset \mathbb{C}^{N_2}$ such that $D = D_1 \times D_2$.

Remark 2.24. If $D \subset \mathbb{C}^N$ is a bounded domain, then we have that (see [Sat80, Chap. II, §4, Rmk. 2] and the references therein)

(i) the group $\text{Hol}(D)$ admits a (unique) structure of Lie group compatible with the open-compact topology;
(ii) D is symmetric if and only if it is homogeneous and $\text{Hol}(D)$ is semisimple.

Moreover, É. Cartan [Car35] showed that, up to dimension three, every homogeneous bounded domain is also symmetric and he asked if this was true in every dimension. Counterexamples were found later, starting from dimension four, by Pyateskii–Shapiro (see [PS69]).

The image of the Harish–Chandra embedding is a bounded symmetric domain inside the complex vector space $\mathfrak{p}_+$.

Theorem 2.25 (Hermann, Harish–Chandra). *For any HSM of non-compact type M together with a fixed base point $o \in M$, the image of the Harish–Chandra embedding $i_{HC} : M \hookrightarrow \mathfrak{p}_+$ is a bounded symmetric domain.*

Proof. Since i_{HC} is a holomorphic open embedding and M is connected, the image $i_{HC}(M)$ is a domain inside $\mathfrak{p}_+$.

The group $\text{Aut}(M)$ of biholomorphic isometries of M is a subgroup of finite index of the group of biholomorphisms $\text{Hol}(M) = \text{Hol}(i_{HC}(M))$ (see [Mil05, Prop. 1.6]). Therefore, the group $\text{Aut}(M)^o = \text{Hol}(i_{HC}(M))^o$ acts transitively on $i_{HC}(M)$ by Theorem 2.11(ii); hence $i_{HC}(M)$ is a homogeneous domain. Moreover, the fact that every point of M has a symmetry in the sense of Definition 2.4(ii)) implies that every point of $i_{HC}(M)$ has a symmetry in the sense of Definition 2.23; hence $i_{HC}(M)$ is a symmetric domain.

Finally, in order to prove that $i_{HC}(M)$ is a bounded domain, we need to recall an explicit description of the image of i_{HC}. Consider the Hermitian SLA of non-compact type $(\mathfrak{g}, \theta, H)$ associated to M as in Sect. 2.2 and the induced decomposition $\mathfrak{g}_\mathbb{C} = \mathfrak{k}_\mathbb{C} \oplus \mathfrak{p}_+ \oplus \mathfrak{p}_-$ as in (6). Denote by τ the complex conjugation on $\mathfrak{g}_\mathbb{C}$ corresponding to the real form $\mathfrak{g}^*$ of $\mathfrak{g}_\mathbb{C}$ introduced in Sect. 2.3. Since $(\mathfrak{g}, \theta, H)$ is of non-compact type, the algebra $\mathfrak{g}^*$ is compact by Theorem 2.19(ii). Therefore, τ is a Cartan involution of $\mathfrak{g}_\mathbb{C}$ (see [Kna96, Prop. 6.14]), which implies that (see [Kna96, Chap. VI, §2])

$$B_\tau : \mathfrak{g}_\mathbb{C} \times \mathfrak{g}_\mathbb{C} \longrightarrow \mathbb{C}$$
$$(X, Y) \mapsto B_\tau(X, Y) := -B(X, \tau Y)$$

is a positive definite Hermitian form, where B denotes as usual the Killing form of $\mathfrak{g}_\mathbb{C}$. For any $X \in \mathfrak{p}_+$, define the linear operator

$$T(X) : \mathfrak{p}_- \longrightarrow \mathfrak{k}_{\mathbb{C}}$$
$$Y \mapsto [Y, X]$$

and denote by $T(X)^* : \mathfrak{k}_{\mathbb{C}} \to \mathfrak{p}_-$ the adjoint of $T(X)$ with respect to B_τ. With these notations, the image of M via the Harish–Chandra embedding can be described as (see [AMRT10, Chap. III, Thm. 2.9]):

$$i_{HC}(M) = \{X \in \mathfrak{p}_+ : T(X)^* \circ T(X) < 2\,\mathrm{id}_{\mathfrak{p}_-}\}. \tag{8}$$

From (8), it follows that $i_{HC}(M)$ is a bounded domain, as required. □

Remark 2.26. Let M be a HSM of non-compact type of dimension n and fix a base point $o \in M$.

(i) The Harish–Chandra embedding $i_{HC} : M \hookrightarrow \mathfrak{p}_+$ (with respect to the base point $o \in M$) can be characterized as the unique open holomorphic embedding of M inside $\mathbb{C}^n$, up to linear complex isomorphisms, such that $i_{HC}(o) = 0$ and $i_{HC}(M)$ is a circular domain, i.e. it is stable under multiplication by $\mathbb{S}^1 \subset \mathbb{C}^*$. See [Sat80, Chap. II, §4, Rmk. 1] and the references therein.

(ii) From the description (8), it follows easily that $i_{HC}(M)$ is a convex bounded domain (Hermann's convexity theorem). Conversely, Mok–Tsai proved that, if the rank of M is greater than one, then the Harish–Chandra embedding is the unique embedding of M inside $\mathbb{C}^n$, up to complex affine transformations, as a bounded convex domain; see [Mok89, Chap. 5, §2] and the references therein.

We are now going to show that, conversely, any bounded symmetric domain can be endowed with a canonical Hermitian metric with respect to which it becomes a HMS of non-compact type.

Let $D \subset \mathbb{C}^N$ be any bounded domain. Let $\mathcal{H}^2(D)$ be the separable Hilbert space consisting of all holomorphic functions on D that are square integrable with respect to the Euclidean measure $d\mu$ on D (see [Hel78, Chap. VIII, Cor. 3.2]). Choose an orthonormal basis $\{e_n(z)\}_{n \in \mathbb{N}}$ of $\mathcal{H}(D)$ and set

$$K_D : D \times D \longrightarrow \mathbb{C},$$
$$(z, w) \mapsto K_D(z, w) := \sum_{n \in \mathbb{N}} e_n(z) \cdot \overline{e_m(w)}, \tag{9}$$

where the right hand side converges absolutely and uniformly on any compact subset of $D \times D$ (see [Hel78, Chap. VIII, Thm. 3.3]). The function K_D, known as the *Bergman kernel function* of D, is independent of the choice of the orthonormal basis $\{e_n(z)\}$ (see [Hel78, Chap. VIII, Thm. 3.3]) and it can be intrinsically characterized (see [Sat80, Chap. II, §6]) as the unique function $K_D : D \times D \to \mathbb{C}$ such that

(i) $K_D(z, w) = \overline{K_D(w, z)}$ for any $z, w \in D$.
(ii) For any $w \in D$, the function $z \mapsto K_D(z, w)$ belongs to $\mathcal{H}^2(D)$.

(iii) For any $f \in \mathcal{H}^2(D)$, we have that

$$f(z) = \int_D K_D(z,w) f(w) d\mu(w).$$

Fix now coordinates $z = (z^1, \cdots, z^N)$ of $\mathbb{C}^N$ and consider the smooth tensor of type $(0, 2)$ defined by

$$h_D = \sum_{1 \leq i,j \leq N} \frac{\partial^2}{\partial z^i \partial \bar{z}^j} \log K_D(z,z) dz^i d\bar{z}^j. \tag{10}$$

Theorem 2.27. *Let $D \subset \mathbb{C}^N$ be a bounded domain and consider the complex structure J_D on D inherited from $\mathbb{C}^N$.*

(i) *The tensor h_D defines a Hermitian metric (called the* Bergman metric *of D) on the complex manifold (D, J_D), which is invariant under $\mathrm{Hol}(D)$. In particular, $\mathrm{Aut}(D, J_D, h_D) = \mathrm{Hol}(D)$.*
(ii) *If D is a bounded symmetric domain, then (D, J_D, h_D) is a HSM of non-compact type.*

Proof. Part (i) is proved in [Hel78, Chap. VIII, Prop. 3.4, Prop. 3.5].

Part (ii): by assumption, $\mathrm{Hol}(D)$ acts transitively on D and each point $p \in D$ has a symmetry $s_p \in \mathrm{Hol}(D)$. Since $\mathrm{Hol}(D) = \mathrm{Aut}(D, J_D, h_D)$ by part (i), it follows that $\mathrm{Aut}(D, J_D, h_D)$ acts transitively on D and that the symmetry s_p at the point $p \in D$ belongs to $\mathrm{Aut}(D, J_D, h_D)$. Therefore, (D, J_D, h_D) is a Hermitian symmetric manifold. The fact that (D, J_D, h_D) is of non-compact type is proved in [Hel78, Chap. VIII, Thm. 7.1(i)]. □

Remark 2.28. (i) The Bergman metric h_D of a bounded domain $D \subset \mathbb{C}^N$ is Kähler, i.e. $\mathrm{Im}\, h_D$ is a closed 2-form (see [Hel78, Chap. VIII, Prop. 3.4]).
(ii) If $D \subset \mathbb{C}^N$ is a homogeneous bounded domain, then h_D is Kähler–Einstein, i.e. its Ricci curvature is proportional to the associated Riemannian metric $\mathrm{Re}\, h_D$ (see [Hel78, Chap. VIII, Prop. 3.6]).

Example 2.29. Consider the open unitary disk

$$\Delta := \{z \in \mathbb{C} : |z| < 1\} \subset\subset \mathbb{C}.$$

Clearly, Δ is a bounded domain. Its Bergman Hermitian metric is equal to (see [FK94, Chap. IX, §2])

$$h_\Delta = \frac{4}{1 - |z|^2} dz d\bar{z}.$$

The unitary disk Δ is biholomorphic to the upper half space $\mathcal{H}$ of Example 2.6(2) via the Cayley transforms

$$\phi : \mathcal{H} \xrightarrow{\cong} \Delta$$
$$\tau \mapsto \frac{\tau - i}{\tau + i}. \tag{11}$$

The pull-back via ϕ of the Riemannian metric $\operatorname{Re} h_\Delta$ on Δ is the hyperbolic metric on $\mathcal{H}$ introduced in Example 2.6(2). Therefore, Δ is a bounded symmetric domain.

Putting together Theorems 2.25 and 2.27, we obtain the following correspondence between HSM of non-compact type and bounded symmetric domains.

Theorem 2.30. *The maps*

$$\{\text{HSMs of non-compact type}\} \longrightarrow \{\text{Bounded symmetric domains}\}$$
$$(M, J, h) \longrightarrow i_{HC}(M) \subset \mathfrak{p}_+ \tag{12}$$
$$(D, J_D, h_D) \longleftarrow D \subset \mathbb{C}^N$$

are bijections which are inverses of each other. Moreover, the above bijections send irreducible HSMs of non-compact type into irreducible bounded symmetric domains and conversely.

Proof. It follows from Theorems 2.25 and 2.27 that the above maps are well-defined. The fact that they are inverses of each other can be extracted from the proof of [Hel78, Chap. VIII, Thm. 7.1]. The last assertion is obvious. □

The rank of a bounded symmetric domain can be characterized in the following way.

Theorem 2.31 (Polydisk theorem). *Let D be a bounded symmetric domain with its Bergman metric h_D and fix a base point $o \in D$. If the rank of D is equal to r then there exists a totally geodesic polydisk $\Delta^r \subseteq D$ of dimension r such that the restriction of h_D to Δ^r is equal to the Bergman metric of Δ^r and $D = \operatorname{Stab}(o) \cdot \Delta^r$.*

Proof. See [Mok89, Chap. 5, (1.1), Thm. 1]. □

The correspondence in Theorem 2.30 together with the classification of irreducible HSMs of non-compact type recalled in Sect. 2.1 gives a classification of irreducible bounded symmetric domains. We record the irreducible bounded symmetric domains in their Harish–Chandra embeddings (together with their complex dimensions and their ranks) into Table 4, referring to Sect. 3 for more details.

Remark 2.32. Bounded symmetric domains play a crucial role in Hodge theory. Indeed, on one hand if a period domain D is such that its universal family of Hodge structures satisfies Griffiths transversality then D is a bounded symmetric domain. On the other hand, Deligne has shown that every bounded symmetric domain can be realized as the subdomain of a period domain on which certain tensors for the universal family are of Hodge type. In particular, every HSM of non-compact type

Table 4 Irreducible bounded symmetric domains in their Harish–Chandra embeddings

Type	Bounded symmetric domain	dim$_\mathbb{C}$	Rank
$I_{p,q}$	$\{Z \in M_{p,q}(\mathbb{C}) : Z^t \cdot \overline{Z} < I_q\}$	pq	q
II_n	$\{Z \in M_{n,n}^{\text{skew}}(\mathbb{C}) : Z^t \cdot \overline{Z} < I_n\}$	$\binom{n}{2}$	$\lfloor \frac{n}{2} \rfloor$
III_n	$\{Z \in M_{n,n}^{\text{sym}}(\mathbb{C}) : Z^t \cdot \overline{Z} < I_n\}$	$\binom{n+1}{2}$	n
IV_n	$\{Z \in \mathbb{C}^n : 2\overline{Z}^t Z < 1 + |Z^t Z|^2, \overline{Z}^t Z < 1\}$	n	$\min\{2,n\}$
V	$\mathcal{D}_V \subset \mathbb{O}_\mathbb{C}^2$	16	2
VI	$\mathcal{D}_{VI} \subset H_3(\mathbb{O}_\mathbb{C})$	27	3

can be realized as a moduli space for Hodge structures plus tensors. We refer the reader to [Mil12, Sec. 7] and the references therein. HSM of non-compact type embedded (equivariantly and horizontally) inside period domains have been recently characterized by R. Friedmann and R. Laza [FL13].

2.6 HSMs of Compact Type as Cominuscle Homogeneous Varieties

As a consequence of the Borel embedding (see Theorem 2.22), we can describe HSMs of compact type as cominuscle homogeneous (projective) varieties.

Definition 2.33. A rational homogeneous projective variety H/Q, where H is a semisimple complex Lie group (or, equivalently, algebraic group) and Q is a parabolic subgroup of H, is said to be a **cominuscle homogeneous variety** if the unipotent radical of Q is abelian.

Remark 2.34. If H is a simple complex algebraic group, then H/Q is a cominuscle homogeneous variety if and only if Q is, up to conjugation, a standard maximal parabolic subgroup associated to a **cominuscle (or special) simple root**, i.e. a simple root occurring with coefficient 1 in the simple root decomposition of the highest positive root (see [RRS92, Lemma 2.2] for a proof). In this case, H/Q is called an *irreducible* cominuscle homogeneous variety and clearly any cominuscle homogenous variety can be written uniquely as a product of irreducible ones.

In Sect. 2.4, we have seen that any HSM $M^* = G_c/K$ of compact type is isomorphic to $G_\mathbb{C}/(P_- \rtimes K_\mathbb{C})$ (see Theorem 2.22), which is a cominuscle homogeneous variety since $G_\mathbb{C}$ is a semisimple complex Lie group and $P_- \rtimes K_\mathbb{C}$ is a parabolic subgroup whose unipotent radical P_- is abelian.

Theorem 2.35. *The map*

$$\{\text{HSMs of compact type}\} \longrightarrow \{\text{Cominuscle homogeneous varieties}\} \quad (13)$$
$$G_c/K \mapsto G_\mathbb{C}/(P_- \rtimes K_\mathbb{C})$$

is a bijection sending irreducible HSMs of compact type into irreducible cominuscle homogeneous varieties.

Proof. See e.g. [RRS92, 5.5]. □

The rank of a cominuscle homogeneous variety can be characterized in the following way.

Theorem 2.36 (Polysphere theorem). *Let X be a cominuscle homogeneous variety with a base point o and let h the Hermitian metric coming from the bijection of Theorem 2.35. If the rank of X is equal to r then there exists a totally geodesic polysphere $(\mathbb{P}^1)^r \subseteq X$ of dimension r such that the restriction of h to $(\mathbb{P}^1)^r$ is equal to the product of the Fubini–Studi metrics on each factor and $X = \mathrm{Stab}(o) \cdot (\mathbb{P}^1)^r$.*

Proof. See [Mok89, Chap. 5, (1.1), Thm. 1]. □

The correspondence in Theorem 2.35 together with the classification of irreducible HSMs of compact type recalled in Sect. 2.1 gives a classification of irreducible cominuscle homogeneous varieties (see also [LM02, Sec. 2.1, 3.1] and [LM03, Sec. 3]). We record the irreducible cominuscle homogeneous varieties together with their associated cominuscle simple roots (see Remark 2.34) into Table 5, referring to Sect. 3 for more details.

Note that each of the varieties $\mathrm{Gr}(q, p+q)$, $\mathrm{Gr}_{\mathrm{ort}}(n, 2n)$ and $\mathbb{P}^2(\mathbb{O})$ corresponds to two cominuscle simple roots; this is due to the fact that in each of the above cases there is an automorphism of the Dinkin diagram that exchanges the two roots, and hence inducing an outer automorphism of the associated simple complex algebraic group that establishes an isomorphism of their associated cominuscle homogeneous varieties.

Remark 2.37. Theorem 2.35 and Remark 2.34, together with the correspondence between (irreducible) bounded symmetric domains and (irreducible) cominuscle varieties (see Corollary 2.21 and Theorem 2.30), provide an explicit bijection between irreducible bounded symmetric domain and cominuscle simple roots of Dinkin diagrams, up to the action of the automorphism group of the Dinkin diagram. This correspondence is described explicitly in [Mil05, Sec. 1] and [Mil12, Sec. 2] (following Deligne), without passing to the compact dual HSM.

2.7 Classifying HSMs of Non-compact Type Via Jordan Theory

The aim of this subsection is to explain an alternative approach to the classification of HSMs of non-compact type which is based on the Jordan theory, rather than the Lie theory as in Sect. 2.2.

Table 5 Irreducible cominuscle homogeneous varieties and their associated cominuscle simple roots

Cominuscle homogeneous variety		Cominuscle simple roots	
$I_{p,q}$:	$\mathrm{Gr}(q, p+q)$	(diagram with nodes $1, \ldots, q, \ldots, p, \ldots, p+q-1$)	A_{p+q-1}
II_n :	$\mathrm{Gr}_{\mathrm{ort}}(n, 2n)^o$	(diagram with nodes $1, \ldots, n-2, n-1, n$)	D_n
III_n :	$\mathrm{Gr}_{\mathrm{sym}}(n, 2n)$	(diagram with nodes $1, \ldots, n$)	C_n
IV_n :	$\mathcal{Q}^n$	(diagram with nodes $1, \ldots, \frac{n}{2}-1, \frac{n}{2}, \frac{n}{2}+1$)	$D_{\frac{n}{2}+1}$ if n is even
		(diagram with nodes $1, \ldots, \frac{n-1}{2}, \frac{n+1}{2}$)	$B_{\frac{n+1}{2}}$ if n is odd
V :	$\mathbb{P}_{\mathbb{O}}^2$ Cayley plane	(diagram)	E_6
VI :	$\mathcal{F} = \mathrm{Gr}_\omega(\mathbb{O}^3, \mathbb{O}^6)$ Freudenthal variety	(diagram)	E_7

We start by giving the definition of Jordan triple systems, referring the interested reader to [FKKLR00, Part V] for an excellent survey on Jordan triple systems, with special emphasis on Hermitian positive JTSs.

Definition 2.38. A *Jordan triple system* over a field F is a pair $(V, \{.,.,.\})$ consisting of a (finite-dimensional) F-vector space V together with a F-multilinear triple product $V \times V \times V \to V$, which satisfies the following properties:

(JT1) $\{x, y, z\} = \{z, y, x\}$ for any $x, y, z \in V$;
(JT2) $[a \square x, b \square y] = ((a \square b)x) \square y - x \square ((b \square a)y)$ for any $a, b, x, y \in V$, where $a \square b$ is the endomorphism of V defined by

$$(a \square b)x := \{a, b, x\},$$

and $[]$ is the usual bracket among endomorphisms of V.

A Jordan triple system $(V, \{.,.,.\})$ over F is said to be:

(i) *semisimple* if the trace form

$$\tau : V \times V \longrightarrow F, \qquad (14)$$
$$(x, y) \mapsto \tau(x, y) := \operatorname{tr}(x \square y),$$

is non-degenerate.

(ii) *simple* if τ is not identically zero and $(V, \{.,.,.\})$ does not have proper ideals, i.e. proper subvector spaces $I \subset V$ such that $\{I, V, V\} \subseteq I$ and $\{V, I, V\} \subseteq I$.

(iii) *Hermitian* if $F = \mathbb{R}$ and V has a complex structure with respect to which $\{.,.,.\}$ is $\mathbb{C}$-linear in the first and third factor and $\mathbb{C}$-antilinear with respect to the second factor.

(iv) *Hermitian positive* (or, for short, a **Hermitian positive JTS**) if it is Hermitian and

(JTp) the trace form τ is positive definite.

Note that property (JTp) makes sense since the trace form is a Hermitian form on V (see [Sat80, p. 55]).

Proposition 2.39. *Any Hermitian positive (resp. complex semisimple) JTS decomposes uniquely as a product of simple Hermitian positive (resp. complex simple) JTSs.*

Proof. See [FKKLR00, Part V, Prop. IV.1.4] for the case of Hermitian positive JTSs and [Loo75, Thm. 10.14] for the case of complex semisimple JTSs. □

Remark 2.40. There is a natural bijection

$$\{\text{Hermitian positive JTSs}\} \xrightarrow{\cong} \{\text{Complex semisimple JTSs}\}$$
$$(V, \{.,.,.\}) \mapsto (V, \{x, y, x\}' := \{x, \overline{y}, z\})$$

which preserves the decomposition into the product of simple Jordan algebras.

Simple Hermitian positive JTSs (or equivalently simple complex JTSs by Remark 2.40) can be classified (see [Loo75, Sec. 17.4]).

Theorem 2.41. *Every simple Hermitian positive JTS is isomorphic to one of the following:*

(i) $M_{p,q}(\mathbb{C})$ with Jordan triple product $\{M_1, M_2, M_3\} = \frac{1}{2}(M_1 \overline{M_2}^t M_3 + M_3 \overline{M_2}^t M_1)$.

(ii) $M_{n,n}^{\text{skew}}(\mathbb{C}) := \{M \in M_{n,n}(\mathbb{C}) : M^t = -M\}$ with the same Jordan triple product of (i).

(iii) $M_{n,n}^{\text{sym}}(\mathbb{C}) := \{M \in M_{n,n}(\mathbb{C}) : M^t = M\}$ with the same Jordan triple product of (i).

(iv) $\mathbb{C}^n$ with Jordan triple product $\{X, Y, Z\} = (X^t \cdot Z)\overline{Y} - (Z^t \cdot \overline{Y})X - (X^t \cdot \overline{Y})Z$.

Table 6 Simple Hermitian positive JTSs

Type	Complex vector space $\mathfrak{p}_+$	Jordan triple product $\{.,.,.\}$
$I_{p,q}$	$M_{p,q}(\mathbb{C})$	$\{M_1, M_2, M_3\} = \frac{1}{2}(M_1\overline{M_2}^t M_3 + M_3\overline{M_2}^t M_1)$
II_n	$M_{n,n}^{\text{skew}}(\mathbb{C})$	$\{M_1, M_2, M_3\} = \frac{1}{2}(M_1\overline{M_2}^t M_3 + M_3\overline{M_2}^t M_1)$
III_n	$M_{n,n}^{\text{sym}}(\mathbb{C})$	$\{M_1, M_2, M_3\} = \frac{1}{2}(M_1\overline{M_2}^t M_3 + M_3\overline{M_2}^t M_1)$
IV_n	$\mathbb{C}^n$	$\{X, Y, Z\} = (X^t \cdot Z)\overline{Y} - (Z^t \cdot \overline{Y})X - (X^t \cdot \overline{Y})Z$
V	$\mathbb{O}_\mathbb{C}^2$	$\left\{\begin{pmatrix}a_1\\a_2\end{pmatrix}, \begin{pmatrix}b_1\\b_2\end{pmatrix}, \begin{pmatrix}c_1\\c_2\end{pmatrix}\right\} = \begin{pmatrix}(a_1\widetilde{\overline{b_1}})c_1 + (c_1\widetilde{\overline{b_1}})a_1 + (a_1\overline{b_2})\widetilde{c_2} + (c_1\overline{b_2})\widetilde{a_2}\\ \widetilde{a_1}(\overline{b_1}c_2) + \widetilde{c_1}(\overline{b_1}a_2) + \widetilde{a_2}(\overline{b_2}c_2) + \widetilde{c_2}(\overline{b_2}a_2)\end{pmatrix}$
VI	$H_3(\mathbb{O}_\mathbb{C})$	$\{a, b, c\} = (a|b)c + (c|b)a - (a \times c) \times \overline{b}$

(v) $\mathbb{O}_\mathbb{C}^2$ with Jordan triple product $\left\{\begin{pmatrix}a_1\\a_2\end{pmatrix}, \begin{pmatrix}b_1\\b_2\end{pmatrix}, \begin{pmatrix}c_1\\c_2\end{pmatrix}\right\} = \begin{pmatrix}(a_1\widetilde{\overline{b_1}})c_1 + (c_1\widetilde{\overline{b_1}})a_1 + (a_1\overline{b_2})\widetilde{c_2} + (c_1\overline{b_2})\widetilde{a_2}\\ \widetilde{a_1}(\overline{b_1}c_2) + \widetilde{c_1}(\overline{b_1}a_2) + \widetilde{a_2}(\overline{b_2}c_2) + \widetilde{c_2}(\overline{b_2}a_2)\end{pmatrix}$.

(vi) $H_3(\mathbb{O}_\mathbb{C}) := \{M \in M_{3,3}(\mathbb{O}_\mathbb{C}) : \tilde{M}^t = M\}$ with Jordan triple product $\{a, b, c\} = (a|b)c + (c|b)a - (a \times c) \times \overline{b}$.

We record the simple Hermitian positive JTSs into Table 6, referring to Sect. 3 for more details. Among the different types of simple Hermitian positive JTSs, there are the same isomorphisms specified in Table 7.

There is a way to construct a Hermitian positive JTS starting from a Hermitian SLA of non-compact type. Indeed, given a Hermitian SLA of non-compact type $(\mathfrak{g}, \theta, H)$, consider the decomposition

$$\mathfrak{g}_\mathbb{C} = \mathfrak{k}_\mathbb{C} \oplus \mathfrak{p}_+ \oplus \mathfrak{p}_-$$

given in (6) and denote by $x \to \overline{x}$ the complex conjugation on $\mathfrak{g}_\mathbb{C}$ with respect to the real form $\mathfrak{g}$ of $\mathfrak{g}_\mathbb{C}$. Then the complex vector space $\mathfrak{p}_+$ endowed with the triple product

$$\{.,.,.\} : \mathfrak{p}_+ \times \mathfrak{p}_+ \times \mathfrak{p}_+ \longrightarrow \mathfrak{p}_+,$$
$$(x, y, z) \mapsto \{x, y, z\} := \frac{1}{2}[[x, \overline{y}], z], \quad (15)$$

is a Hermitian positive JTS (see [Sat80, p. 55]).

Theorem 2.42. *The map*

$$\{\text{Hermitian SLAs of non-compact type}\} \longrightarrow \{\text{Hermitian positive JTSs}\}$$
$$(\mathfrak{g}, \theta, H) \mapsto (\mathfrak{p}_+, \{.,.,.\}) \quad (16)$$

Hermitian Symmetric Manifolds

Table 7 Isomorphic types

Isomorphic types	$\mathbb{R}$-dimension	Rank
$I_{1,1} = II_2 = III_1 = IV_1$	2	1
$I_{3,1} = II_3$	6	1
$III_2 = IV_3$	6	2
$I_{2,2} = IV_4$	8	2
$II_4 = IV_6$	12	2

is a bijection sending irreducible Hermitian SLAs of non-compact type into simple Hermitian positive JTSs.

Proof. We limit ourself to defining a map in the other direction, referring to [Sat80, Chap. II, Prop. 3.3] for the verification that it is the inverse of the map (16).

Let $(V, \{.,.,.\})$ be a Hermitian positive JTS. Consider the graded complex vector space

$$\mathfrak{S} = \mathfrak{S}_{-1} \oplus \mathfrak{S}_0 \oplus \mathfrak{S}_1 := V \oplus (V \square V) \oplus \overline{V},$$

where $V \square V := \{x \square y : x, y \in V\} \subseteq \text{End}(V)$ and $\overline{V}$ is the complex conjugate vector space of V, i.e. the complex vector space whose underlying real vector space is equal to the one of V and such that $i \cdot \overline{v} = -i \cdot v$ for any $v \in V$.

Observe that (JT2) of Definition 2.38 implies that $V \square V$ is a complex Lie subalgebra of $\text{End}(V)$ with respect to the usual Lie bracket $[\,]$ on $\text{End}(V)$. Moreover, if we denote by T^* the adjoint of an endomorphism $T \in \text{End}(V)$ with respect to the Hermitian positive definite trace form τ (see (JTp) of Definition 2.38), then (JT2) implies that $(x \square y)^* = y \square x$. In particular, $V \square V \subseteq \text{End}(V)$ is closed under the adjoint operator. Using these observations, we can define a Lie bracket on $\mathfrak{S} = V \oplus (V \square V) \oplus \overline{V}$ as it follows:

$$[(a, T, \overline{b}), (a', T', \overline{b'})] := (Ta' - T'a, 2a' \square \overline{b} + [T, T'] - 2a \square \overline{b'}, (T')^* \overline{b} - T^* \overline{b'}).$$

It turns out that $(\mathfrak{S}, [\,])$ is a graded semisimple complex Lie algebra (see [Sat80, Chap. I, Prop. 7.1])

Consider now the map

$$\sigma : \mathfrak{S} \longrightarrow \mathfrak{S},$$

$$(a, T, b) \mapsto (b, -T^*, a).$$

It is easily checked that σ is a complex conjugation on the graded complex Lie algebra $\mathfrak{S}$, i.e. it is a $\mathbb{C}$-antilinear involution such that $[\sigma(X), \sigma(Y)] = [X, Y]$ for any $X, Y \in \mathfrak{S}$ and $\sigma(\mathfrak{S}_i) = \mathfrak{S}_{-i}$ for $i = -1, 0, 1$. Moreover, the fact that the trace form τ is positive definite (by (JTp) of Definition 2.38) implies that σ is a Cartan involution of the complex semisimple Lie algebra $\mathfrak{S}$ (see [Sat80, Chap. I, §9]), or, equivalently, that the real form $\mathfrak{G}_\sigma := \{X \in \mathfrak{G} : \sigma(X) = X\}$ of the complex Lie algebra $\mathfrak{G}$ defined by σ is a compact real form of $\mathfrak{G}$ (see [Kna96, Prop. 6.14]).

Finally, consider the $\mathbb{C}$-linear involution on $\mathfrak{G}$ which preserves the grading on $\mathfrak{G}$ and such that

$$\theta_{|\mathfrak{G}_i} = \begin{cases} \text{id} & \text{if } i = 0, \\ -\text{id} & \text{if } i = -1, 1. \end{cases}$$

Then the complex conjugation $X \mapsto \theta(\sigma(X))$ on $\mathfrak{G}$ defines another real form $\mathfrak{g} := \{X \in \mathfrak{G} : \theta(\sigma(X)) = X\}$ of $\mathfrak{G}$, on which θ induces a Cartan involution (see e.g. [Sat80, Chap. I, §4]). If $\mathfrak{g} = \mathfrak{k} \oplus \mathfrak{p}$ is the Cartan decomposition relative to θ (as in (1)), then by construction it follows that $\mathfrak{k}_\mathbb{C} = \mathfrak{G}_0$ and $\mathfrak{p} \otimes_\mathbb{R} \mathbb{C} = \mathfrak{G}_{-1} \oplus \mathfrak{G}_1$. Therefore, arguing as in Sect. 2.2, there exists a unique element $H \in Z(\mathfrak{k})$ such that $\text{ad}(H)^2_{|\mathfrak{p}} = -\text{id}_\mathfrak{p}$ and such that $\mathfrak{p} \otimes_\mathbb{R} \mathbb{C} = \mathfrak{G}_{-1} \oplus \mathfrak{G}_1$ is the decomposition into eigenspaces for $\text{ad}(H)_{|\mathfrak{p}}$ relative to the eigenvalues $+1$ and -1 (as in Sect. 2.4).

Summing up, starting with a Hermitian positive JTS $(V, \{., ., .\})$, we have constructed a Hermitian SLA of non-compact type $(\mathfrak{g}, \theta, H)$ and this defines the inverse of the map (16) (see [Sat80, Chap. II, Prop. 3.3]). □

Note that the correspondence in Theorem 2.42 together with Theorem 2.16 and the classification of simple Hermitian positive JTSs in Theorem 2.41 gives a new approach to the classification of HSMs of non-compact type, as recalled in Sect. 2.1.

Remark 2.43. The bijection (16) becomes an equivalence of categories if the left hand set is endowed with morphisms of Hermitian SLAs as in Remark 2.17 and the right hand side is endowed with morphisms of Hermitian positive JTSs defined as it follows: a morphism of Hermitian positive JTSs between two Hermitian positive JTSs $(V, \{., ., .\})$ and $(V', \{., ., .\}')$ is a $\mathbb{C}$-linear map $\psi : V \to V'$ such that

$$\psi(\{x, y, z\}) = \{\psi(x), \psi(y), \psi(z)\}' \text{ for any } x, y, z \in V.$$

See [Sat80, Chap. II, §8].

Remark 2.44. Let M be a HSM of non-compact type and consider its associated Hermitian positive JTS $(\mathfrak{p}_+, \{., ., .\})$ (via the bijections of Theorems 2.16 and 2.42). Then the image of the Harish–Chandra embedding $i_{HC} : M \hookrightarrow \mathfrak{p}_+$ (as in Theorem 2.22) is equal to

$$i_{HC}(M) = \{z \in \mathfrak{p}_+ : z \square z < \text{id}_{\mathfrak{p}_+}\}. \tag{17}$$

See [Sat80, Chap. II, Thm. 5.9] and the references therein.

Using the above presentation of HSMs, it is possible to give a Jordan theoretic description of the rank of a HSM of non-compact type (as in Definition 2.10). Recall that a *tripotent* (or idempotent) of a Jordan triple system $(V, \{., ., .\})$ is an element $e \in V$ such that $\{e, e, e\} = e$. Two tripotents e_1 and e_2 are said to be orthogonal if $\{e_1, e_2, x\} = 0$ for any $x \in V$. A *Jordan frame* of $(V, \{.,.,.\})$ is a maximal collection $\{e_1, \ldots, e_n\}$ of pairwise orthogonal distinct tripotents.

Proposition 2.45. *Let M be a HSM of non-compact type and let $(\mathfrak{p}_+, \{.,.,.\})$ the corresponding Hermitian positive JTS according to Theorems 2.16 and 2.42. Then the rank of M is equal to the cardinality of every Jordan frame of $(\mathfrak{p}_+, \{.,.,.\})$.*

Proof. See [Loo77]. □

3 Irreducible Hermitian Symmetric Manifolds

Irreducible HSMs of non-compact type (resp. of compact type) have been classified by Cartan in [Car35] (based upon [Car26-27]) and they are divided in six **types** which according to the nowadays standard Siegel's notation[1] are called: $I_{p,q}$ (with $p \geq q \geq 1$), II_n (with $n \geq 2$), III_n (with $n \geq 1$), IV_n (with $2 \neq n \geq 1$), V and VI. The first four families are called *classical HSMs* while the last two are called *exceptional HSMs*. For small values of the parameters there are some isomorphisms between the above types as shown in Table 7.

For a modern proof of the above classification, the reader is referred to [Hel78, Chap. X, §6], where such a result is deduced as a corollary of the more general classification of irreducible Riemannian symmetric spaces. The notation used by Helgason differs from the one used by Siegel and the translation between the two different notations is given as it follows: $I_{p,q} = A_{III}$, $II_n = D_{III}$, $III_n = CI$, $IV_n = BDI(q=2)$, $V = EVI$ and $VI = EVII$. A direct proof which avoids the classification of Riemannian symmetric manifolds appears in [Wol64].

In the subsections below, we describe in detail each of the above types following mainly [Wol72] and [Mok89, Chap. 4, §2] for classical HSMs and [Roo08] for exceptional HSMs.

3.1 Type $I_{p,q}$

The **bounded symmetric domain** of type $I_{p,q}$ (with $p \geq q \geq 1$) in its *Harish–Chandra embedding* is given by

$$\mathcal{D}_{I_{p,q}} := \{Z \in M_{p,q}(\mathbb{C}) : Z^t \cdot \overline{Z} < I_q\} \subset M_{p,q}(\mathbb{C}). \tag{18}$$

Note that, in the special case $q = 1$, $\mathcal{D}_{I_{p,1}}$ is the unitary ball $B^p := \{(z_1, \cdots, z_p) \in \mathbb{C}^p : \sum_i |z_i|^2 < 1\} \subset \mathbb{C}^p$.

---

[1] Cartan's original notation permutes Type III and Type IV.

Let $SU(p, q)$ be the connected simple non-compact Lie subgroup of $SL(p + q, \mathbb{C})$ that leaves invariant the bilinear Hermitian form on $\mathbb{C}^{p+q} \times \mathbb{C}^{p+q}$ given by $-x_1 \overline{y}_1 - \cdots - x_p \overline{y}_p + x_{p+1} \overline{y}_{p+1} + \cdots + x_{p+q} \overline{y}_{p+q}$. More explicitly

$$SU(p, q) = \left\{ g \in SL(p + q, \mathbb{C}) : \overline{g}^t \begin{pmatrix} -I_p & 0 \\ 0 & I_q \end{pmatrix} g = \begin{pmatrix} -I_p & 0 \\ 0 & I_q \end{pmatrix} \right\}$$

$$= \left\{ \begin{pmatrix} A & B \\ C & D \end{pmatrix} \in SL(p + q, \mathbb{C}) : \begin{array}{l} \overline{A}^t A - \overline{C}^t C = I_p \\ \overline{D}^t D - \overline{B}^t B = I_q \\ \overline{A}^t B = \overline{C}^t D \end{array} \right\}.$$

The Lie group $SU(p, q)$ acts transitively on $\mathcal{D}_{I_{p,q}}$ via generalized Möbius transformations

$$SU(p, q) \times \mathcal{D}_{I_{p,q}} \longrightarrow \mathcal{D}_{I_{p,q}}$$

$$\left(\begin{pmatrix} A & B \\ C & D \end{pmatrix}, Z \right) \mapsto (AZ + B)(CZ + D)^{-1}. \tag{19}$$

Notice that the center $Z(SU(p, q)) = \{\lambda I_{p+q} : \lambda^{p+q} = 1\}$ of $SU(p, q)$ acts trivially on $\mathcal{D}_{I_{p,q}}$; indeed, it turns out that the connected component of the group of biholomorphisms of $\mathcal{D}_{I_{p,q}}$ is given by

$$\mathrm{Hol}(\mathcal{D}_{I_{p,q}})^o = SU(p, q)/Z(SU(p, q)) := PSU(p, q),$$

which is the connected non-compact adjoint simple Lie group of type A_{p+q-1}.

The symmetry of $\mathcal{D}_{I_{p,q}}$ at the base point 0 is given by the element

$$s_0 = \left[\begin{pmatrix} -I_p & 0 \\ 0 & I_q \end{pmatrix} \right] \in PSU(p, q), \tag{20}$$

which acts on $\mathcal{D}_{I_{p,q}}$ by sending Z into $-Z$. The symmetry s_0 induces an involution on $SU(p, q)$

$$\sigma : SU(p, q) \longrightarrow SU(p, q),$$

$$\begin{pmatrix} A & B \\ C & D \end{pmatrix} \mapsto \begin{pmatrix} -I_p & 0 \\ 0 & I_q \end{pmatrix} \begin{pmatrix} A & B \\ C & D \end{pmatrix} \begin{pmatrix} -I_p & 0 \\ 0 & I_q \end{pmatrix}^{-1} = \begin{pmatrix} A & -B \\ -C & D \end{pmatrix}, \tag{21}$$

whose fixed Lie subgroup is equal to the maximal compact Lie subgroup

$$\left\{ \begin{pmatrix} A & 0 \\ 0 & D \end{pmatrix} \in SU(p, q) \right\} = \left\{ \begin{pmatrix} A & 0 \\ 0 & D \end{pmatrix} : \begin{array}{l} \overline{A}^t A = I_p, \overline{D}^t D = I_q \\ \det(A) \det(D) = 1 \end{array} \right\} =: S(U_p \times U_q),$$

which is also equal to the stabilizer of $0 \in \mathcal{D}_{I_{p,q}}$. In particular, the pair $(SU(p,q), S(U_p \times U_q))$ is a Riemannian symmetric pair. Notice that the involution σ descends to an involution of $PSU(p,q)$ whose fixed locus is the maximal compact Lie subgroup $\overline{S(U_p \times U_q)} := S(U_p \times U_q)/Z(SU(p,q))$ of $PSU(p,q)$. Therefore also the pair $(PSU(p,q), \overline{S(U_p \times U_q)})$ is a Riemannian symmetric pair.

By the above discussion, we get the following presentation of $\mathcal{D}_{I_{p,q}}$ as an irreducible *HSM of non-compact type*

$$\mathcal{D}_{I_{p,q}} \cong SU(p,q)/S(U_p \times U_q) = PSU(p,q)/\overline{S(U_p \times U_q)}, \tag{22}$$

associated to the Riemannian symmetric pair $(SU(p,q), S(U_p \times U_q))$ (resp. to $(PSU(p,q), \overline{S(U_p \times U_q)})$). Notice that the last description of $\mathcal{D}_{I_{p,q}}$ is the one appearing in Theorem 2.12(i).

The irreducible *Hermitian SLA of non-compact type* associated to the Riemannian symmetric pair $(SU(p,q), S(U_p \times U_q))$ (or equivalently to $(PSU(p,q), \overline{S(U_p \times U_q)})$) is given by the Lie algebra

$$\text{Lie } SU(p,q) = \mathfrak{su}(p,q) = \left\{ M \in \mathfrak{sl}(p+q, \mathbb{C}) : \overline{M}^t \begin{pmatrix} -I_p & 0 \\ 0 & I_q \end{pmatrix} = -\begin{pmatrix} -I_p & 0 \\ 0 & I_q \end{pmatrix} M \right\} \tag{23}$$

$$= \left\{ \begin{pmatrix} Z_1 & Z_2 \\ \overline{Z_2}^t & Z_3 \end{pmatrix} \in \mathfrak{gl}(p+q, \mathbb{C}) : \begin{array}{l} \overline{Z_1}^t = -Z_1, \ \overline{Z_3}^t = -Z_3 \\ Z_2 \in M_{p,q}(\mathbb{C}), \text{Tr}(Z_1) + \text{Tr}(Z_3) = 0 \end{array} \right\}$$

endowed with the Cartan involution $\theta = d\sigma$ given by

$$\theta \begin{pmatrix} Z_1 & Z_2 \\ \overline{Z_2}^t & Z_3 \end{pmatrix} = \begin{pmatrix} Z_1 & -Z_2 \\ -\overline{Z_2}^t & Z_3 \end{pmatrix}$$

and with the element

$$H = i \begin{pmatrix} \frac{q}{p+q} I_p & 0 \\ 0 & \frac{-p}{p+q} I_q \end{pmatrix} \in \text{Fix}(\theta) = \text{Lie } S(U_p \times U_q) = \left\{ \begin{pmatrix} Z_1 & 0 \\ 0 & Z_3 \end{pmatrix} \in \mathfrak{su}(p,q) \right\}.$$

The **cominuscle homogeneous variety** of type $I_{p,q}$ is the Grassmannian $Gr(q, p+q)$ parametrizing q-dimensional subspaces of $\mathbb{C}^{p+q}$:

$$Gr(q, p+q) := \{[W \subset \mathbb{C}^{p+q}] : \dim W = q\}. \tag{24}$$

The *Borel embedding* of $\mathcal{D}_{I_{p,q}}$ into $Gr(q, p+q)$ is given by

$$\mathcal{D}_{I_{p,q}} \subset M_{p,q}(\mathbb{C}) \hookrightarrow \mathrm{Gr}(q, p+q),$$

$$Z \mapsto \left\{ \langle v_1, \cdots v_q \rangle \ : \ \{v_1, \cdots, v_q\} \text{ are the column vectors of } \begin{pmatrix} Z \\ I_q \end{pmatrix} \right\}. \tag{25}$$

The complex algebraic simple group $\mathrm{SL}(p+q, \mathbb{C})$ of type A_{p+q-1} acts transitively on $\mathrm{Gr}(q, p+q)$ via

$$\mathrm{SL}(p+q, \mathbb{C}) \times \mathrm{Gr}(q, p+q) \longrightarrow \mathrm{Gr}(q, p+q)$$

$$(g, [W \subset \mathbb{C}^{p+q}]) \mapsto [g(W) \subset \mathbb{C}^{p+q}].$$

Note that the center $Z(\mathrm{SL}(p+q, \mathbb{C})) = \{\lambda I_{p+q} : \lambda^{p+q} = 1\}$ of $\mathrm{SL}(p+q, \mathbb{C})$ acts trivially on $\mathrm{Gr}(q, p+q)$; indeed, it turns out that the group of automorphisms of the algebraic variety $\mathrm{Gr}(q, p+q)$ is equal to

$$\mathrm{PSL}(p+q, \mathbb{C}) := \mathrm{SL}(p+q, \mathbb{C}) / Z(\mathrm{SL}(p+q, \mathbb{C})),$$

which is the connected simple adjoint complex algebraic group of type A_{p+q-1} and it is the complexification of the Lie group $\mathrm{PSU}(p, q)$.

Consider now the base point $W_o := \langle e_{p+1}, \cdots, e_{p+q} \rangle \in \mathrm{Gr}(q, p+q)$ with respect to the standard basis $\{e_1, \cdots, e_{p+q}\}$ of $\mathbb{C}^{p+q}$. The stabilizer of W_o is the maximal parabolic subgroup associated to the q-th simple root of the Dinkin diagram A_{p+q-1} (which is cominuscle, see Table 5)

$$Q_q := \left\{ \begin{pmatrix} A & 0 \\ C & D \end{pmatrix} \in \mathrm{SL}(p+q, \mathbb{C}) \right\} \subset \mathrm{SL}(p+q, \mathbb{C}),$$

where $A \in M_{p,p}(\mathbb{C})$, $C \in M_{q,p}(\mathbb{C})$ and $D \in M_{q,q}(\mathbb{C})$. The parabolic group Q_q admits the following Levi decomposition

$$Q_q = R_u(Q_q) \rtimes L(Q_q) := \left\{ \begin{pmatrix} I_p & 0 \\ C & I_q \end{pmatrix} \right\} \rtimes \left\{ \begin{pmatrix} A & 0 \\ 0 & D \end{pmatrix} \in \mathrm{SL}(p+q, \mathbb{C}) \right\},$$

which coincides with the Levi decomposition appearing in Theorem 2.35.

From the above discussion, we obtain the following explicit presentation of $\mathrm{Gr}(q, p+q)$ as a cominuscle homogeneous variety (as in Definition 2.33)

$$\mathrm{Gr}(q, p+q) \cong \mathrm{SL}(p+q, \mathbb{C})/Q_q = \mathrm{PSL}(p+q, \mathbb{C})/\overline{Q_q}, \tag{26}$$

where $\overline{Q_q} := Q_q / \{\lambda I_{p+q} : \lambda^{p+q} = 1\}$.

Consider now the compact real form of $\mathrm{SL}(p+q, \mathbb{C})$, which is the Lie subgroup $\mathrm{SU}(p+q) \subset \mathrm{SL}(p+q, \mathbb{C})$ that leaves invariant the positive definite Hermitian form $x_1 \overline{y}_1 + \cdots + x_{p+q} \overline{y}_{p+q}$ on $\mathbb{C}^{p+q} \times \mathbb{C}^{p+q}$. More explicitly

$$SU(p+q) = \{g \in SL(p+q, \mathbb{C}) : \overline{g}^t g = I_{p+q}\}$$

$$= \left\{ \begin{pmatrix} A & B \\ C & D \end{pmatrix} \in SL(p+q, \mathbb{C}) : \begin{array}{l} \overline{A}^t A + \overline{C}^t C = I_p \\ \overline{D}^t D + \overline{B}^t B = I_q \\ \overline{A}^t B = -\overline{C}^t D \end{array} \right\}.$$

Similarly, the quotient of $SU(p+q)$ by its center

$$PSU(p+q) := SU(p+q)/Z(SU(p+q)) = SU(p+q)/\{\lambda I_{p+q} : \lambda^{p+q} = 1\}$$

is the compact real form of $PSL(p+q, \mathbb{C})$.

The restriction of the action of $SL(p+q, \mathbb{C})$ on $Gr(q, p+q)$ to the subgroup $SU(p+q) \subset SL(p+q, \mathbb{C})$ is still transitive and the stabilizer of W_o is the maximal proper connected and compact subgroup

$$SU(p+q) \cap Q_q = \left\{ \begin{pmatrix} A & 0 \\ 0 & D \end{pmatrix} \in SU(p+q) \right\} \cong S(U_p \times U_q).$$

The action of $SU(p+q)$ on $Gr(q, p+q)$ factors through a transitive action of $PSU(p+q)$ in such a way that the stabilizer of W_o is equal to the maximal proper connected and compact subgroup

$$PSU(p+q) \cap \overline{Q_q} = \left\{ \left[\begin{pmatrix} A & 0 \\ 0 & D \end{pmatrix} \right] \in PSU(p,q) \right\} \cong \overline{S(U_p \times U_q)}.$$

The pair $(SU(p+q), S(U_p \times U_q))$ is a Riemannian symmetric pair since $S(U_p \times U_q)$ is the fixed subgroup of the involution

$$\sigma^* : SU(p+q) \longrightarrow SU(p+q)$$

$$\begin{pmatrix} A & B \\ C & D \end{pmatrix} \mapsto \begin{pmatrix} A & -B \\ -C & D \end{pmatrix},$$

and similarly for the pair $(PSU(p+q), \overline{S(U_p \times U_q)})$.

By the above discussion, we get the following presentation of $Gr(q, p+q)$ as the irreducible *HSM of compact type*

$$Gr(q, p+q) \cong SU(p+q)/S(U_p \times U_q) = PSU(p+q)/\overline{S(U_p \times U_q)}, \quad (27)$$

associated to the Riemannian symmetric pair $(SU(p+q), S(U_p \times U_q))$ (resp. to $(PSU(p+q), \overline{S(U_p \times U_q)})$. In particular, the last description of $Gr(q, p+q)$ is the one appearing in Theorem 2.12(ii). Notice that the symmetry at the base point W_o of $\mathrm{Gr}(q, p+q)$, seen as a Hermitian symmetric manifold, is given by the element

$$s_{W_o} = \left[\begin{pmatrix} -I_p & 0 \\ 0 & I_q \end{pmatrix}\right] \in \mathrm{PSU}(p+q).$$

The irreducible *Hermitian SLA of compact type* associated to the Riemann symmetric pair $(\mathrm{SU}(p+q), S(U_p \times U_q))$ is given by the Lie algebra

$$\mathrm{Lie}\,\mathrm{SU}(p+q) = \mathfrak{su}(p+q) = \left\{ \begin{pmatrix} Z_1 & Z_2 \\ -\overline{Z_2}^t & Z_3 \end{pmatrix} : \begin{array}{l} \overline{Z_1}^t = -Z_1,\ \overline{Z_3}^t = -Z_3 \\ Z_2 \in M_{p,q}(\mathbb{C}),\ \mathrm{Tr}(Z_1) + \mathrm{Tr}(Z_3) = 0 \end{array} \right\} \tag{28}$$

endowed with the involution $\theta^* = d\sigma^*$

$$\theta^* \begin{pmatrix} Z_1 & Z_2 \\ -\overline{Z_2}^t & Z_3 \end{pmatrix} = \begin{pmatrix} Z_1 & -Z_2 \\ \overline{Z_2}^t & Z_3 \end{pmatrix}$$

and with the element

$$H^* = i \begin{pmatrix} \frac{q}{p+q} I_p & 0 \\ 0 & \frac{-p}{p+q} I_q \end{pmatrix} \in \mathrm{Fix}(\theta^*) = \left\{ \begin{pmatrix} Z_1 & 0 \\ 0 & Z_3 \end{pmatrix} \in \mathfrak{su}(p+q) \right\} \cong \mathrm{Lie}\,S(U_p \times U_q).$$

Notice that the Hermitian SLA $(\mathfrak{su}(p+q), \theta^*, H^*)$ is the dual of the Hermitian SLA $(\mathfrak{su}(p,q), \theta, H)$ in the sense of Sect. 2.3.

The complexification of the Lie algebras $\mathfrak{su}(p,q)$ and $\mathfrak{su}(p+q)$ is the complex simple Lie algebra of type A_{p+q-1}

$$\mathrm{Lie}\,\mathrm{SL}(p+q,\mathbb{C}) = \mathfrak{sl}(p+q,\mathbb{C}) = \left\{ \begin{pmatrix} Z_1 & Z_2 \\ Z_4 & Z_3 \end{pmatrix} \in \mathfrak{gl}(p+q,\mathbb{C}) : \mathrm{Tr}(Z_1) + \mathrm{Tr}(Z_3) = 0 \right\}.$$

The decomposition (6) of $\mathfrak{sl}(p+q,\mathbb{C})$ is given by

$$\mathfrak{sl}(p+q,\mathbb{C}) = \left\{ \begin{pmatrix} Z_1 & 0 \\ 0 & Z_3 \end{pmatrix} : \mathrm{Tr}(Z_1) + \mathrm{Tr}(Z_3) = 0 \right\} \oplus \left\{ \begin{pmatrix} 0 & Z_2 \\ 0 & 0 \end{pmatrix} \right\} \oplus \left\{ \begin{pmatrix} 0 & 0 \\ Z_4 & 0 \end{pmatrix} \right\}.$$

In particular, we have the identification

$$M_{p,q}(\mathbb{C}) \xrightarrow{\cong} \mathfrak{p}_+$$
$$M \mapsto \begin{pmatrix} 0 & M \\ 0 & 0 \end{pmatrix}. \tag{29}$$

Using the above identification and the formula (15), $M_{p,q}(\mathbb{C})$ becomes a *Hermitian positive JTS* with respect to the triple product

$$\{M_1, M_2, M_3\} = \frac{1}{2}\left(M_1\overline{M_2}^t M_3 + M_3\overline{M_2}^t M_1\right). \tag{30}$$

3.2 Type II_n

The **bounded symmetric domain** of type II_n ($n \geq 2$) in its *Harish–Chandra embedding* is given by

$$\mathcal{D}_{II_n} := \{Z \in M_{n,n}^{\text{skew}}(\mathbb{C}) : Z^t \cdot \overline{Z} < I_n\} \subset M_{n,n}^{\text{skew}}(\mathbb{C}) := \{Z \in M_{n,n}(\mathbb{C}) : Z^t = -Z\}. \tag{31}$$

Let $SO(2n, \mathbb{C})$ be the connected complex Lie subgroup of $SL(2n, \mathbb{C})$ that leaves invariant the bilinear symmetric form on $\mathbb{C}^{2n} \times \mathbb{C}^{2n}$ given by $S(\underline{x}, \underline{y}) = x_1 y_{n+1} + \cdots + x_n y_{2n} + \cdots + x_{n+1} y_1 + \cdots + x_{2n} y_n$.[2] The group $SO(2n.\mathbb{C})$ is simple of type D_n and it is explicitly given in $n \times n$ block notation as

$$SO(2n, \mathbb{C}) = \{g \in SL(2n, \mathbb{C}) : g^t S_n g = S_n\}$$

$$= \left\{\begin{pmatrix} A & B \\ C & D \end{pmatrix} \in SL(2n, \mathbb{C}) : \begin{array}{l} A^t C = -C^t A \\ B^t D = -D^t B \\ A^t D + C^t B = I_n \end{array}\right\}.$$

Consider the non-compact real form $SO^{nc}(2n)$ of $SO(2n, \mathbb{C})$ consisting of all the elements of $SO(2n, \mathbb{C})$ that leave invariant the bilinear Hermitian form on $\mathbb{C}^{2n} \times \mathbb{C}^{2n}$ given by $-x_1\overline{y}_1 - \cdots - x_n\overline{y}_n + x_{n+1}\overline{y}_{n+1} + \cdots + x_{2n}\overline{y}_{2n}$. Explicitly,

$$SO^{nc}(2n) = SO(2n, \mathbb{C}) \cap SU(n, n) = \left\{g \in SO(2n, \mathbb{C}) : \overline{g}^t \begin{pmatrix} I_n & 0 \\ 0 & -I_n \end{pmatrix} g = \begin{pmatrix} I_n & 0 \\ 0 & -I_n \end{pmatrix}\right\}$$

$$= \left\{\begin{pmatrix} A & B \\ -\overline{B} & \overline{A} \end{pmatrix} \in SL(2n, \mathbb{C}) : \begin{array}{l} \overline{A}^t A - B^t \overline{B} = I_n \\ \overline{A}^t B + B^t \overline{A} = 0 \end{array}\right\}.$$

Note that the group $SO^{nc}(2n)$ is isomorphic to the classical real Lie group $SO^*(2n)$ via the conjugation inside $SO(2n, \mathbb{C})$ given by (see [Mok89, p. 74])

---

[2] Usually, one defines $SO(2n, \mathbb{C})$ with respect to the standard bilinear symmetric form on $\mathbb{C}^{2n} \times \mathbb{C}^{2n}$ given by $x_1 y_1 + \cdots + x_{2n} y_{2n}$. However, for our purposes it will be more convenient to use this alternative presentation.

$$\text{SO}^{nc}(2n) \xrightarrow{\cong} \text{SO}^*(2n) := \{g \in \text{SL}(2n, \mathbb{C}) : g^t g = I_{2n} \text{ and } \overline{g}^t J_n g = J_n\}$$

$$h \mapsto \begin{pmatrix} I_n & iI_n \\ iI_n & I_n \end{pmatrix} h \begin{pmatrix} I_n & iI_n \\ iI_n & I_n \end{pmatrix}^{-1}. \tag{32}$$

The Lie group $\text{SO}^{nc}(2n)$ acts transitively on $\mathcal{D}_{II_n}$ via generalized Möbius transformations, as in (19). Notice that the center $Z(\text{SO}^{nc}(2n)) = \{\pm I_{2n}\}$ of $\text{SO}^{nc}(2n)$ acts trivially on $\mathcal{D}_{II_n}$; indeed, it turns out that the connected component of the group of biholomorphisms of $\mathcal{D}_{II_n}$ is given by

$$\text{Hol}(\mathcal{D}_{II_n})^o = \text{SO}^{nc}(2n)/Z(\text{SO}^{nc}(2n)) := \text{PSO}^{nc}(2n),$$

which is the connected non-compact adjoint simple Lie group of type D_n.

The symmetry of $\mathcal{D}_{II_n}$ at the base point 0 is given by the element

$$s_0 = \left[\begin{pmatrix} iI_n & 0 \\ 0 & -iI_n \end{pmatrix}\right] \in \text{PSO}^{nc}(2n), \tag{33}$$

which acts on $\mathcal{D}_{II_n}$ by sending Z into $-Z$. The symmetry s_0 induces an involution on $\text{SO}^{nc}(2n)$

$$\sigma : \text{SO}^{nc}(2n) \longrightarrow \text{SO}^{nc}(2n)$$

$$\begin{pmatrix} A & B \\ C & D \end{pmatrix} \mapsto \begin{pmatrix} iI_n & 0 \\ 0 & -iI_n \end{pmatrix} \begin{pmatrix} A & B \\ C & D \end{pmatrix} \begin{pmatrix} iI_n & 0 \\ 0 & -iI_n \end{pmatrix}^{-1} = \begin{pmatrix} A & -B \\ -C & D \end{pmatrix},$$

whose fixed Lie subgroup is equal to the maximal compact Lie subgroup

$$\left\{\begin{pmatrix} A & 0 \\ 0 & D \end{pmatrix} \in \text{SO}^{nc}(2n)\right\} = \left\{\begin{pmatrix} A & 0 \\ 0 & \overline{A} \end{pmatrix} : \overline{A}^t A = I_n\right\} =: \text{U}(n),$$

which is also equal to the stabilizer of $0 \in \mathcal{D}_{II_n}$. In particular, the pair $(\text{SO}^{nc}(2n), \text{U}(n))$ is a Riemannian symmetric pair. Notice that the involution σ descends to an involution of $\text{PSO}^{nc}(2n)$ whose fixed locus is the maximal compact Lie subgroup $\overline{\text{U}(n)} := \text{U}(n)/\{\pm I_n\}$ of $\text{PSO}^{nc}(2n)$. Therefore, also the pair $(\text{PSO}^{nc}(2n), \overline{\text{U}(n)})$ is a Riemannian symmetric pair.

By the above discussion, we get the following presentation of $\mathcal{D}_{II_n}$ as an irreducible *HSM of non-compact type*

$$\mathcal{D}_{II_n} \cong \text{SO}^{nc}(2n)/\text{U}(n) = \text{PSO}^{nc}(2n)/\overline{\text{U}(n)}, \tag{34}$$

associated to the Riemannian symmetric pair $(\text{SO}^{nc}(2n), \text{U}(n))$ (resp. to $(\text{PSO}^{nc}(2n), \overline{\text{U}(n)})$). Notice that the last description of $\mathcal{D}_{II_n}$ is the one appearing in Theorem 2.12(i).

The irreducible *Hermitian SLA of non-compact type* associated to the Riemannian symmetric pair $(\mathrm{SO}^{nc}(2n), \mathrm{U}(n))$ (or equivalently to $(\mathrm{PSO}^{nc}(2n), \overline{\mathrm{U}(n)})$) is given by the Lie algebra

$$\mathfrak{so}^{nc}(2n) := \operatorname{Lie} \mathrm{SO}^{nc}(2n) = \left\{ M \in \mathfrak{sl}(2n, \mathbb{C}) : \begin{array}{c} M^t S_n = -S_n M \\ \overline{M}^t \begin{pmatrix} -I_n & 0 \\ 0 & I_n \end{pmatrix} = -\begin{pmatrix} -I_n & 0 \\ 0 & I_n \end{pmatrix} M \end{array} \right\}$$
(35)

$$= \left\{ \begin{pmatrix} Z_1 & Z_2 \\ Z_2^t & -Z_1^t \end{pmatrix} \in \mathfrak{gl}(2n, \mathbb{C}) : \overline{Z_1}^t = -Z_1, \ Z_2^t = -Z_2 \right\}^3$$

endowed with the Cartan involution $\theta = d\sigma$ given by

$$\theta \begin{pmatrix} Z_1 & Z_2 \\ Z_2^t & -Z_1^t \end{pmatrix} = \begin{pmatrix} Z_1 & -Z_2 \\ -Z_2^t & -Z_1^t \end{pmatrix}$$

and with the element

$$H = \frac{i}{2} \begin{pmatrix} I_n & 0 \\ 0 & -I_n \end{pmatrix} \in \operatorname{Fix}(\theta) = \left\{ \begin{pmatrix} Z_1 & 0 \\ 0 & -Z_1^t \end{pmatrix} : \overline{Z_1}^t = -Z_1 \right\} \cong \operatorname{Lie} U(n).$$

The **cominuscle homogeneous variety** of type II_n is the connected component $\operatorname{Gr}_{\mathrm{ort}}(n, 2n)^o$ containing $[W_o := \langle e_{n+1}, \cdots, e_{2n}\rangle \subset \mathbb{C}^{2n}]$ of the orthogonal Grassmannian $\operatorname{Gr}_{\mathrm{ort}}(n, 2n)$ parametrizing Lagrangian n-dimensional subspaces of $\mathbb{C}^{2n}$:

$$\operatorname{Gr}_{\mathrm{ort}}(n, 2n) := \{[W \subset \mathbb{C}^{2n}] : \dim W = p, \ S_{|W \times W} \equiv 0\},\tag{36}$$

where S is the bilinear symmetric non-degenerate form on $\mathbb{C}^{2n}$ which is represented by the matrix $S_n = \begin{pmatrix} 0 & I_n \\ I_n & 0 \end{pmatrix}$ in the standard basis of $\mathbb{C}^{2n}$.

The *Borel embedding* of $\mathcal{D}_{II_n}$ into $\operatorname{Gr}_{\mathrm{ort}}(n, 2n)^o$ is given by

$$\mathcal{D}_{II_n} \subset M_{n,n}^{\mathrm{skew}}(\mathbb{C}) \hookrightarrow \operatorname{Gr}_{\mathrm{ort}}(n, 2n)^o$$

$$Z \mapsto \left\{ \langle v_1, \cdots v_n\rangle : \{v_1, \cdots, v_n\} \text{ are the column vectors of } \begin{pmatrix} Z \\ I_n \end{pmatrix} \right\}.$$
(37)

---

[3] Note that $\mathfrak{so}^{nc}(2n)$ is isomorphic to the classical real Lie algebra $\mathfrak{so}^*(2n) = \operatorname{Lie} \mathrm{SO}^*(2n)$ via the same conjugation map as in (32).

The complex algebraic simple group $SO(2n, \mathbb{C})$ acts transitively on $\mathrm{Gr}_{\mathrm{ort}}(n, 2n)^o$ via

$$SO(2n, \mathbb{C}) \times \mathrm{Gr}_{\mathrm{ort}}(n, 2n)^o \longrightarrow \mathrm{Gr}_{\mathrm{ort}}(n, 2n)^o$$
$$(g, [W \subset \mathbb{C}^{2n}]) \mapsto [g(W) \subset \mathbb{C}^{2n}].$$

Note that the center $Z(SO(2n, \mathbb{C})) = \{\pm I_{2n}\}$ of $SO(2n, \mathbb{C})$ acts trivially on $\mathrm{Gr}_{\mathrm{ort}}(n, 2n)^o$; indeed, it turns out that the group of automorphisms of the algebraic variety $\mathrm{Gr}_{\mathrm{ort}}(n, 2n)^o$ is equal to

$$PSO(2n, \mathbb{C}) := PSO(2n, \mathbb{C})/\{\pm I_{2n}\},$$

which is the connected simple adjoint complex algebraic group of type D_n and it is the complexification of the Lie group $PSO^{nc}(2n)$.

The stabilizer of $[W_o = \langle e_{n+1}, \cdots, e_{2n}\rangle \subset \mathbb{C}^{2n}] \in \mathrm{Gr}_{\mathrm{ort}}(n, 2n)^o$ is the maximal parabolic subgroup associated to the n-th simple root of the Dinkin diagram D_n (which is a cominuscle simple root of D_n, see Table 5)[4]

$$Q_n := \left\{ \begin{pmatrix} A & 0 \\ C & D \end{pmatrix} \in SO(2n, \mathbb{C}) \right\} \subset SO(2n, \mathbb{C}),$$

where $A \in M_{n,n}(\mathbb{C})$, $C \in M_{n,n}(\mathbb{C})$ and $D \in M_{n,n}(\mathbb{C})$. The parabolic group Q_n admits the following Levi decomposition

$$Q_n = R_u(Q_n) \rtimes L(Q_n) := \left\{ \begin{pmatrix} I_p & 0 \\ C & I_q \end{pmatrix} : C^t = -C \right\} \rtimes \left\{ \begin{pmatrix} A & 0 \\ 0 & D \end{pmatrix} \in SO(2n, \mathbb{C}) \right\},$$

which coincides with the Levi decomposition appearing in Theorem 2.35.

From the above discussion, we obtain the following explicit presentation of $\mathrm{Gr}_{\mathrm{ort}}(n, 2n)^o$ as a cominuscle homogeneous variety (as in Definition 2.33)

$$\mathrm{Gr}_{\mathrm{ort}}(n, 2n)^o \cong SO(2n, \mathbb{C})/Q_n = PSO(n, \mathbb{C})/\overline{Q_n}, \tag{38}$$

where $\overline{Q_n} := Q_n/\{\pm I_{2n}\}$.

Consider now the compact real form $SO^c(2n, \mathbb{C})$ of $SO(2n, \mathbb{C})$ consisting of all the elements of $SO(2n, \mathbb{C})$ that leaves invariant the positive definite Hermitian form $x_1 \overline{y}_1 + \cdots + x_{2n} \overline{y}_{2n}$ on $\mathbb{C}^{2n}$. More explicitly

---

[4] As it is seen from Table 5, we could have chosen the $(n-1)$-th simple root and we would have gotten an isomorphic (although non conjugate) parabolic subgroup.

$$\mathrm{SO}^c(2n) := \mathrm{SO}(2n,\mathbb{C}) \cap \mathrm{SU}(2n) = \{g \in \mathrm{SO}(2n,\mathbb{C}) : \overline{g}^t g = I_{2n}\}$$

$$= \left\{ \begin{pmatrix} A & B \\ \overline{B} & \overline{A} \end{pmatrix} \in \mathrm{SL}(2n,\mathbb{C}) : \begin{array}{l} A^t \overline{A} + \overline{B}^t B = I_n \\ A^t \overline{B} + \overline{B}^t A = 0 \end{array} \right\}.$$

Note that the group $\mathrm{SO}^c(2n)$ is isomorphic to the real orthogonal group $\mathrm{SO}(2n)$ via the same conjugation map as in (32). Similarly, the quotient of $\mathrm{SO}^c(2n)$ by its center

$$\mathrm{PSO}^c(2n) := \mathrm{SO}^c(2n)/Z(\mathrm{SO}^c(2n)) = \mathrm{SO}^c(2n)/\{\pm I_{2n}\}$$

is a compact real form of $\mathrm{PSO}(2n,\mathbb{C})$.

The restriction of the action of $\mathrm{SO}(2n,\mathbb{C})$ on $\mathrm{Gr}_{\mathrm{ort}}(n,2n)^o$ to the subgroup $\mathrm{SO}^c(2n) \subset \mathrm{SO}(2n,\mathbb{C})$ is still transitive and the stabilizer of W_o is the maximal proper connected and compact subgroup

$$\mathrm{SO}^c(2n) \cap Q_n = \left\{ \begin{pmatrix} A & 0 \\ 0 & D \end{pmatrix} \in \mathrm{SO}^c(2n) \right\} = \left\{ \begin{pmatrix} A & 0 \\ 0 & \overline{A} \end{pmatrix} : \overline{A}^t A = I_{2n} \right\} \cong U(n).$$

The action of $\mathrm{SO}^c(2n)$ on $\mathrm{Gr}_{\mathrm{ort}}(n,2n)^o$ factors through a transitive action of $\mathrm{PSO}^c(2n)$ in such a way that the stabilizer of W_o is equal to the maximal proper connected and compact subgroup

$$\mathrm{PSO}^c(2n) \cap \overline{Q_n} = \left\{ \left[\begin{pmatrix} A & 0 \\ 0 & D \end{pmatrix} \right] \in \mathrm{PSO}^c(2n) \right\} = \left\{ \left[\begin{pmatrix} A & 0 \\ 0 & \overline{A} \end{pmatrix} \right] : \overline{A}^t A = I_{2n} \right\} \cong \overline{U(n)}.$$

The pair $(\mathrm{SO}^c(2n), U(n))$ is a Riemannian symmetric pair since $U(n)$ is the fixed subgroup of the involution

$$\sigma^* : \mathrm{SO}^c(2n) \longrightarrow \mathrm{SO}^c(2n)$$

$$\begin{pmatrix} A & B \\ C & D \end{pmatrix} \mapsto \begin{pmatrix} A & -B \\ -C & D \end{pmatrix},$$

and similarly for the pair $(\mathrm{PSO}^c(2n), \overline{U(2n)})$.

By the above discussion, we get the following presentation of $\mathrm{Gr}_{\mathrm{ort}}(n,2n)^o$ as the irreducible *HSM of compact type*

$$\mathrm{Gr}_{\mathrm{ort}}(n,2n)^o \cong \mathrm{SO}^c(2n)/U(n) = \mathrm{PSO}^c(2n)/\overline{U(n)}, \tag{39}$$

associated to the Riemannian symmetric pair $(\mathrm{SO}^c(2n), U(n))$ (resp. to $(\mathrm{PSO}^c(2n), \overline{U(n)})$). In particular, the last description of $\mathrm{Gr}_{\mathrm{ort}}(n,2n)^o$ is the one appearing in

Theorem 2.12(ii). Notice that the symmetry at the base point W_o of $\mathrm{Gr}_{\mathrm{ort}}(n, 2n)^o$, seen as a Hermitian symmetric manifold, is given by the element

$$s_{W_o} = \left[\begin{pmatrix} -I_p & 0 \\ 0 & I_q \end{pmatrix}\right] \in \mathrm{PSO}^c(2n).$$

The irreducible *Hermitian SLA of compact type* associated to the Riemann symmetric pair $(\mathrm{SO}^c(2n), U(2n))$ is given by the Lie algebra

$$\mathfrak{so}^c(2n) := \mathrm{Lie}\,\mathrm{SO}^c(2n) = \left\{ M \in \mathfrak{sl}(p+q, \mathbb{C}) : \begin{matrix} M^t S_n = -S_n M \\ \overline{M}^t = -M \end{matrix} \right\} \quad (40)$$

$$= \left\{ \begin{pmatrix} Z_1 & Z_2 \\ -\overline{Z}_2^t & -\overline{Z}_1^t \end{pmatrix} \in \mathfrak{gl}(p+q, \mathbb{C}) : \overline{Z_1}^t = -Z_1,\ Z_2^t = -Z_2 \right\}^5$$

endowed with the involution $\theta^* = d\sigma^*$

$$\theta^* \begin{pmatrix} Z_1 & Z_2 \\ -\overline{Z}_2^t & -\overline{Z}_1^t \end{pmatrix} = \begin{pmatrix} Z_1 & -Z_2 \\ \overline{Z}_2^t & -\overline{Z}_1^t \end{pmatrix}$$

and with the element

$$H^* = \frac{i}{2} \begin{pmatrix} I_n & 0 \\ 0 & -I_n \end{pmatrix} \in \mathrm{Fix}(\theta^*) = \left\{ \begin{pmatrix} Z_1 & 0 \\ 0 & -\overline{Z}_1^t \end{pmatrix} : \overline{Z_1}^t = -Z_1 \right\} \cong \mathrm{Lie}\,U(n).$$

Notice that the Hermitian SLA $(\mathfrak{so}^c(2n), \theta^*, H^*)$ is the dual of the Hermitian SLA $(\mathfrak{so}^{nc}(2n), \theta, H)$ in the sense of Sect. 2.3.

The complexification of the Lie algebras $\mathfrak{so}^{nc}(2n)$ and $\mathfrak{so}^c(2n)$ is the complex simple Lie algebra of type D_n

$$\mathrm{Lie}\,\mathrm{SO}(2n, \mathbb{C}) = \mathfrak{so}(2n, \mathbb{C}) = \left\{ \begin{pmatrix} Z_1 & Z_2 \\ Z_3 & -Z_1^t \end{pmatrix} \in \mathfrak{gl}(p+q, \mathbb{C}) : Z_2^t = -Z_2,\ Z_3^t = -Z_3 \right\}.$$

The decomposition (6) of $\mathfrak{so}(p+q, \mathbb{C})$ is given by

$$\mathfrak{sl}(p+q, \mathbb{C}) = \left\{ \begin{pmatrix} Z_1 & 0 \\ 0 & -Z_1^t \end{pmatrix} \right\} \oplus \left\{ \begin{pmatrix} 0 & Z_2 \\ 0 & 0 \end{pmatrix} : Z_2^t = -Z_2 \right\} \oplus \left\{ \begin{pmatrix} 0 & 0 \\ Z_3 & 0 \end{pmatrix} : Z_3^t = -Z_3 \right\}.$$

---

[5]Note that $\mathfrak{so}^c(2n)$ is isomorphic to the classical real Lie algebra $\mathfrak{so}(2n) = \mathrm{Lie}\,\mathrm{SO}(2n)$ via the same conjugation map as in (32).

In particular, we have the identification

$$M_{n,n}^{\text{skew}}(\mathbb{C}) \xrightarrow{\cong} \mathfrak{p}_+$$
$$M \mapsto \begin{pmatrix} 0 & M \\ 0 & 0 \end{pmatrix}. \tag{41}$$

Using the above identification and the formula (15), $M_{n,n}^{\text{skew}}(\mathbb{C})$ becomes a *Hermitian positive JTS* with respect to the triple product

$$\{M_1, M_2, M_3\} = \frac{1}{2}\left(M_1 \overline{M_2}^t M_3 + M_3 \overline{M_2}^t M_1\right). \tag{42}$$

3.3 Type III_n

The **bounded symmetric domain** of type III_n ($n \geq 1$) in its *Harish–Chandra embedding* is given by

$$\mathcal{D}_{III_n} := \{Z \in M_{n,n}^{\text{sym}}(\mathbb{C}) : Z^t \cdot \overline{Z} < I_n\} \subset M_{n,n}^{\text{sym}}(\mathbb{C}) := \{Z \in M_{n,n}(\mathbb{C}) : Z^t = Z\}. \tag{43}$$

Let $\text{Sp}(n, \mathbb{C})$ be the connected complex Lie subgroup of $\text{SL}(2n, \mathbb{C})$ that leaves invariant the bilinear alternating form on $\mathbb{C}^{2n} \times \mathbb{C}^{2n}$ given by $S(\underline{x}, \underline{y}) = x_1 y_{n+1} + \cdots + x_n y_{2n} + \cdots - x_{n+1} y_1 - \cdots - x_{2n} y_n$. The group $\text{Sp}(n.\mathbb{C})$ is simple of type C_n and it is explicitly given in $n \times n$ block notation as

$$\text{Sp}(n, \mathbb{C}) = \{g \in \text{SL}(2n, \mathbb{C}) : g^t J_n g = J_n\}$$
$$= \left\{ \begin{pmatrix} A & B \\ C & D \end{pmatrix} \in \text{SL}(2n, \mathbb{C}) : \begin{array}{l} A^t C = C^t A \\ B^t D = D^t B \\ A^t D - C^t B = I_n \end{array} \right\}.$$

Consider the non-compact real form $\text{Sp}^{\text{nc}}(n)$ of $\text{Sp}(n, \mathbb{C})$ consisting of all the elements of $\text{Sp}(n, \mathbb{C})$ that leave invariant the bilinear Hermitian form on $\mathbb{C}^{2n} \times \mathbb{C}^{2n}$ given by $-x_1 \overline{y}_1 - \cdots - x_n \overline{y}_n + x_{n+1} \overline{y}_{n+1} + \cdots + x_{2n} \overline{y}_{2n}$. Explicitly,

$$\text{Sp}^{\text{nc}}(n) = \text{Sp}(n, \mathbb{C}) \cap \text{SU}(n, n) = \left\{ g \in \text{Sp}(n, \mathbb{C}) : \overline{g}^t \begin{pmatrix} I_n & 0 \\ 0 & -I_n \end{pmatrix} g = \begin{pmatrix} I_n & 0 \\ 0 & -I_n \end{pmatrix} \right\}$$
$$= \left\{ \begin{pmatrix} A & B \\ \overline{B} & \overline{A} \end{pmatrix} \in \text{SL}(2n, \mathbb{C}) : \begin{array}{l} \overline{A}^t A - B^t \overline{B} = I_n \\ \overline{A}^t B - B^t \overline{A} = 0 \end{array} \right\}.$$

Note that the Lie group $\mathrm{Sp}^{nc}(n)$ is isomorphic to the real symplectic group $\mathrm{Sp}(n, \mathbb{R})$ via the conjugation inside $\mathrm{Sp}(n, \mathbb{C})$ given by (see [Mok89, p. 71])

$$\mathrm{Sp}^{nc}(n) \xrightarrow{\cong} \mathrm{Sp}(n, \mathbb{R}) := \{g \in \mathrm{GL}(2n, \mathbb{R}) : g^t J_n g = J_n\}$$

$$h \mapsto \begin{pmatrix} I_n & iI_n \\ iI_n & I_n \end{pmatrix} h \begin{pmatrix} I_n & iI_n \\ iI_n & I_n \end{pmatrix}^{-1}. \tag{44}$$

The Lie group $\mathrm{Sp}^{nc}(n)$ acts transitively on $\mathcal{D}_{III_n}$ via generalized Möbius transformations, as in (19). Notice that the center $Z(\mathrm{Sp}^{nc}(n)) = \{\pm I_{2n}\}$ of $\mathrm{Sp}^{nc}(n)$ acts trivially on $\mathcal{D}_{III_n}$; indeed, it turns out that the connected component of the group of biholomorphisms of $\mathcal{D}_{III_n}$ is given by

$$\mathrm{Hol}(\mathcal{D}_{III_n})^o = \mathrm{Sp}^{nc}(n)/Z(\mathrm{Sp}^{nc}(n)) := \mathrm{PSp}^{nc}(n),$$

which is the connected non-compact adjoint simple Lie group of type C_n.

The symmetry of $\mathcal{D}_{III_n}$ at the base point 0 is given by the element

$$s_0 = \left[\begin{pmatrix} iI_n & 0 \\ 0 & -iI_n \end{pmatrix}\right] \in \mathrm{PSp}^{nc}(n), \tag{45}$$

which acts on $\mathcal{D}_{III_n}$ by sending Z into $-Z$. The symmetry s_0 induces an involution on $\mathrm{Sp}^{nc}(n)$

$$\sigma : \mathrm{Sp}^{nc}(n) \longrightarrow \mathrm{Sp}^{nc}(n)$$

$$\begin{pmatrix} A & B \\ C & D \end{pmatrix} \mapsto \begin{pmatrix} iI_n & 0 \\ 0 & -iI_n \end{pmatrix} \begin{pmatrix} A & B \\ C & D \end{pmatrix} \begin{pmatrix} iI_n & 0 \\ 0 & -iI_n \end{pmatrix}^{-1} = \begin{pmatrix} A & -B \\ -C & D \end{pmatrix},$$

whose fixed Lie subgroup is equal to the maximal compact Lie subgroup

$$\left\{\begin{pmatrix} A & 0 \\ 0 & D \end{pmatrix} \in \mathrm{Sp}^{nc}(2n)\right\} = \left\{\begin{pmatrix} A & 0 \\ 0 & \overline{A} \end{pmatrix} : \overline{A}^t A = I_n\right\} =: \mathrm{U}(n),$$

which is also equal to the stabilizer of $0 \in \mathcal{D}_{III_n}$. In particular, the pair $(\mathrm{Sp}^{nc}(n), \mathrm{U}(n))$ is a Riemannian symmetric pair. Notice that the involution σ descends to an involution of $\mathrm{PSp}^{nc}(n)$ whose fixed locus is the maximal compact Lie subgroup $\overline{\mathrm{U}(n)} := \mathrm{U}(n)/\{\pm I_n\}$ of $\mathrm{PSp}^{nc}(n)$. Therefore, also the pair $(\mathrm{PSp}^{nc}(n), \overline{\mathrm{U}(n)})$ is a Riemannian symmetric pair.

By the above discussion, we get the following presentation of $\mathcal{D}_{III_n}$ as an irreducible *HSM of non-compact type*

$$\mathcal{D}_{III_n} \cong \mathrm{Sp}^{nc}(n)/\mathrm{U}(n) = \mathrm{PSp}^{nc}(n)/\overline{\mathrm{U}(n)}, \tag{46}$$

associated to the Riemannian symmetric pair $(\operatorname{Sp}^{nc}(n), \operatorname{U}(n))$ (resp. to $(\operatorname{PSp}^{nc}(n), \overline{\operatorname{U}(n)})$). Notice that the last description of $\mathcal{D}_{III_n}$ is the one appearing in Theorem 2.12(i).

The irreducible *Hermitian SLA of non-compact type* associated to the Riemannian symmetric pair $(\operatorname{Sp}^{nc}(n), \operatorname{U}(n))$ (or equivalently to $(\operatorname{PSp}^{nc}(n), \overline{\operatorname{U}(n)})$) is given by the Lie algebra

$$\mathfrak{sp}^{nc}(n) := \operatorname{Lie} \operatorname{Sp}^{nc}(n) = \left\{ M \in \mathfrak{sl}(2n, \mathbb{C}) : \begin{matrix} M^t S_n = -S_n M \\ \overline{M}^t \begin{pmatrix} -I_n & 0 \\ 0 & I_n \end{pmatrix} = - \begin{pmatrix} -I_n & 0 \\ 0 & I_n \end{pmatrix} M \end{matrix} \right\} \tag{47}$$

$$= \left\{ \begin{pmatrix} Z_1 & Z_2 \\ Z_2^t & -Z_1^t \end{pmatrix} \in \mathfrak{gl}(2n, \mathbb{C}) : \overline{Z_1}^t = -Z_1, \ Z_2^t = Z_2 \right\}^{6}.$$

endowed with the Cartan involution $\theta = d\sigma$ given by

$$\theta \begin{pmatrix} Z_1 & Z_2 \\ Z_2^t & -Z_1^t \end{pmatrix} = \begin{pmatrix} Z_1 & -Z_2 \\ -\overline{Z}_2^t & -Z_1^t \end{pmatrix}$$

and with the element

$$H = \frac{i}{2} \begin{pmatrix} I_n & 0 \\ 0 & -I_n \end{pmatrix} \in \operatorname{Fix}(\theta) = \left\{ \begin{pmatrix} Z_1 & 0 \\ 0 & -Z_1^t \end{pmatrix} : \overline{Z}_1^t = -Z_1 \right\} \cong \operatorname{Lie} U(n).$$

The **cominuscle homogeneous variety** of type III_n is the symplectic Grassmannian $\operatorname{Gr}_{\operatorname{sym}}(n, 2n)$ parametrizing Lagrangian n-dimensional subspaces of $\mathbb{C}^{2n}$:

$$\operatorname{Gr}_{\operatorname{sym}}(n, 2n) := \{[W \subset \mathbb{C}^{2n}] : \dim W = p, J_{|W \times W} \equiv 0\}, \tag{48}$$

where J is the standard symplectic form on $\mathbb{C}^{2n}$ which is represented by the matrix $J_n = \begin{pmatrix} 0 & I_n \\ -I_n & 0 \end{pmatrix}$ in the standard basis of $\mathbb{C}^{2n}$.

The *Borel embedding* of $\mathcal{D}_{III_n}$ into $\operatorname{Gr}_{\operatorname{sym}}(n, 2n)$ is given by

$$\mathcal{D}_{III_n} \subset M_{n,n}^{\operatorname{sym}}(\mathbb{C}) \hookrightarrow \operatorname{Gr}_{\operatorname{sym}}(n, 2n),$$

$$Z \mapsto \left\{ \langle v_1, \cdots v_n \rangle : \{v_1, \cdots, v_n\} \text{ are the column vectors of } \begin{pmatrix} Z \\ I_n \end{pmatrix} \right\}. \tag{49}$$

---

[6] Note that $\mathfrak{sp}^{nc}(n)$ is isomorphic to the classical real Lie algebra $\mathfrak{sp}(n, \mathbb{R})$ via the same conjugation given in formula (44).

The complex simple algebraic group $\mathrm{Sp}(n, \mathbb{C})$ acts transitively on $\mathrm{Gr}_{\mathrm{sym}}(n, 2n)$ via

$$\mathrm{Sp}(n, \mathbb{C}) \times \mathrm{Gr}_{\mathrm{sym}}(n, 2n) \longrightarrow \mathrm{Gr}_{\mathrm{sym}}(n, 2n)$$
$$(g, [W \subset \mathbb{C}^{2n}]) \mapsto [g(W) \subset \mathbb{C}^{2n}].$$

Note that the center $Z(\mathrm{Sp}(n, \mathbb{C})) = \{\pm I_{2n}\}$ of $\mathrm{Sp}(n, \mathbb{C})$ acts trivially on $\mathrm{Gr}_{\mathrm{sym}}(n, 2n)$; indeed, it turns out that the group of automorphisms of the algebraic variety $\mathrm{Gr}_{\mathrm{sym}}(n, 2n)$ is equal to

$$\mathrm{PSp}(n, \mathbb{C}) := \mathrm{Sp}(n, \mathbb{C})/\{\pm I_{2n}\},$$

which is a connected semisimple complex algebraic group of adjoint type and it is the complexification of the Lie group $\mathrm{PSp}(n, \mathbb{R})$.

Consider now the base point $W_o := \langle e_{n+1}, \cdots, e_{2n} \rangle \in \mathrm{Gr}_{\mathrm{sym}}(n, 2n)$ with respect to the standard basis $\{e_1, \cdots, e_{2n}\}$ of $\mathbb{C}^{2n}$ (recall that we have normalized J so that it is represented by the standard symplectic matrix J_n with respect to this basis). The stabilizer of W_o is the maximal parabolic subgroup associated to the n-th simple root of the Dinkin diagram C_n (which is the unique cominuscle simple root of C_n, see Table 5)

$$Q_n := \left\{ \begin{pmatrix} A & 0 \\ C & D \end{pmatrix} \in \mathrm{Sp}(n, \mathbb{C}) \right\} \subset \mathrm{Sp}(n, \mathbb{C}),$$

where $A \in M_{n,n}(\mathbb{C})$, $C \in M_{n,n}(\mathbb{C})$ and $D \in M_{n,n}(\mathbb{C})$. The parabolic group Q_n admits the following Levi decomposition

$$Q_n = R_u(Q_n) \rtimes L(Q_n) := \left\{ \begin{pmatrix} I_n & 0 \\ C & I_n \end{pmatrix} : C^t = C \right\} \rtimes \left\{ \begin{pmatrix} A & 0 \\ 0 & D \end{pmatrix} \in \mathrm{Sp}(n, \mathbb{C}) \right\},$$

which coincides with the Levi decomposition appearing in Theorem 2.35.

From the above discussion, we obtain the following explicit presentation of $\mathrm{Gr}_{\mathrm{sym}}(n, 2n)$ as a cominuscle homogeneous variety (as in Definition 2.33)

$$\mathrm{Gr}_{\mathrm{sym}}(n, 2n) \cong \mathrm{Sp}(n, \mathbb{C})/Q_n = \mathrm{PSp}(n, \mathbb{C})/\overline{Q_n}, \tag{50}$$

where $\overline{Q_n} := Q_n/\{\pm I_{2n}\}$.

Consider now the compact real form of $\mathrm{Sp}(n, \mathbb{C})$, which is the Lie subgroup $\mathrm{Sp}(n) := \mathrm{Sp}(n, \mathbb{C}) \cap \mathrm{SU}(2n) \subset \mathrm{Sp}(n, \mathbb{C})$ that leaves invariant the positive definite Hermitian form $x_1 \overline{y}_1 + \cdots + x_{2n} \overline{y}_{2n}$ on $\mathbb{C}^{2n}$. More explicitly[7]

---

[7]The Lie group $\mathrm{Sp}(n)$ admits another natural description in terms of matrices with coefficients in $\mathbb{H}$. Namely, there an isomorphism of Lie group

Hermitian Symmetric Manifolds

$$\mathrm{Sp}(n) = \{g \in \mathrm{Sp}(n, \mathbb{C}) : \overline{g}^t g = I_{2n}\}$$

$$= \left\{ \begin{pmatrix} A & B \\ -\overline{B} & \overline{A} \end{pmatrix} \in \mathrm{SL}(2n, \mathbb{C}) : \begin{array}{l} \overline{A}^t A + \overline{B}^t B = I_n \\ \overline{A}^t B = -B^t \overline{A} \end{array} \right\}.$$

Similarly, the quotient of $\mathrm{Sp}(n)$ by its center

$$\mathrm{PSp}(n) := \mathrm{Sp}(n)/Z(\mathrm{Sp}(n)) = \mathrm{Sp}(n)/\{\pm I_{2n}\}$$

is the compact real form of $\mathrm{PSp}(n, \mathbb{C})$.

The restriction of the action of $\mathrm{Sp}(n, \mathbb{C})$ on $\mathrm{Gr}(q, p+q)$ to the subgroup $\mathrm{Sp}(n) \subset \mathrm{Sp}(n, \mathbb{C})$ is still transitive and the stabilizer of W_o is the maximal proper connected and compact subgroup

$$\mathrm{Sp}(n) \cap Q_n = \left\{ \begin{pmatrix} A & 0 \\ 0 & \overline{A} \end{pmatrix} : \overline{A}^t A = I_n \right\} \cong \mathrm{U}(n).$$

The action of $\mathrm{Sp}(n)$ on $\mathrm{Gr}(q, p+q)$ factors through a transitive action of $\mathrm{PSp}(n)$ in such a way that the stabilizer of W_o is equal to the maximal proper connected and compact subgroup

$$\mathrm{PSp}(n) \cap \overline{Q_n} = \left\{ \left[\begin{pmatrix} A & 0 \\ 0 & \overline{A} \end{pmatrix} \right] : \overline{A}^t A = I_n \right\} \cong \overline{\mathrm{U}(n)} = \mathrm{U}(n)/\{\pm I_n\}.$$

The pair $(\mathrm{Sp}(n), \mathrm{U}(n))$ is a Riemannian symmetric pair since $\mathrm{U}(n)$ is the fixed subgroup of the involution

$$\sigma^* : \mathrm{Sp}(n) \longrightarrow \mathrm{Sp}(n)$$

$$\begin{pmatrix} A & B \\ -\overline{B} & \overline{A} \end{pmatrix} \mapsto \begin{pmatrix} A & -B \\ \overline{B} & \overline{A} \end{pmatrix},$$

and similarly for the pair $(\mathrm{PSp}(n), \overline{\mathrm{U}(n)})$.

By the above discussion, we get the following presentation of $\mathrm{Gr}_{\mathrm{sym}}(n, 2n)$ as the irreducible *HSM of compact type*

$$\mathrm{Gr}_{\mathrm{sym}}(n, 2n) \cong \mathrm{Sp}(n)/\mathrm{U}(n) = \mathrm{PSp}(n)/\overline{\mathrm{U}(n)}, \tag{51}$$

$$\mathrm{Sp}(n) \xrightarrow{\cong} \mathrm{U}(n, \mathbb{H}) := \{g \in \mathrm{GL}(n, \mathbb{H}) : \overline{g}^t g = I_n\}$$

$$\begin{pmatrix} A & B \\ -\overline{B} & \overline{A} \end{pmatrix} \mapsto A - j\overline{B}.$$

associated to the Riemannian symmetric pair $(\mathrm{Sp}(n), \mathrm{U}(n))$ (resp. to $(\mathrm{PSp}(n), \overline{\mathrm{U}(n)})$). In particular, the last description of $\mathcal{D}_{III_n}$ is the one appearing in Theorem 2.12(ii). Notice that the symmetry at the base point W_o of $\mathrm{Gr}_{\mathrm{sym}}(n, 2n)$, seen as a Hermitian symmetric manifold, is given by the element

$$s_{W_o} = \left[\begin{pmatrix} iI_n & 0 \\ 0 & -iI_n \end{pmatrix}\right] \in \mathrm{PSp}(n).$$

The irreducible *Hermitian SLA of compact type* associated to the Riemann symmetric pair $(\mathrm{Sp}(n), \mathrm{U}(n))$ is given by the Lie algebra

$$\mathrm{Lie}\,\mathrm{Sp}(n) = \mathfrak{sp}(n) = \left\{ \begin{pmatrix} Z_1 & Z_2 \\ -\overline{Z_2}^t & -\overline{Z_1}^t \end{pmatrix} : \overline{Z_1}^t = -Z_1,\ Z_2^t = Z_2 \right\} \tag{52}$$

endowed with the involution $\theta^* = d\sigma^*$

$$\theta^* \begin{pmatrix} Z_1 & Z_2 \\ -\overline{Z_2}^t & -\overline{Z_1}^t \end{pmatrix} = \begin{pmatrix} Z_1 & -Z_2 \\ \overline{Z_2}^t & -\overline{Z_1}^t \end{pmatrix}$$

and with the element

$$H^* = \frac{1}{2}\begin{pmatrix} I_n & 0 \\ 0 & -I_n \end{pmatrix} \in \mathrm{Fix}(\theta^*) = \left\{ \begin{pmatrix} Z_1 & 0 \\ 0 & -\overline{Z_1}^t \end{pmatrix} : \overline{Z_1}^t = -Z_1 \right\} \cong \mathfrak{u}(n) = \mathrm{Lie}\,\mathrm{U}(n)$$

Notice that the Hermitian SLA $(\mathfrak{sp}(n), \theta^*, H^*)$ is the dual of the Hermitian SLA $(\mathfrak{sp}^{nc}(n), \theta, H)$ in the sense of Sect. 2.3.

The complexification of the Lie algebras $\mathfrak{sp}^{nc}(n)$ and $\mathfrak{sp}(n)$ is the complex simple Lie algebra of type C_n

$$\mathrm{Lie}\,\mathrm{Sp}(n, \mathbb{C}) = \mathfrak{sp}(n, \mathbb{C}) = \{M \in \mathfrak{gl}(2n, \mathbb{C}) : M^t J_n = -J_n M\}$$

$$= \left\{ \begin{pmatrix} Z_1 & Z_2 \\ Z_3 & -Z_1^t \end{pmatrix} \in \mathfrak{gl}(2n, \mathbb{C}) : Z_2^t = Z_2 \text{ and } Z_3^t = Z_3 \right\}.$$

The decomposition (6) of $\mathfrak{sp}(n, \mathbb{C})$ is given by

$$\mathfrak{sp}(n, \mathbb{C}) = \left\{ \begin{pmatrix} Z_1 & 0 \\ 0 & -Z_1^t \end{pmatrix} \right\} \oplus \left\{ \begin{pmatrix} 0 & Z_2 \\ 0 & 0 \end{pmatrix} : Z_2^t = Z_2 \right\} \oplus \left\{ \begin{pmatrix} 0 & 0 \\ Z_3 & 0 \end{pmatrix} : Z_3^t = Z_3 \right\}.$$

In particular, we have the identification

$$M_{n,n}^{\text{sym}}(\mathbb{C}) \xrightarrow{\cong} \mathfrak{p}_+$$
$$M \mapsto \begin{pmatrix} 0 & M \\ 0 & 0 \end{pmatrix}. \tag{53}$$

Using the above identification and the formula (15), $M_{n,n}^{\text{sym}}(\mathbb{C})$ becomes a *Hermitian positive JTS* with respect to the triple product

$$\{M_1, M_2, M_3\} = \frac{1}{2}\left(M_1 \overline{M_2}^t M_3 + M_3 \overline{M_2}^t M_1\right). \tag{54}$$

3.4 Type IV_n

The **bounded symmetric domain** of type IV_n ($1 \leq n \neq 2$) in its *Harish–Chandra embedding* is given by

$$\mathcal{D}_{IV_n} := \{Z \in \mathbb{C}^n : 2\overline{Z}^t Z < 1 + |Z^t Z|^2, \ \overline{Z}^t Z < 1\} \subset \mathbb{C}^n. \tag{55}$$

Note that the first inequality, together with the fact that $|Z^t Z|^2 \leq (\overline{Z}^t Z)^2$, implies that

$$2\overline{Z}^t Z < 1 + |Z^t Z|^2 \implies 2\overline{Z}^t Z < 1 + (\overline{Z}^t Z)^2 \iff 0 < (1 - \overline{Z}^t Z)^2.$$

Therefore the open subset $\{Z \in \mathbb{C}^n : 2\overline{Z}^t Z < 1 + |Z^t Z|^2\} \subset \mathbb{C}^n$ is the disjoint union of two connected components defined by, respectively, $\overline{Z}^t Z < 1$ and by $\overline{Z}^t Z > 1$. The first connected component is the one that contains the origin $0 \in \mathbb{C}^n$ and it coincides with the domain $\mathcal{D}_{IV_n}$.

The domain $\mathcal{D}_{IV_n}$ (which is also called the Lie ball) admits another real analytic incarnation in terms of $2 \times n$ real matrices, namely we have a real analytic diffeomorphism (see [Hua46, Sec. 12 and 13])

$$\mathcal{D}_{IV_n} \xrightarrow{\cong} \{M \in M_{2,n}(\mathbb{R}) : M \cdot M^t < I_2\}$$
$$Z \mapsto 2\begin{pmatrix} Z^t Z + 1 & i(Z^t Z - 1) \\ \overline{Z}^t \overline{Z} + 1 & -i(\overline{Z}^t \overline{Z} - 1) \end{pmatrix}^{-1} \cdot \begin{pmatrix} Z \\ \overline{Z} \end{pmatrix}. \tag{56}$$

Consider the subgroup $SO^{nc}(n, 2)$ of $SO(n+2, \mathbb{C})$ consisting of all the elements of $SO(2+n, \mathbb{C})$ that leave invariant the bilinear Hermitian form on $\mathbb{C}^{n+2} \times \mathbb{C}^{2+n}$ given by $-x_1 \overline{y}_1 - \cdots - x_n \overline{y}_n + x_{n+1} \overline{y}_{n+1} + x_{n+2} \overline{y}_{n+2}$. Explicitly,

$$\mathrm{SO}^{nc}(n,2) = \mathrm{SO}(n+2,\mathbb{C}) \cap \mathrm{U}(n,2)$$

$$= \left\{ g \in \mathrm{SL}(n+2,\mathbb{C}) : g^t g = I_{n+2}, \overline{g}^t \begin{pmatrix} I_n & 0 \\ 0 & -I_2 \end{pmatrix} g = \begin{pmatrix} I_n & 0 \\ 0 & -I_2 \end{pmatrix} \right\}$$

$$= \left\{ \begin{pmatrix} A & B \\ C & D \end{pmatrix} \in \mathrm{SL}(n+2,\mathbb{C}) : \begin{array}{ll} A^t A + C^t C = I_n, & \overline{A}^t A - \overline{C}^t C = I_n \\ D^t D + B^t B = I_2, & \overline{D}^t D - \overline{B}^t B = I_2 \\ A^t B = -C^t D, & \overline{A}^t B = \overline{C}^t D \end{array} \right\}$$

The Lie group $\mathrm{SO}^{nc}(n,2)$ acts transitively on $\mathcal{D}_{IV_n}$ via

$$\mathrm{SO}^{nc}(n,2) \times \mathcal{D}_{IV_n} \longrightarrow \mathcal{D}_{IV_n}$$

$$\left(\begin{pmatrix} A & B \\ C & D \end{pmatrix}, Z \right) \mapsto \frac{2iAZ + B \begin{pmatrix} 1 + Z^t \cdot Z \\ i - iZ^t \cdot Z \end{pmatrix}}{(1,i) \cdot \left(2iCZ + D \begin{pmatrix} 1 + Z^t \cdot Z \\ i - iZ^t \cdot Z \end{pmatrix} \right)}.$$

Notice that the center $Z(\mathrm{SO}^{nc}(n,2)) = \{\pm I_{n+2}\}$ of $\mathrm{SO}^{nc}(n,2)$ acts trivially on $\mathcal{D}_{IV_n}$; indeed, it turns out that the connected component of the group of biholomorphisms of $\mathcal{D}_{IV_n}$ is given by

$$\mathrm{Hol}(\mathcal{D}_{IV_n})^o = \mathrm{SO}^{nc}(n,2)/Z(\mathrm{SO}^{nc}(n,2)) := \mathrm{PSO}^{nc}(n,2),$$

which is the connected non-compact adjoint simple Lie group of type $D_{n/2+1}$ if n is even and $B_{(n+1)/2}$ if n is odd.

The symmetry of $\mathcal{D}_{IV_n}$ at the base point 0 is given by the element

$$s_0 = \left[\begin{pmatrix} I_n & 0 \\ 0 & -I_2 \end{pmatrix} \right] \in \mathrm{PSO}^{nc}(n,2), \tag{57}$$

which acts on $\mathcal{D}_{IV_n}$ by sending Z into $-Z$. The symmetry s_0 induces an involution on $\mathrm{SO}^{nc}(n,2)$

$$\sigma : \mathrm{SO}^{nc}(n,2) \longrightarrow \mathrm{SO}^{nc}(n,2)$$

$$\begin{pmatrix} A & B \\ C & D \end{pmatrix} \mapsto \begin{pmatrix} I_n & 0 \\ 0 & -I_2 \end{pmatrix} \begin{pmatrix} A & B \\ C & D \end{pmatrix} \begin{pmatrix} I_n & 0 \\ 0 & -I_2 \end{pmatrix}^{-1} = \begin{pmatrix} A & -B \\ -C & D \end{pmatrix},$$

whose fixed Lie subgroup is equal to the maximal compact Lie subgroup

$$\left\{ \begin{pmatrix} A & 0 \\ 0 & D \end{pmatrix} \in \mathrm{SO}^{nc}(n,2) \right\} = \left\{ \begin{pmatrix} A & 0 \\ 0 & D \end{pmatrix} : \begin{array}{l} \overline{A}^t A = A^t A = I_n \\ \overline{D}^t D = D^t D = I_2 \end{array} \right\} = \mathrm{SO}(n) \times \mathrm{SO}(2),$$

which is also equal to the stabilizer of $0 \in \mathcal{D}_{IV_n}$. In particular, the pair $(\mathrm{SO}^{nc}(n,2), \mathrm{SO}(n) \times \mathrm{SO}(2))$ is a Riemannian symmetric pair. Notice that the involution σ descends to an involution of $\mathrm{PSO}^{nc}(n,2)$ whose fixed locus is the maximal compact Lie subgroup $\overline{\mathrm{SO}(n) \times \mathrm{SO}(2)} := \mathrm{SO}(n) \times \mathrm{SO}(2)/Z(\mathrm{SO}^{nc}(n,2))$ of $\mathrm{PSO}^{nc}(n,2)$. Therefore, also the pair $(\mathrm{PSO}^{nc}(n,2), \overline{\mathrm{SO}(n) \times \mathrm{SO}(2)})$ is a Riemannian symmetric pair.

By the above discussion, we get the following presentation of $\mathcal{D}_{IV_n}$ as an irreducible *HSM of non-compact type*

$$\mathcal{D}_{IV_n} \cong \mathrm{SO}^{nc}(n,2)/(\mathrm{SO}(n) \times \mathrm{SO}(2)) = \mathrm{PSO}^{nc}(n,2)/\overline{\mathrm{SO}(n) \times \mathrm{SO}(2)}, \quad (58)$$

associated to the Riemannian symmetric pair $(\mathrm{SO}^{nc}(n,2), \mathrm{SO}(n) \times \mathrm{SO}(2))$ (resp. to $(\mathrm{PSO}^{nc}(n,2), \overline{\mathrm{SO}(n) \times \mathrm{SO}(2)})$. Notice that the last description of $\mathcal{D}_{IV_n}$ is the one appearing in Theorem 2.12(i).

The irreducible *Hermitian SLA of non-compact type* associated to the Riemannian symmetric pair $(\mathrm{SO}^{nc}(n,2), \mathrm{SO}(n) \times \mathrm{SO}(2))$ (or equivalently to $(\mathrm{PSO}^{nc}(n,2), \overline{\mathrm{SO}(n) \times \mathrm{SO}(2)})$ is given by the Lie algebra

$$\mathfrak{so}^{nc}(n,2) := \mathrm{Lie}\,\mathrm{SO}^{nc}(n,2) = \left\{ M \in \mathfrak{sl}(n+2,\mathbb{C}) : \begin{array}{l} M^t = -M \\ \overline{M}^t \begin{pmatrix} I_n & 0 \\ 0 & -I_2 \end{pmatrix} = -\begin{pmatrix} I_n & 0 \\ 0 & -I_2 \end{pmatrix} M \end{array} \right\} \quad (59)$$

$$= \left\{ \begin{pmatrix} X_1 & iX_2 \\ -iX_2^t & X_3 \end{pmatrix} \in \mathfrak{gl}(n+2,\mathbb{C}) : \begin{array}{l} \overline{X}_1 = X_1,\ \overline{X}_2 = X_2,\ \overline{X}_3 = X_3 \\ X_1^t = -X_1,\ X_3^t = -X_3 \end{array} \right\}$$

endowed with the involution $\theta = d\sigma$

$$\theta \begin{pmatrix} X_1 & iX_2 \\ -iX_2^t & X_3 \end{pmatrix} = \begin{pmatrix} X_1 & -iX_2 \\ iX_2^t & X_3 \end{pmatrix}$$

and with the element

$$H = \begin{pmatrix} 0 & 0 \\ 0 & J_1 \end{pmatrix} \in \mathrm{Fix}(\theta) = \left\{ \begin{pmatrix} X_1 & 0 \\ 0 & X_3 \end{pmatrix} \in \mathfrak{gl}(n+2,\mathbb{R}) : X_1^t = -X_1,\ X_3^t = -X_3 \right\}$$

$$\cong \mathrm{Lie}(\mathrm{SO}(n) \times \mathrm{SO}(2)).$$

The **cominuscle homogeneous variety** of type IV_n is the complex quadric hypersurface of dimension n:

$$\mathcal{Q}^n := \{[v] \in \mathbb{P}^{n+1} : Q(v,v) = 0\} \subset \mathbb{P}^{n+1} \tag{60}$$

where Q is the bilinear symmetric non-degenerate form on $\mathbb{C}^{n+2}$ given by $Q(v) = v_1^2 + \ldots + v_{n+2}^2$.

Observe that the complex quadric hypersurface $\mathcal{Q}^n \subset \mathbb{P}^{n+1}$ admits another real analytic incarnation. Namely, $\mathcal{Q}^n$ is real analytic diffeomorphic to the oriented real Grassmannian $\mathrm{Gr}_\mathbb{R}^+(2, n+2)$ parametrizing two-dimensional oriented subspaces of $\mathbb{R}^{n+2}$ via the map (see [Sat80, Appendix §6])

$$\mathrm{Gr}_\mathbb{R}^+(2, n+2) \xrightarrow{\cong} \mathcal{Q}^n \tag{61}$$
$$\langle v_1, v_2 \rangle \mapsto [v_1 + i v_2].$$

The *Borel embedding* of $\mathcal{D}_{IV_n}$ into $\mathcal{Q}^n$ is given by

$$\mathcal{D}_{IV_n} \subset \mathbb{C}^n \hookrightarrow \mathcal{Q}^n$$
$$Z \mapsto \left[\begin{pmatrix} 2iZ \\ 1 + Z^t \cdot Z \\ i - i Z^t \cdot Z \end{pmatrix}\right]. \tag{62}$$

The complex algebraic simple group $SO(n+2, \mathbb{C})$ acts transitively on $\mathcal{Q}^n$ via

$$SO(n+2, \mathbb{C}) \times \mathcal{Q}^n \longrightarrow \mathcal{Q}^n$$
$$(g, [v]) \mapsto [g(v)].$$

Note that the center

$$Z(SO(n+2, \mathbb{C})) = \begin{cases} \{\pm I_{n+2}\} & \text{if } n \text{ is even,} \\ \{I_{n+2}\} & \text{if } n \text{ is odd,} \end{cases} \tag{63}$$

acts trivially on $\mathcal{Q}^n$; indeed, it turns out that the group of automorphisms of the algebraic variety $\mathcal{Q}^n$ is equal to

$$PSO(2n, \mathbb{C}) := PSO(2n, \mathbb{C})/Z(SO(n+2, \mathbb{C})),$$

which is the connected simple adjoint complex algebraic group of type $D_{n/2+1}$ if n is even and $B_{(n+1)/2}$ if n is odd.

The stabilizer of $v_o = [(0, \cdots, 0, 1, i)] \in \mathcal{Q}^n$ is the maximal parabolic subgroup associated to the first simple root of the Dinkin diagram $D_{n/2+1}$ if n is even and of the Dinkin diagram $B_{(n+1)/2}$ if n is odd (which are cominuscle simple roots, see Table 5)

$$Q_1 := \left\{ \begin{pmatrix} A & B \\ C & D \end{pmatrix} \in \mathrm{SO}(n+2,\mathbb{C}) : \begin{array}{l} B = (B', iB') \text{ for some } B' \in M_{n,1}(\mathbb{C}) \\ D = \begin{pmatrix} a & b \\ c & d \end{pmatrix} \text{ such that } ia - b = c + id \end{array} \right\},$$

where $A \in M_{n,n}(\mathbb{C})$, $B \in M_{n,2}(\mathbb{C})$, $C \in M_{2,n}(\mathbb{C})$ and $D \in M_{2,2}(\mathbb{C})$.

From the above discussion, we obtain the following explicit presentation of Q^n as a cominuscle homogeneous variety (as in Definition 2.33)

$$Q^n \cong \mathrm{SO}(n+2,\mathbb{C})/Q_1 = \mathrm{PSO}(n,\mathbb{C})/\overline{P}_1, \tag{64}$$

where $\overline{P}_1 := Q_1/Z(\mathrm{SO}(n+2,\mathbb{C}))$.

Consider now the compact real form $\mathrm{SO}(n+2)$ of $\mathrm{SO}(n+2,\mathbb{C})$ consisting of all the real matrices in $\mathrm{SO}(n+2,\mathbb{C})$ or, equivalently, of all the elements in $\mathrm{SO}(n+2,\mathbb{C})$ that leave invariant the positive definite Hermitian form $x_1 \overline{y}_1 + \cdots + x_{n+2} \overline{y}_{n+2}$ on $\mathbb{C}^{n+2}$. More explicitly

$$\mathrm{SO}(n+2) := \mathrm{SO}(n+2,\mathbb{C}) \cap \mathrm{SU}(n+2) = \{g \in \mathrm{SO}(n+2,\mathbb{C}) : \overline{g} = g\}$$

$$= \left\{ \begin{pmatrix} A & B \\ C & D \end{pmatrix} \in \mathrm{SL}(n+2,\mathbb{R}) : \begin{array}{l} A^t A + C^t C = I_n \\ D^t D + B^t B = I_2 \\ A^t B = -C^t D \end{array} \right\}.$$

Similarly, the quotient of $\mathrm{SO}(n+2)$ by its center (which is given by (63))

$$\mathrm{PSO}(n+2) := \mathrm{SO}(n+2)/Z(\mathrm{SO}(n+2))$$

is a compact real form of $\mathrm{PSO}(2n,\mathbb{C})$.

The restriction of the action of $\mathrm{SO}(n+2,\mathbb{C})$ on Q^n to the subgroup $\mathrm{SO}(n+2) \subset \mathrm{SO}(n+2,\mathbb{C})$ is still transitive and the stabilizer of v_o is the maximal proper connected and compact subgroup

$$\mathrm{SO}(n+2) \cap Q_1 = \left\{ \begin{pmatrix} A & 0 \\ 0 & D \end{pmatrix} : \begin{array}{l} A^t A = I_n, \ \det(A) = 1 \\ D^t D = I_2, \ \det(D) = 1 \end{array} \right\} \cong \mathrm{SO}(n) \times \mathrm{SO}(2).$$

The pair $(\mathrm{SO}(n+2), \mathrm{SO}(n) \times \mathrm{SO}(2))$ is a Riemannian symmetric pair since $\mathrm{SO}(n) \times \mathrm{SO}(2)$ is the connected component of the fixed subgroup of the involution

$$\sigma^* : \mathrm{SO}(n+2) \longrightarrow \mathrm{SO}(n+2)$$

$$\begin{pmatrix} A & B \\ C & D \end{pmatrix} \mapsto \begin{pmatrix} A & -B \\ -C & D \end{pmatrix}.$$

and similarly for the pair $(\mathrm{PSO}(n+2), \overline{\mathrm{SO}(n)\times \mathrm{SO}(2)})$, where $\overline{\mathrm{SO}(n)\times \mathrm{SO}(2)}$ is the image of $\mathrm{SO}(n)\times SO(2)$ in $\mathrm{PSO}(n+2)$.

By the above discussion, we get the following presentation of $\mathcal{Q}^n$ as the irreducible *HSM of compact type*

$$\mathcal{Q}^n \cong \mathrm{SO}(n+2)/\mathrm{SO}(n)\times \mathrm{SO}(2) = \mathrm{PSO}(n+2)/\overline{\mathrm{SO}(n)\times \mathrm{SO}(2)}, \qquad (65)$$

associated to the $\overline{\text{Riemannian}}$ symmetric pair $(\mathrm{SO}(n+2), \mathrm{SO}(n)\times \mathrm{SO}(2))$ (resp. to $(\mathrm{PSO}(n+2), \overline{\mathrm{SO}(n)\times \mathrm{SO}(2)})$). In particular, the last description of $\mathcal{Q}^n$ is the one appearing in Theorem 2.12(ii). Notice that the symmetry at the base point v_o of $\mathcal{Q}^n$, seen as a Hermitian symmetric manifold, is given by the element

$$s_{v_o} = \left[\begin{pmatrix} I_n & 0 \\ 0 & -I_2 \end{pmatrix}\right] \in \mathrm{PSO}(n+2).$$

The irreducible *Hermitian SLA of compact type* associated to the Riemann symmetric pair $(\mathrm{SO}(n+2), \mathrm{SO}(n)\times \mathrm{SO}(2))$ is given by the Lie algebra

$$\mathfrak{so}(n+2) := \mathrm{Lie}\,\mathrm{SO}(n+2) = \{M \in \mathfrak{sl}(p+q, \mathbb{R}) : M^t = -M\} \qquad (66)$$

$$= \left\{ \begin{pmatrix} X_1 & X_2 \\ -X_2^t & X_3 \end{pmatrix} \in \mathfrak{gl}(p+q, \mathbb{R}) : X_1^t = -X_1,\ X_3^t = -X_3 \right\}$$

endowed with the involution $\theta^* = d\sigma^*$

$$\theta^* \begin{pmatrix} X_1 & X_2 \\ -X_2^t & X_3 \end{pmatrix} = \begin{pmatrix} X_1 & -X_2 \\ X_2^t & X_3 \end{pmatrix}$$

and with the element

$$H^* = \begin{pmatrix} 0 & 0 \\ 0 & J_1 \end{pmatrix} \in \mathrm{Fix}(\theta^*) = \left\{ \begin{pmatrix} X_1 & 0 \\ 0 & X_3 \end{pmatrix} : X_1^t = -X_1,\ X_3^t = -X_3 \right\}$$

$$\cong \mathrm{Lie}(\mathrm{SO}(n)\times \mathrm{SO}(2)).$$

Notice that the Hermitian SLA $(\mathfrak{so}(n+2), \theta^*, H^*)$ is the dual of the Hermitian SLA $(\mathfrak{so}^{nc}(n,2), \theta, H)$ in the sense of Sect. 2.3.

The complexification of the Lie algebras $\mathfrak{so}^{nc}(n,2)$ and $\mathfrak{so}(n+2)$ is the complex simple Lie algebra of type $D_{n/2+1}$ if n is even and $B_{(n+1)/2}$ if n is odd:

$$\mathrm{Lie}\,\mathrm{SO}(n+2, \mathbb{C}) = \mathfrak{so}(n+2, \mathbb{C}) = \left\{ \begin{pmatrix} Z_1 & Z_2 \\ -Z_2^t & Z_3 \end{pmatrix} \in \mathfrak{gl}(n+2, \mathbb{C}) : Z_1^t = -Z_1,\ Z_3^t = -Z_3 \right\}.$$

The decomposition (6) of $\mathfrak{so}(n+2,\mathbb{C})$ is given by

$$\mathfrak{so}(n+2,\mathbb{C}) = \left\{ \begin{pmatrix} Z_1 & 0 \\ 0 & Z_3 \end{pmatrix} : \begin{array}{l} Z_1^t = -Z_1 \\ Z_3^t = -Z_3 \end{array} \right\} \oplus \left\{ \begin{pmatrix} 0 & (iZ', Z') \\ -(iZ', Z')^t & 0 \end{pmatrix} \right\}$$

$$\oplus \left\{ \begin{pmatrix} 0 & (Z'', iZ'') \\ -(Z'', iZ'')^t & 0 \end{pmatrix} \right\}.$$

In particular, we have the identification

$$\mathbb{C}^n \xrightarrow{\cong} \mathfrak{p}_+$$

$$Z \mapsto \begin{pmatrix} 0 & (iZ, Z) \\ -(iZ, Z)^t & 0 \end{pmatrix}. \tag{67}$$

Using the above identification and the formula (15), $\mathbb{C}^n$ becomes a *Hermitian positive JTS* with respect to the triple product

$$\{X, Y, Z\} = (X^t \cdot Z)\overline{Y} - (Z^t \cdot \overline{Y})X - (X^t \cdot \overline{Y})Z. \tag{68}$$

3.5 Type VI

Let $\mathbb{O}$ be the $\mathbb{R}$-algebra of octonions or Cayley algebra (we refer the reader to [Bae02] for a beautiful introduction to the octonions). Recall that $\mathbb{O}$ is the alternative $\mathbb{R}$-algebra (neither associative nor commutative) of dimension 8 whose underlying vector space is equal to $\mathbb{H} \times \mathbb{H}$ and whose multiplication is equal to

$$(a_1, b_1) \cdot (a_2, b_2) := (a_1 a_2 - b_2 \widetilde{b_1}, \widetilde{a_1} b_2 + a_2 b_1),$$

where $\mathbb{H}$ is the division $\mathbb{R}$-algebra of quaternions and $\widetilde{}$ denotes its involution

$$\widetilde{} : \mathbb{H} \longrightarrow \mathbb{H}$$

$$x_0 + ix_1 + jx_2 + kx_3 \mapsto x_0 - ix_1 - jx_2 - kx_3.$$

The algebra $\mathbb{O}$ is endowed with the unity element $e = (1, 0)$ and with an involutive anti-automorphism

$$(a, b) \mapsto \widetilde{(a, b)} := (\tilde{a}, -b).$$

The above involution gives rise to a norm

$$|.|^2 : \mathbb{O} \longrightarrow \mathbb{R}$$
$$(a,b) \mapsto |(a,b)|^2 := (a,b) \cdot \widetilde{(a,b)} = a\tilde{a} + b\tilde{b},$$

which is a positive define quadratic form and it is multiplicative (i.e. $|(a_1,b_1) \cdot (a_2,b_2)|^2 = |(a_1,b_1)|^2|(a_2,b_2)|^2$). Therefore the pair $(\mathbb{O}, |.|^2)$ is a Euclidean composition algebra of dimension 8 and indeed it is the unique such algebra. We will denote by $\langle , \rangle$ the bilinear form associated to the quadratic form $|.|^2$, i.e.

$$\langle x, y \rangle := |x+y|^2 - |x|^2 - |y|^2,$$

for any $x, y \in \mathbb{O}$.

Let $\mathbb{O}_\mathbb{C} := \mathbb{O} \otimes_\mathbb{R} \mathbb{C}$ be the complexification of $\mathbb{O}$ (it is called the complex Cayley algebra). The involution $\widetilde{}$ and the quadratic form $|.|^2$ on $\mathbb{O}$ extend naturally on $\mathbb{O}_\mathbb{C}$ (by a slight abuse of notation, we will continue to denote them by the same symbols). Moreover, $\mathbb{O}_\mathbb{C}$ is endowed with a complex conjugation with respect to its real form $\mathbb{O}$:

$$\lambda \otimes x \mapsto \overline{\lambda \otimes x} := \overline{\lambda} \otimes x,$$

where $\lambda \in \mathbb{C}$ and $x \in \mathbb{O}$.

Consider the complex vector space $H_3(\mathbb{O}_\mathbb{C})$ consisting of Hermitian 3×3-matrices with entries in $\mathbb{O}_\mathbb{C}$

$$H_3(\mathbb{O}_\mathbb{C}) := \{a \in M_{3,3}(\mathbb{O}_\mathbb{C}) : \tilde{a}^t = a\} = \tag{69}$$

$$= \left\{ \begin{pmatrix} \alpha_1 & a_3 & \tilde{a}_2 \\ \tilde{a}_3 & \alpha_2 & a_1 \\ a_2 & \tilde{a}_1 & \alpha_3 \end{pmatrix} : \alpha_1, \alpha_2, \alpha_3 \in \mathbb{C}; a_1, a_2, a_3 \in \mathbb{O}_\mathbb{C} \right\}.$$

The complex vector space $H_3(\mathbb{O}_\mathbb{C})$ is endowed with a product (called the Freudenthal product) defined by

$$a \times b := \begin{pmatrix} \alpha_1 & a_3 & \tilde{a}_2 \\ \tilde{a}_3 & \alpha_2 & a_1 \\ a_2 & \tilde{a}_1 & \alpha_3 \end{pmatrix} \times \begin{pmatrix} \beta_1 & b_3 & \tilde{b}_2 \\ \tilde{b}_3 & \beta_2 & b_1 \\ b_2 & \tilde{b}_1 & \beta_3 \end{pmatrix} := \tag{70}$$

$$= \begin{pmatrix} \alpha_2\beta_3 + \alpha_3\beta_2 - \langle a_1, b_1 \rangle & a_1 b_2 + b_1 a_2 - \alpha_3 \tilde{b}_3 - \beta_3 \tilde{a}_3 & \tilde{b}_1 \tilde{a}_3 + \tilde{a}_1 \tilde{b}_3 - \alpha_2 b_2 - \beta_2 a_2 \\ \tilde{b}_2 \tilde{a}_1 + \tilde{a}_2 \tilde{b}_1 - \alpha_3 b_3 - \beta_3 a_3 & \alpha_3\beta_1 + \alpha_1\beta_3 - \langle a_2, b_2 \rangle & a_2 b_3 + b_2 a_3 - \alpha_1 \tilde{b}_1 - \beta_1 \tilde{a}_1 \\ a_3 b_1 + b_3 a_1 - \alpha_2 \tilde{b}_2 - \beta_2 \tilde{a}_2 & \tilde{b}_3 \tilde{a}_2 + \tilde{a}_3 \tilde{b}_2 - \alpha_1 b_1 - \beta_1 a_1 & \alpha_1\beta_2 + \alpha_2\beta_1 - \langle a_3, b_3 \rangle \end{pmatrix}.$$

Moreover, $H_3(\mathbb{O}_\mathbb{C})$ is endowed with a positive definite Hermitian form defined by

$$(a|b) := \sum_{i=1}^{3} \alpha_i \overline{\beta_i} + \sum_{j=1}^{3} \langle a_j, \overline{b_j} \rangle, \tag{71}$$

where $a, b \in H_3(\mathbb{O}_\mathbb{C})$ are written as in (70). Using the Freudenthal product and the above positive define Hermitian form, we can define a Jordan triple product on $H_3(\mathbb{O}_\mathbb{C})$ via

$$\{a, b, c\} := (a|b)c + (c|b)a - (a \times c) \times \overline{b}, \tag{72}$$

where $\overline{b}$ is the element of $H_3(\mathbb{O}_\mathbb{C})$ obtained by conjugating all the entries of b with respect to the complex conjugation of $\mathbb{O}_\mathbb{C}$. The pair $(H_3(\mathbb{O}_\mathbb{C}), \{.,.,.\})$ is an irreducible Hermitian positive JTS of dimension 27 (see [Roo08, Sec. 2.2]), called sometimes the exceptional Hermitian positive JTS of dimension 27 or the **Hermitian positive JTS of type VI**.

From the above explicit description of the Hermitian positive JTS $(H_3(\mathbb{O}_\mathbb{C}), \{.,..,.\})$ and formula (17), we can deduce an explicit expression of the associated bounded symmetric domain in its Harish–Chandra embedding. In order to do that, we need to introduce the determinant and the adjoint of an element of $H_3(\mathbb{O}_\mathbb{C})$. The determinant is defined by

$$\det : H_3(\mathbb{O}_\mathbb{C}) \longrightarrow \mathbb{O}_\mathbb{C},$$

$$a \mapsto \frac{1}{3!}(a \times a|\overline{a}) = \alpha_1 \alpha_2 \alpha_3 - \sum_{i=1}^{3} \alpha_i |a_i|^2 + a_1(a_2 a_3) + (\widetilde{a_3}\widetilde{a_2})a_1, \tag{73}$$

where $a \in H_3(\mathbb{O}_\mathbb{C})$ is written as in (70). The adjoint of $a \in H_3(\mathbb{O}_\mathbb{C})$ is defined by

$$(a)^\sharp := \frac{a \times a}{2}. \tag{74}$$

The relation between the determinant and the adjoint is given by the following formulas (see [Roo08, Sec. 2.1])

$$\begin{cases} (x^\sharp | \overline{x}) = 3 \det(x), \\ (x^\sharp)^\sharp = \det(x)x. \end{cases} \tag{75}$$

The **bounded symmetric domain** of type *VI* in its *Harish–Chandra embedding* is given by (see [Roo08, Sec. 3.1])

$$\mathcal{D}_{VI} := \left\{ a \in H_3(\mathbb{O}_\mathbb{C}) : \begin{array}{c} 1 - (a|a) + (a^\#|a^\#) - |\det(a)|^2 > 0 \\ 3 - 2(a|a) + (a^\#|a^\#) > 0 \\ 3 - (a|a) > 0 \end{array} \right\} \subset H_3(\mathbb{O}_\mathbb{C}).$$
(76)

The **cominuscle homogeneous variety** of type *VI* is the Freudenthal variety

$$\mathcal{F} := \{[\lambda, x, y, \mu] \in \mathbb{P}(\mathbb{C} \oplus H_3(\mathbb{O}_\mathbb{C}) \oplus H_3(\mathbb{O}_\mathbb{C}) \oplus \mathbb{C}) : y^\# = \mu x, \, x^\# = \lambda y, \, (x|\overline{y}) = 3\lambda\mu\}.$$
(77)

The *Borel embedding* of $\mathcal{D}_{VI}$ into $\mathcal{F}$ is given by

$$\mathcal{D}_{VI} \subset H_3(\mathbb{O}_\mathbb{C}) \xhookrightarrow{j} \mathcal{F}$$
$$x \mapsto \left[(1, x, x^\#, \det(x))\right].$$
(78)

Using the relations (75), it is easy to see that j is an open embedding and that $j(H_3(\mathbb{O}_\mathbb{C}))$ is the Zariski open subset of $\mathcal{F}$ defined by $\{\lambda \neq 0\}$.

3.6 Type V

In this subsection, we are going to use the notation introduced in Sect. 3.5.

The **Hermitian positive JTS of type V** (sometimes also called the exceptional Hermitian positive JTS of dimension 16) is the simple Hermitian positive JTS $(\mathbb{O}_\mathbb{C}^2, \{.,.,.\})$ where the Jordan triple product $\{.,.,.\}$ is defined by

$$\left\{ \begin{pmatrix} a_1 \\ a_2 \end{pmatrix}, \begin{pmatrix} b_1 \\ b_2 \end{pmatrix}, \begin{pmatrix} c_1 \\ c_2 \end{pmatrix} \right\} := \begin{pmatrix} (a_1\widetilde{\overline{b_1}})c_1 + (c_1\widetilde{\overline{b_1}})a_1 + (a_1\overline{b_2})\widetilde{c_2} + (c_1\overline{b_2})\widetilde{a_2} \\ \widetilde{a_1}(\overline{b_1}c_2) + \widetilde{c_1}(\overline{b_1}a_2) + \widetilde{a_2}(\overline{b_2}c_2) + \widetilde{c_2}(\overline{b_2}a_2) \end{pmatrix}.$$
(79)

The **bounded symmetric domain** of type V in its *Harish–Chandra embedding* is given by (see [Roo08, Sec. 3.1])

$$\mathcal{D}_V := \left\{ x = \begin{pmatrix} x_1 \\ x_2 \end{pmatrix} \in \mathbb{O}_\mathbb{C}^2 : \begin{array}{c} 1 - \sum_{i=1}^{2} \langle x_i, \overline{x_i} \rangle + \sum_{i=1}^{2} (|x_i|^2)^2 + \langle x_2 x_3, \overline{x_2 x_3} \rangle > 0 \\ 2 - \sum_{i=1}^{2} \langle x_i, \overline{x_i} \rangle > 0 \end{array} \right\} \subset \mathbb{O}_\mathbb{C}^2.$$
(80)

The **cominuscle homogeneous variety** of type V is the Cayley plane

$$\mathbb{P}_\mathbb{O}^2 := \{[a] \in \mathbb{P}(H_3(\mathbb{O}_\mathbb{C})) : a^\# = 0\} = \qquad (81)$$

$$\left\{ [a] \in \mathbb{P}(H_3(\mathbb{O}_\mathbb{C})) : \begin{array}{c} \alpha_2\alpha_3 = |a_1|^2, \ \alpha_3\alpha_1 = |a_2|^2, \ \alpha_1\alpha_2 = |a_3|^2 \\ a_1 a_2 = \alpha_3 \widetilde{a_3}, \ a_2 a_3 = \alpha_1 \widetilde{a_1}, \ a_3 a_1 = \alpha_2 \widetilde{a_2} \end{array} \right\},$$

where $a \in H_3(\mathbb{O}_\mathbb{C})$ is written as in (69).

The Cayley plane $\mathbb{P}_\mathbb{O}^2$ is homogeneous with respect to the natural action of the subgroup $\mathrm{SL}_3(\mathbb{O}_\mathbb{C}) \subset \mathrm{GL}(H_3(\mathbb{O}_\mathbb{C})) = \mathrm{GL}_{27}(\mathbb{C})$ consisting of the elements preserving the determinant (73) (see [LM03, Sec. 6.2]). The group $\mathrm{SL}_3(\mathbb{O}_\mathbb{C})$ is a complex simple Lie group of type E_6. Moreover, the stabilizer of any element is isomorphic to the maximal parabolic subgroup P_6 corresponding to the 6-th simple root of the diagram E_6 (which is a cominuscle simple root, see Table 5). Therefore, we obtain the following explicit presentation of $\mathbb{P}_\mathbb{O}^2$ as a cominuscle homogeneous variety (as in Definition 2.33)

$$\mathbb{P}_\mathbb{O}^2 \cong \mathrm{SL}_3(\mathbb{O}_\mathbb{C})/P_6. \qquad (82)$$

The *Borel embedding* of $\mathcal{D}_V$ into $\mathbb{P}_\mathbb{O}^2$ is given by

$$\mathcal{D}_V \subset \mathbb{O}_\mathbb{C}^2 \xhookrightarrow{j} \mathbb{P}_\mathbb{O}^2$$

$$\begin{pmatrix} x_1 \\ x_2 \end{pmatrix} \mapsto \left[\begin{pmatrix} 1 & x_2 & \widetilde{x_1} \\ \widetilde{x_2} & |x_2|^2 & \widetilde{x_2}\widetilde{x_1} \\ x_1 & x_1 x_2 & |x_1|^2 \end{pmatrix} \right]. \qquad (83)$$

Indeed, it is easily checked that $j(\mathbb{O}_\mathbb{C}^2)$ is the Zariski open subset of $\mathbb{P}_\mathbb{O}^2$ consisting of all the matrices $[a] \in \mathbb{P}_\mathbb{O}^2$ whose $(1, 1)$-entry is non-zero.

4 Boundary Components

The aim of this section is to define and study the boundary components of a Hermitian symmetric manifold of non-compact type, or, equivalently, of a bounded symmetric domain, see Sect. 2.5.

Let $D \xhookrightarrow{i_{HC}} \mathbb{C}^N$ be a bounded symmetric domain in its Harish–Chandra embedding and let $D \xhookrightarrow{i_{HC}} \mathbb{C}^N \xhookrightarrow{j} D^c$ be the Borel embedding into the compact dual D^c of D (see Theorem 2.22). Denote by $\overline{D}$ the closure of D inside $\mathbb{C}^N$ with respect to the Euclidean topology.

Definition 4.1. Consider the following equivalence relation $\sim$ on $\overline{D}$: $p \sim q$ if and only if there exist holomorphic maps $\lambda_1, \cdots, \lambda_m : \Delta = \{z \in \mathbb{C} : |z| < 1\} \to \overline{D}$ (for some $m \in \mathbb{N}$) such that

- $\lambda_1(0) = p$ and $\lambda_m(0) = q$;
- $\text{Im}\,\lambda_i \cup \text{Im}\,\lambda_{i+1} \neq \emptyset$ for any $1 \leq i \leq m-1$.

A *boundary component* F of D is an equivalence class for the above equivalence relation $\sim$ on $\overline{D}$.

Given two boundary components F_1 and F_2 of D, we say that F_1 dominates F_2 (and we write $F_2 \leq F_1$) if $F_2 \subseteq \overline{F_1}$.

In other words, two points p and q of $\overline{D}$ belong to the same boundary component if they can be connected by a finite chain of holomorphic disks contained in $\overline{D}$. Note that D is always a boundary component of itself and that for every boundary component F of D it holds that $F \leq D$.

Theorem 4.2. *Let $D \subset \mathbb{C}^N$ be a bounded symmetric domain in its Harish–Chandra embedding. Then*

(i) $\overline{D} = \coprod_{F \leq D} F$ *and* $G = \text{Hol}(D)^o$ *preserves this decomposition.*
(ii) *Let $F \leq D$ and denote by $\langle F \rangle$ be the smallest linear subspace of $\mathbb{C}^N$ containing F, by $\overline{F}$ be the Euclidean closure of F inside $\mathbb{C}^N$ (or equivalently inside $\langle F \rangle$) and by F^c the Zariski closure of F inside D^c. Then F is a Hermitian symmetric manifold of non-compact type such that*

- $F \subset \langle F \rangle$ *is the Harish–Chandra embedding of F;*
- $F \subset \langle F \rangle \subset \overline{F}$ *is the Borel embedding of F.*

Moreover the following diagram of inclusions is Cartesian

$$\begin{array}{ccccccc} F & \hookrightarrow & \overline{F} & \hookrightarrow & \langle F \rangle & \hookrightarrow & F^c \\ \downarrow & & \downarrow & & \downarrow & & \downarrow \\ D & \hookrightarrow & \overline{D} & \hookrightarrow & \mathbb{C}^N & \hookrightarrow & D^c \end{array} \qquad (84)$$

(iii) *If $F \leq D$ and $F' \leq F$ then $F' \leq D$.*
(iv) *If $D = D_1 \times \cdots \times D_r$ is the decomposition of D into irreducible bounded symmetric domains, then the boundary components of D are the product of the boundary components of the D_i's.*

Proof. See [AMRT10, Chap. III, Thm. 3.3]. □

Remark 4.3. It has been proved by Bott–Korányi (see [KW65, §3]) that the union of the zero-dimensional boundary components of a bounded symmetric domain $D \subset \mathbb{C}^N$ is the Bergman–Silov boundary of D, i.e. the smallest closed subset of the boundary $\partial D := \overline{D} \setminus D$ on which the absolute value of any function continuous on $\overline{D}$ and holomorphic on D achieves its maximum.

Hermitian Symmetric Manifolds

4.1 The Normalizer Subgroup of a Boundary Component

The aim of this subsection is to study the normalizer subgroup of a boundary component F of D.

Definition 4.4. The *normalizer* subgroup of a boundary component $F \leq D$ is the subgroup
$$N(F) := \{g \in G = \mathrm{Hol}(D)^o : gF = F\} \subseteq G.$$

We can classify the boundary components of D in terms of their normalizer subgroups.

Theorem 4.5. *Let $D = D_1 \times \cdots \times D_s$ the decomposition of D into its irreducible bounded symmetric domains and let $\mathrm{Hol}(D)^o = G = G_1 \times \cdots \times G_s = \mathrm{Hol}(D_1)^o \times \cdots \times \mathrm{Hol}(D_s)^o$ the associated decomposition of G into its simple factors. Then there is a bijection*

$$\{\text{Boundary components } F \leq D\} \xrightarrow{\cong} \left\{ \begin{array}{l} \text{Subgroups } P_1 \times \cdots \times P_s \subseteq G_1 \times \cdots \times G_s \text{ such that} \\ P_i = G_i \text{ or } P_i \text{ is a maximal parabolic subgroup of } G_i \end{array} \right\}$$

$$F \mapsto N(F).$$

Proof. See [AMRT10, Chap. III, Prop. 3.9]. □

We want now to take a closer look at the structure of the normalizer subgroup associated to a boundary component of D. We will need the following technical result.

Lemma 4.6. *Let F be a boundary component of D and fix a base point $o \in D$. Then there exists a unique pair*

$$\begin{aligned} f_F &: \Delta \to D, \\ \phi_F &: \mathbb{S}^1 \times \mathrm{SL}_2(\mathbb{R}) \to G = \mathrm{Hol}(D)^o, \end{aligned} \tag{85}$$

such that

(i) *f_F is a symmetric morphism (in the sense of Remark 2.17) such that $f_F(0) = o \in D$ and $o_F := f_F(1) := \lim_{z \to 1} f_F(z) \in F$;*
(ii) *ϕ_F is a morphism of Lie groups such that*

$$h_o(e^{i\theta}) := \phi_F\left(e^{i\theta}, \begin{pmatrix} \cos\theta & \sin\theta \\ -\sin\theta & \cos\theta \end{pmatrix}\right)$$

belongs to $K = \mathrm{Stab}(o) \subset G$ and it acts on $T_o D$ as multiplication by $e^{2i\theta}$.

(iii) f_F is equivariant with respect to the morphism ϕ_F and the natural actions of G on D and of $\mathbb{S}^1 \times \mathrm{SL}_2(\mathbb{R})$ on Δ via

$$[\mathbb{S}^1 \times SL_2(\mathbb{R})] \times \Delta \longrightarrow \Delta$$

$$\left(\left[e^{i\theta}, \begin{pmatrix} a & b \\ c & d \end{pmatrix}\right], z\right) \mapsto \frac{[i(a+d) - (b-c)]z + [i(a-d) + (b+c)]}{[i(a-d) - (b+c)]z + [i(a+d) + (b-c)]}.$$

Proof. See [AMRT10, Chap. III, Thm. 3.3(v) and Thm. 3.7]. □

Remark 4.7. (i) Since $f_F : \Delta \to D$ is a symmetric morphism, it extends uniquely to a morphism between their Harish–Chandra and Borel embeddings (see [AMRT10, Chap. III, Sec. 2.2])

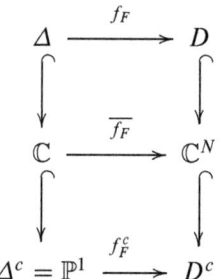

In particular $f_F(1) = \overline{f_F}(1) \in \overline{D}$.

(ii) For any non-Euclidean Hermitian symmetric domain M with a fixed base point o there exists a unique morphism $u_o : \mathbb{S}^1 \to G = \mathrm{Aut}(M)^o$ such that $\mathrm{Im}\, u_o \subset K = \mathrm{Stab}(o) \subset G$ and $u_o(e^{i\theta})$ induces the multiplication by $e^{i\theta}$ on $T_o M$ (see [AMRT10, Chap. III, Sec. 2.1]). Therefore, part (ii) is equivalent to saying that $h_o^2 = u_o$.

(iii) The action of $\mathrm{SL}_2(\mathbb{R})$ on Δ in part (iii) is equivalent to the action of $\mathrm{SL}_2(\mathbb{R})$ on the upper half space $\mathcal{H}$ via Moëbius transformations (see Example 2.6(2)) using the Cayley isomorphism $\mathcal{H} \cong \Delta$ of (11).

The connected component of the normalizer subgroup of a boundary component of D admits the following decomposition, know as the *5-term decomposition*.

Theorem 4.8. *Let F be a boundary component of D and let $N(F)$ its associated normalizer subgroup. Consider the one-parameter subgroup of $G = \mathrm{Hol}(D)^o$*

$$w_F : \mathbb{G}_m \longrightarrow G$$

$$t \mapsto \phi_F\left(1, \begin{pmatrix} t & 0 \\ 0 & t^{-1} \end{pmatrix}\right). \tag{86}$$

(i) The normalizer subgroup $N(F)$ of F is equal to

$$N(F) = \{g \in G : \lim_{t \to 0} w_F(t) \cdot g \cdot w_F(t)^{-1} \text{ exists}\}.$$

(ii) The connected component $N(F)^o$ of $N(F)$ is equal to the semidirect product

$$N(F)^o = Z(w_F)^o \ltimes W(F),$$

where:

- $W(F)$ is the unipotent radical of $N(F)^o$ and it is equal to

$$W(F) := \{g \in G : \lim_{t \to 0} w_F(t) \cdot g \cdot w_F(t)^{-1} = 1 \in G\};$$

- $Z(w_F)^o$ is a Levi subgroup of $N(F)^o$ and it is the connected component of the centralizer $Z(w_F)$ of w_F, i.e.

$$Z(w_F) := \{g \in G : w_F(t) \cdot g \cdot w_F(t)^{-1} = g \text{ for any } t \in \mathbb{G}_m\}.$$

(iii) $W(F)$ is a two-step unipotent group which is given as an extension of two abelian groups

$$0 \to U(F) \to W(F) \to V(F) \to 0,$$

where $U(F)$ is the center of $W(F)$ and $V(F) := W(F)/U(F)$.

(iv) $Z(w_F)^o$ is a reductive group which is equal to the product modulo finite subgroups

$$Z(w_F)^o = G_h(F) \cdot G_l(F) \cdot M(F),$$

where

- $M(F)$ is compact and semisimple;
- $G_h(F)$ is semisimple and it satisfies $G_h(F)/Z_{G_h(F)} = \operatorname{Aut}(F)^o$;
- $G_l(F)$ is reductive without compact factors.

Proof. See [AMRT10, Chap. III. Thm. 3.7, Thm. 3.10, §4.1]. □

4.2 The Decomposition of D Along a Boundary Component

The aim of this subsection is to deduce from the 5-term decomposition of the normalizer $N(F)$ of a boundary component F of D (see Theorem 4.8) a decomposition of D along F.

Proposition 4.9. *Let D be a bounded symmetric domain and fix a boundary component $F \leq D$. Then $N(F)^o$ acts transitively on D.*

Proof. See [AMRT10, Chap. III, §4.3]. □

Recall that, fixing a base point $o \in D$, D is diffeomorphic to G/K where $G = \text{Hol}(D)^o$ and $K = \text{Stab}(o) \subset G$ is a maximal compact subgroup (see Theorem 2.11). Using this and the above Proposition 4.9, we have a diffeomorphism

$$D \cong N(F)^o/(K \cap N(F)^o). \tag{87}$$

From the 5-term decomposition of $N(F)^o$ (see Theorem 4.8), it follows that

$$K \cap N(F)^o = K \cap Z(w_F) = K_l(F) \cdot K_h(F) \cdot M(F) \subset G_l(F) \cdot G_h(F) \cdot M(F) = Z(w_F), \tag{88}$$

where $K_l(F) \subset G_l(F)$ and $K_h(F) \subset G_h(F)$ are maximal compact subgroups. Substituting (88) into (87), we get the diffeomorphism

$$D \xrightarrow{\cong} \frac{N(F)^o}{K \cap N(F)^o} = \frac{[G_h(F) \cdot G_l(F) \cdot M(F)] \ltimes W(F)}{K_h(F) \cdot K_l(F) \cdot M(F)} = \frac{G_h(F)}{K_h(F)} \times \frac{G_l(F)}{K_l(F)} \times W(F). \tag{89}$$

In order to get a better description of the above diffeomorphism, we will now describe more geometrically the right hand side of (89).

Theorem 4.10. *Notations as above.*

(i) *Under the natural action of $G_h(F) \subset G$ on $\overline{D} \subset \mathbb{C}^N$, the orbit of the point $o_F := f_F(1)$ is equal to $F \subset \overline{D}$ and its stabilizer is equal to $K_h(F)$. Therefore, we have a diffeomorphism*

$$\frac{G_h(F)}{K_h(F)} \cong F. \tag{90}$$

(ii) *Under the action of $G_l(F) \subset N(F)^o$ on $U(F)$ by conjugation, the orbit of the point $\Omega_F := \phi_F\left(1, \begin{pmatrix} 1 & 1 \\ 0 & 1 \end{pmatrix}\right) \in U(F)$ is an open cone $C(F) \subset U(F)$ and its stabilizer is equal to $K_l(F)$. Therefore, we have a diffeomorphism*

$$\frac{G_l(F)}{K_l(F)} \cong C(F). \tag{91}$$

Proof. Part (i) follows from [AMRT10, Chap. III, Thm. 3.10 and Lemma 4.6]. Part (ii) follows from [AMRT10, Chap. III, Thm. 4.1]. □

Remark 4.11. Let $o \in D \subseteq \mathfrak{p}_+$ be a bounded symmetric domain (with a fixed base point o) in its Harish–Chandra embedding (see Theorem 2.22). Consider the Jordan triple product $\{.,.,.\}$ of (15) with respect to which $(\mathfrak{p}_+,\{.,.,.\})$ is a Hermitian positive JTS (see Sect. 2.7). Then there is a bijection (see [Sat80, Chap. III, §8, Rmk. 2])

$$\{\text{Boundary components of } D\} \xrightarrow{\cong} \{\text{Tripotents of } (\mathfrak{p}_+,\{.,.,.\})\} \qquad (92)$$
$$F \longmapsto o_F$$

where a tripotent (or idempotent) of $(\mathfrak{p}_+,\{.,.,.\})$ is an element $e \in \mathfrak{p}_+$ such that $\{e,e,e\} = e$. For a detailed study of the tripotents of a Jordan triple system, we refer the reader to [FKKLR00, Chap. V, Part V].

Using the above Theorem 4.10, the diffeomorphism (89) can be written as

$$D \xrightarrow{\cong} F \times C(F) \times W(F) \qquad (93)$$
$$x \mapsto (\pi_F(x), \Phi_F(x), w(x)).$$

The above smooth maps w, π_F and Φ_F can be described explicitly as it follows. By (87) and (88), we can write $x \in D$ as

$$x = g_h g_l w \cdot o,$$

for some unique $w \in W(F)$ and for some $g_h \in G_h(F)$ (resp. $g_l \in G_l(F)$) which is unique up to multiplication by an element of $K_h(F)$ (resp. $K_l(F)$). Then we have that (see [AMRT10, Chap. III, §4.3]):

$$\begin{cases} w(x) := w \in W(F), \\ \pi_F(x) := g_h \cdot o_F \in F, \\ \Phi_F(x) := g_l \cdot \Omega_F \in C(F). \end{cases} \qquad (94)$$

The maps π_F and Φ_F are closely related as we are now going to explain. Recall that D embeds as an open subset in its compact dual D^c via the Borel embedding and that D^c is a homogeneous projective variety with respect to the action of the complexification $G_\mathbb{C}$ of $G = \text{Hol}(D)^o$ (see Theorem 2.22). We will denote by $U(F)_\mathbb{C} \subset G_\mathbb{C}$ the complexification of $U(F) \subset N(F)^o \subset G$.

Definition 4.12. Notations as above. Denote by $D(F)$ the analytic open subset of D^c:

$$D \subset D(F) := U(F)_\mathbb{C} \cdot D = \bigcup_{g \in U(F)_\mathbb{C}} g \cdot D \subset D^c.$$

The open subset $D(F)$ admits the following Lie-theoretic description.

Lemma 4.13. *We have a diffeomorphism*

$$D(F) \cong \frac{N(F)^o \cdot U(F)_{\mathbb{C}}}{K_h(F) \cdot G_l(F) \cdot M(F)},$$

where $N(F)^o \cdot U(F)_{\mathbb{C}}$ is the subgroup of $G_{\mathbb{C}}$ generated by $N(F)^o$ and $U(F)_{\mathbb{C}}$.

Proof. The group $N(F)^o \cdot U(F)_{\mathbb{C}}$ acts transitively on $D(F)$ by Proposition 4.9 and Definition 4.12. The stabilizer of the point

$$o_F^o := s_o(o_F) = f_F(-1) \in D(F)$$

is equal to $K_h(F) \cdot G_l(F) \cdot M(F)$ by Ash et al. [AMRT10, Chap. III, Lemma 4.6]. Hence, we get the desired diffeomorphism. □

Using the diffeomorphism in Lemma 4.13, we can define two smooth and surjective maps

$$\begin{cases} \tilde{\pi}_F : D(F) \cong \dfrac{N(F)^o \cdot U(F)_{\mathbb{C}}}{K_h(F) \cdot G_l(F) \cdot M(F)} \longrightarrow \dfrac{G_h(F)}{K_h(F)} \cong F, \\ \tilde{\Phi}_F : D(F) \cong \dfrac{N(F)^o \cdot U(F)_{\mathbb{C}}}{K_h(F) \cdot G_l(F) \cdot M(F)} \longrightarrow \dfrac{N(F)^o \cdot U(F)_{\mathbb{C}}}{N(F)^o} \cong U(F), \end{cases} \quad (95)$$

where the last diffeomorphism is obtained by projecting onto $i \cdot U(F) \subset U(F)_{\mathbb{C}}$.

Theorem 4.14. *Notations as above.*

(i) The smooth maps $\tilde{\pi}_F$ and $\tilde{\Phi}_F$ fit into the following commutative diagram

$$\begin{array}{ccc} C(F) & \hookrightarrow & U(F) \\ \Phi_F \uparrow & \square & \uparrow \tilde{\Phi}_F \\ D & \hookrightarrow & D(F) \\ & \pi_F \searrow & \downarrow \tilde{\pi}_F \\ & & F \end{array}$$

where the upper square is Cartesian.

(ii) The smooth map $\tilde{\pi}_F$ factors as

$$D(F) \xrightarrow{\pi'_F} D(F)' := D(F)/U(F)_{\mathbb{C}} \xrightarrow{p_F} F = D(F)'/V(F),$$

in such a way that π'_F is a trivial holomorphic $U(F)_{\mathbb{C}}$-torsor and p_F is a smooth $V(F)$-torsor and, at the same time, a trivial complex vector bundle. In *particular, we have a diffeomorphism*

$$D(F) \cong U(F)_{\mathbb{C}} \times \mathbb{C}^k \times F, \qquad (96)$$

for some $k \in \mathbb{N}$.
(iii) Under the diffeomorphism (96), the smooth map $\tilde{\Phi}_F$ can be written as

$$\tilde{\Phi}_F : D(F) \cong U(F)_{\mathbb{C}} \times \mathbb{C}^k \times F \longrightarrow U(F)$$
$$(x, y, z) \mapsto \operatorname{Im} x - h_z(y, y),$$

for some bilinear symmetric form $h_z : \mathbb{C}^k \times \mathbb{C}^k \to U(F)$ varying smoothly with $z \in F$.

Proof. See [AMRT10, Chap. III, §4.3]. □

From the above Theorem, we deduce the following presentation of D as a *Siegel domain of the third kind*.

Corollary 4.15. *Notations as above. We have a diffeomorphism*

$$D \cong \{(x, y, z) \in U(F)_{\mathbb{C}} \times \mathbb{C}^k \times F : \operatorname{Im} x - h_z(y, y) \in C(F)\}.$$

4.3 Symmetric Cones and Euclidean Jordan Algebras

The cone $C(F)$ associated to a boundary component $F \leq D$ (see Theorem 4.10(ii)) belongs to a special class of cones, namely the symmetric cones, that we now introduce.

Definition 4.16. Let V be a real (finite-dimensional) vector space endowed with a scalar product $\langle, \rangle$ (i.e. an Euclidean space). An open (pointed and convex) cone $C \subset V$ is said to be:

(i) *homogeneous* if the group of automorphisms of C:

$$\mathbb{G}(C) := \{g \in \operatorname{GL}(V) : g \cdot C = C\} \subset \operatorname{GL}(V)$$

acts transitively on C.
(ii) *symmetric* if it is homogeneous and self-dual, i.e. C is equal to the its dual cone

$$C^* = \{x \in V : \langle x, y \rangle > 0 \text{ for any } y \in \overline{C}\}.$$

Some basic properties of homogeneous and symmetric cones are contained in the following

Theorem 4.17. *Let* $C \subset (V, \langle, \rangle)$ *be a homogeneous cone and fix a base point* $o \in C$.

(i) *The stabilizer subgroup of o*

$$\mathbb{G}(C)_o := \{g \in \mathbb{G}(C) : g(o) = o\} \subset \mathbb{G}(C)$$

is a maximal compact subgroup of $\mathbb{G}(C)$ and, conversely, every maximal compact subgroup of $\mathbb{G}(\Omega)$ is the stabilizer subgroup of some point of Ω

(ii) *We have a diffeomorphism*

$$\frac{\mathbb{G}(C)}{\mathbb{G}(C)_o} \cong C.$$

(iii) *C is symmetric if and only if $\mathbb{G}(C)$ is equal to its dual group*

$$\mathbb{G}(C)^* = \{g^* : g \in \mathbb{G}(C)\},$$

where g^* denote the adjoint of the element $g \in \mathrm{GL}(V)$ with respect to the scalar product $\langle, \rangle$. In particular, in this case, $\mathbb{G}(C)$ is a reductive Lie group.

Proof. For part (i), see [Sat80, Chap. I, Prop. 8.4]. For part (ii), see [FK94, Chap. I, §4]. For part (iii), see [Sat80, Chap. I, Lemma 8.3]. □

Remark 4.18. It can be shown that symmetric cones are Riemannian symmetric manifolds in the sense of Remark 2.5(iv); see [FK94, Chap. I, §4] for a proof.

It turns out (see [FK94, Prop. III.4.5]) that any symmetric cone decomposes uniquely as the product of irreducible symmetric cones, defined as it follows.

Definition 4.19. A symmetric cone $\Omega \subset$ is said to be *irreducible* if and only if there does not exist a non-trivial decomposition $V = V_1 \oplus V_2$ and two symmetric cones $\Omega_1 \subset V_1$ and $\Omega_2 \subset V_2$ such that $\Omega = \Omega_1 + \Omega_2$ (in this case, we say that Ω is the product of Ω_1 and Ω_2).

Symmetric cones can be classified via Euclidean Jordan algebras, which we now introduce.

Definition 4.20. A **Jordan algebra** over a field F is a (finite-dimensional) algebra $(A, \circ)$ over F such that

(J1) $x \circ y = y \circ x$ for any $x, y \in A$;
(J2) T_x and $T_{x \circ x}$ commutes, where for any $x \in A$ we denote by T_x the endomorphism of A given by

$$T_x : A \longrightarrow A,$$
$$y \mapsto T_x(y) := x \circ y.$$

A Jordan F-algebra $(A, \circ)$ is said to be

(i) *semisimple* if the trace form

$$\tau : A \times A \longrightarrow F, \qquad (97)$$
$$(x, y) \mapsto \tau(x, y) := \operatorname{tr}(T_{x \circ y}).$$

is non-degenerate.
(ii) *simple* if τ is not identically zero and A does not contain proper ideals, i.e. proper subvector spaces $I \subset A$ such that for any $x \in I$ and $y \in A$ we have that $x \circ y \in I$.
(iii) *Euclidean* if $F = \mathbb{R}$ and the trace form τ is positive definite.

The following properties of Jordan algebras follow quite easily from the axioms (J1) and (J2).

Lemma 4.21. *Let $(A, \circ)$ be a Jordan F-algebra. Then*

(i) *$(A, \circ)$ is power-associative, i.e. if we define inductively $x^p := x \circ x^{p-1}$ (for any $p \in \mathbb{Z}_{>0}$) then we have that $x^p \circ x^q = x^{p+q}$ for any $p, q \in \mathbb{Z}_{>0}$.*
(ii) *For any $x \in A$ and any $p, q \in \mathbb{Z}_{>0}$ the endomorphisms T_{x^p} and T_{x^q} commute.*
(iii) *The trace form τ is associative, i.e.*

$$\tau(x \circ y, z) = \tau(x, y \circ z),$$

for any $x, y, z \in A$. In particular, if $(A, \circ)$ is semisimple then, for any $y \in A$, the endomorphism T_y is self-adjoint with respect to τ.
(iv) *If $(A, \circ)$ is semisimple then $(A, \circ)$ has a unique unit element $e \in A$, i.e. an element $e \in A$ such that $e \circ x = x$ for any $x \in A$.*

Proof. For (i) and (ii), see [FK94, Prop. II.1.2]. For (iii), see [FK94, Prop. 2.4.3]. Part (iv): since τ is non-degenerate, there exists a unique $e \in A$ such that $\tau(e, x) = \operatorname{tr} T_x$ for any $x \in A$. Using the associativity of τ, we get (for any $x, y \in A$)

$$\tau(x, e \circ y) = \tau(x \circ y, e) = \operatorname{tr} T_{x \circ y} = \tau(x, y),$$

which implies (again by the non-degeneracy of τ) that $e \circ y = y$, q.e.d. □

Example 4.22. (1) Let $(A, \cdot)$ be an associative F-algebra. Then A becomes a Jordan algebra with respect to the Jordan product

$$x \circ y := (x \cdot y + y \cdot x).$$

(2) Let W be a F-vector space and let B be a symmetric bilinear form on $W \times W$. Then $A = F \oplus W$ becomes a Jordan algebra with respect to the Jordan product

$$(\lambda, u) \circ_B (\mu, v) := (\lambda \mu + B(u, v), \lambda v + \mu u). \qquad (98)$$

It is easily checked that the Jordan algebra $(F \oplus W, \circ_B)$ is semisimple if and only if B is non-degenerate and that it is Euclidean if and only if $F = \mathbb{R}$ and B is positive definite.

(3) Let D be equal to $\mathbb{R}, \mathbb{C}$ or $\mathbb{H}$ and denote by $x \mapsto \bar{x}$ the natural involution. For any $n \geq 2$, the real vector space of Hermitian matrices of order n with entries in D

$$\text{Herm}_n(D) := \{M \in M_{n,n}(D) : \overline{M}^t = M\}$$

becomes a Euclidean Jordan algebra with respect to the Jordan product (see [FK94, Chap. 5, §2])

$$M_1 \circ M_2 = \frac{1}{2}(M_1 M_2 + M_2 M_1). \tag{99}$$

If D is equal to the algebra of octonions $\mathbb{O}$, then $\text{Herm}_n(\mathbb{O})$ with the product (99) is an Euclidean Jordan algebra if $m \leq 3$ (see [FK94, Chap. 5, §2]). In particular, $\text{Herm}_3(\mathbb{O})$ is an Euclidean Jordan algebra of dimension 27, known as the Albert algebra.

The Jordan algebras $\text{Herm}_2(D)$ for $D = \mathbb{R}, \mathbb{C}, \mathbb{H}$ or $\mathbb{O}$ are isomorphic to the Jordan algebras associated to a suitable bilinear symmetric form as in Example 4.22; more precisely, we have that

$$\begin{cases} \text{Herm}_2(\mathbb{R}) \cong (\mathbb{R} \oplus \mathbb{R}^2, \circ_Q) \\ \text{Herm}_2(\mathbb{C}) \cong (\mathbb{R} \oplus \mathbb{R}^3, \circ_Q) \\ \text{Herm}_2(\mathbb{H}) \cong (\mathbb{R} \oplus \mathbb{R}^5, \circ_Q) \\ \text{Herm}_2(\mathbb{O}) \cong (\mathbb{R} \oplus \mathbb{R}^9, \circ_Q) \end{cases} \tag{100}$$

where $\circ_Q$ is defined in Example 98 with respect to the positive definite symmetric bilinear form B on the suitable vector space.

(4) Let $(A, \circ)$ be a Jordan algebra over F. Then A becomes a Jordan triple system with respect to the triple product $\{.,.,.\}_\circ$ defined by (see [Sat80, Chap. 1,§6])

$$\{x, y, z\}_\circ := (x \circ y) \circ z + (z \circ y) \circ x - (x \circ z) \circ y.$$

It turns out that $x \square y = T_{x \circ y} + [T_x, T_y]$ for any $x, y \in A$ which implies that the trace form of the Jordan algebra $(A, \circ)$ as defined in (97) coincides with the trace form of the JTS $(A, \{.,.,.\}_\circ)$ as defined in (14). In particular, $(A, \circ)$ is semisimple if and only if $(A, \{.,.,.\}_\circ)$ is semisimple.

If $(A, \circ)$ is a Jordan algebra over $\mathbb{R}$, then $A_\mathbb{C} := A \otimes_\mathbb{R} \mathbb{C}$ becomes a Hermitian JTS with respect to the triple product

$$\{x, y, z\}'_\circ := (x \circ \bar{y}) \circ z + (z \circ \bar{y}) \circ x - (x \circ z) \circ \bar{y},$$

Table 8 Simple Euclidean Jordan algebras

Real vector space V	Jordan product $\circ$
$\mathbb{R} \times \mathbb{R}^n$	$\underline{x} \circ \underline{y} = (x_0 y_0 + \sum_{i=1}^n x_i y_i, x_0 y_1 + y_0 x_1, \cdots, x_0 y_n + y_0 x_n)$
$\mathrm{Herm}_n(\mathbb{R})$ $(n \geq 3)$	$A \circ B = \frac{1}{2}(AB + BA)$
$\mathrm{Herm}_n(\mathbb{C})$ $(n \geq 3)$	$A \circ B = \frac{1}{2}(AB + BA)$
$\mathrm{Herm}_n(\mathbb{H})$ $(n \geq 3)$	$A \circ B = \frac{1}{2}(AB + BA)$
$\mathrm{Herm}_3(\mathbb{O})$	$A \circ B = \frac{1}{2}(AB + BA)$

where $y \mapsto \overline{y}$ is the complex conjugation corresponding to the real form $A \subset A_{\mathbb{C}}$ and the Jordan product $\circ$ is extended linearly to $A_{\mathbb{C}}$. Then $(A, \circ)$ is Euclidean if and only if $(A_{\mathbb{C}}, \{.,.,.\}'_\circ)$ is a positive Hermitian JTS.

Observe that simple Jordan algebras are semisimple. Moreover, the direct sum of simple Jordan algebras is semisimple, where the direct sum of two Jordan algebras $(A_1, \circ_1)$ and $(A_2, \circ_2)$ is the vector space $A := A_1 \oplus A_2$ endowed with the component-wise Jordan product $(x_1, x_2) \circ (y_1, y_2) := (x_1 \circ y_1, x_1 \circ y_2)$. Conversely, we have the following decomposition theorem.

Proposition 4.23. *Any semisimple (resp. Euclidean) Jordan algebra decomposes uniquely as the product of simple (resp. Euclidean and simple) Jordan algebras.*

Sketch of the Proof. Let $(A, \circ)$ be a semisimple (resp. Euclidean) Jordan algebra. If A is not simple, then there exists a proper ideal $I \subset A$. Consider the orthogonal complement of I

$$I^\perp := \{x \in A : \tau(x, y) = 0 \text{ for any } y \in I\}.$$

It is possible to prove (see [FK94, Prop. III.4.4]) that

(i) $I^\perp$ is an ideal of A;
(ii) I and $I^\perp$ are semisimple (resp. Euclidean) Jordan algebras;
(iii) $A = I \oplus I^\perp$ as Jordan algebras.

Iterating this construction for I and $I^\perp$, we get the existence of the decomposition. For the unicity, see loc. cit. □

Simple Euclidean Jordan algebras were classified by Jordan–Neumann–Wigner (see [FK94, Chap. V] and the references therein).

Theorem 4.24. *Every simple Euclidean Jordan algebra is isomorphic to one of the following Jordan algebras (see Table 8)*

(i) $(\mathbb{R} \oplus \mathbb{R}^n, \circ_Q)$ *where Q is the standard scalar product on $\mathbb{R}^n$;*
(ii) $\mathrm{Herm}_n(\mathbb{R})$ *for $n \geq 3$;*
(iii) $\mathrm{Herm}_n(\mathbb{C})$ *for $n \geq 3$;*
(iv) $\mathrm{Herm}_n(\mathbb{H})$ *for $n \geq 3$;*
(v) $\mathrm{Herm}_3(\mathbb{O})$ *(the Albert algebra).*

Remark 4.25. The complexification of a real Jordan algebra $(A, \circ)$ is the complex vector space $A^\mathbb{C} := A \otimes_\mathbb{R} \mathbb{C}$ endowed with the Jordan product $\circ^\mathbb{C}$ obtained by extending linearly the Jordan product $\circ$ on A. The complexification induces a bijection

$$\{\text{Euclidean Jordan algebras}\} \xrightarrow{\cong} \{\text{Semisimple Jordan } \mathbb{C}\text{-algebras}\} \tag{101}$$
$$(A, \circ) \mapsto (A^\mathbb{C}, \circ^\mathbb{C})$$

which preserves the decomposition into the product of simple Jordan algebras (see [FK94, Chap. VIII]).

We now explain the relationship between Euclidean Jordan algebras and symmetric cones, due to the work of Koecher and Vinberg.

Let $(V, \circ)$ be a Euclidean Jordan algebra and denote by $e \in V$ the unit element of V (see Lemma 4.21(iv)). Denote by V^* the set of invertible elements of V, i.e. the elements $x \in V$ for which there exists $y \in V$ such that $x \circ y = e$. Then we define an open cone inside V by

$$\Omega_{(V,\circ)} := \{x^2 : x \in V^*\} = \{x : x \in V^*\}^o = \{x \in V : T_x > 0\}, \tag{102}$$

where $\{.,.,.\}^o$ denotes the connected component containing the identity $e \in V$. Indeed, $\Omega_{(V,\circ)}$ is a symmetric cone with respect to the positive definite form $\langle,\rangle := \tau(,)$ (see [FK94, Chap. III, §2]).

Conversely, let $\Omega \subset V$ be a symmetric cone with respect to a scalar product $\langle,\rangle$ on V and choose a base point $e \in \Omega$. Let $\mathfrak{g}$ be the Lie algebra of $\mathbb{G}(\Omega)$, $\mathfrak{k}$ the Lie algebra of the maximal compact subgroup $\mathbb{G}(\Omega)_e \subset \mathbb{G}(\Omega)$ and $\mathfrak{g} = \mathfrak{k} \oplus \mathfrak{p}$ be the associated Cartan decomposition. The action of $\mathbb{G}(\Omega)$ on V induces an action of $\mathfrak{g}$ on V. Clearly an element $X \in \mathfrak{g}$ belongs to $\mathfrak{k}$ if and only if $X \cdot e = e$. Therefore, the map $\mathfrak{p} \to V$ sending X into $X \cdot e$ is a bijection; hence, for any $x \in V$, there exists a unique $L_x \in \mathfrak{k}$ such that $L_x \cdot e = x$. Define a product $\circ_\Omega$ on V by

$$x \circ_\Omega y := L_x \cdot y. \tag{103}$$

The pair $(V, \circ_\Omega)$ is an Euclidean Jordan algebra with unit element e (see [FK94, Chap. III,§3]).

Theorem 4.26. *There is a bijection*

$$\{\text{Euclidean Jordan algebras}\} \xleftrightarrow{\cong} \{\text{Symmetric cones}\}$$
$$(V, \circ) \longrightarrow \Omega_{(V,\circ)} \subset V \tag{104}$$
$$(V, \circ_\Omega) \longleftarrow \Omega \subset V$$

preserving the decomposition of Euclidean Jordan algebras into simple ones and the decomposition of symmetric cones into irreducible ones.

Proof. See [FK94, Chap. III] or [Sat80, Chap. I, §8]. □

Remark 4.27. The bijection of Theorem 4.26 becomes an equivalence of categories if the two sets are endowed with the following morphisms (see [Sat80, Chap. I, §9]):

(i) A *unital Jordan algebra* homomorphism between two Euclidean Jordan algebras $(V, \circ)$ and $(V', \circ')$ is a linear map $f : V \to V'$ such that

$$f(x \circ y) = f(x) \circ' f(y) \qquad \text{for any } x, y \in V,$$
$$f(e) = e',$$

where e (resp. e') is the unit element of $(V, \circ)$ (resp. $(V', \circ')$).

(ii) An *equivariant* morphism between two symmetric cones $\Omega \subset (V, \langle, \rangle)$ and $\Omega' \subset (V', \langle, \rangle')$ is a linear map $\phi : V \to V'$ sending Ω into Ω' and such that there exists a morphism of Lie algebras $\rho : \text{Lie}\,\mathbb{G}(\Omega) \to \text{Lie}\,\mathbb{G}(\Omega')$ satisfying

$$\phi(T \cdot x) = \rho(T) \cdot \phi(x) \qquad \text{for any } x \in V \text{ and any } T \in \text{Lie}\,\mathbb{G}(\Omega),$$
$$\rho(T^t) = \rho(T)^t$$

where ^{t} denotes the transpose with respect to either $\langle, \rangle$ or $\langle, \rangle'$.

As a consequence of the bijection between symmetric cones and Euclidean Jordan algebras in Theorem 4.26 and the classification of simple Euclidean Jordan algebra given in Theorem 4.24, we get the following classification of irreducible symmetric cones (see [Sat80, Chap. 1, §8]) (see Table 9).

Theorem 4.28. *Every irreducible symmetric cone is isomorphic to one of the following cones*

(i) $\mathcal{P}(1, n) := \{\underline{x} \in \mathbb{R} \oplus \mathbb{R}^n : x_0 > \sqrt{x_1^2 + \cdots + x_n^2}\} \subset \mathbb{R} \oplus \mathbb{R}^n$ *for* $n \geq 1$ *(the Lorentz or light cone);*

(ii) $\mathcal{P}_n(\mathbb{R}) = \text{Herm}_n^{>0}(\mathbb{R}) := \{M \in \text{Herm}_n(\mathbb{R}) : M > 0\} \subset \text{Herm}_n(\mathbb{R})$ *for* $n \geq 3$;

(iii) $\mathcal{P}_n(\mathbb{C}) = \text{Herm}_n^{>0}(\mathbb{C}) := \{M \in \text{Herm}_n(\mathbb{C}) : M > 0\} \subset \text{Herm}_n(\mathbb{C})$ *for* $n \geq 3$;

(iv) $\mathcal{P}_n(\mathbb{H}) = \text{Herm}_n^{>0}(\mathbb{H}) := \{M \in \text{Herm}_n(\mathbb{H}) : M > 0\} \subset \text{Herm}_n(\mathbb{H})$ *for* $n \geq 3$;

(v) $\mathcal{P}_3(\mathbb{O}) = \text{Herm}_3^{>0}(\mathbb{O}) := \{M \in \text{Herm}_3(\mathbb{O}) : M > 0\} \subset \text{Herm}_3(\mathbb{O})$.

The closure $\overline{\Omega}$ of a symmetric cone $\Omega \subset V$ can be decomposed into a disjoint union of boundaries components, which we are now going to define. Recall first that an *idempotent* of an Euclidean Jordan algebra $(V, \circ)$ is an element $e \in V$ such that $e \circ e = e$. The operator T_e (see Definition 4.20) is self-adjoint with respect to the positive definite scalar product τ given by the trace form (see Lemma 4.21(iii)) and its eigenvalues are 0, 1/2 and 1 (see [FK94, Prop. III.1.3]). Therefore, we get an orthogonal decomposition of V into eigenspaces

Table 9 Irreducible symmetric cones

Rank	Dimension	Cone
2	$n+1$	$\mathcal{P}(1,n) = \{\underline{x} \in \mathbb{R} \oplus \mathbb{R}^n : x_0 > \sqrt{x_1^2 + \cdots + x_n^2}\}$
n	$\binom{n+1}{2}$	$\mathcal{P}_n(\mathbb{R}) = \mathrm{Herm}_n^{\geq 0}(\mathbb{R})\ (n \geq 3)$
n	n^2	$\mathcal{P}_n(\mathbb{C}) = \mathrm{Herm}_n^{\geq 0}(\mathbb{C})\ (n \geq 3)$
n	$n(2n-1)$	$\mathcal{P}_n(\mathbb{H}) = \mathrm{Herm}_n^{\geq 0}(\mathbb{H})\ (n \geq 3)$
3	27	$\mathcal{P}_3(\mathbb{O}) = \mathrm{Herm}_3^{\geq 0}(\mathbb{O})$

$$V = V(e,1) \oplus V(c,1/2) \oplus V(c,0), \tag{105}$$

relative to, respectively, the eigenvalues 0, $1/2$ and 1.

Lemma 4.29. *For any idempotent $e \in (V, \circ)$, we have that $V(e,1)$ is an Euclidean Jordan subalgebra of $(V, \circ)$ such that $e \in V(e,1)$ is the identity element.*

Proof. See [FK94, Prop. IV.1.1]. □

Consider now the Euclidean Jordan algebra $(V, \circ_\Omega)$ corresponding to a symmetric cone $\Omega \subset V$ according to Theorem 4.26.

Definition 4.30. For an idempotent $e \in (V, \circ_\Omega)$, we define the *boundary component* of Ω associated to e as the symmetric cone

$$\Omega(e) := \Omega_{(V(c,1), \circ_\Omega)} \subset V(c,1)$$

corresponding to the Euclidean Jordan algebra $(V(c,1), \circ_\Omega)$ according to Theorem 4.26.

The closure $\overline{\Omega}$ of the symmetric cone Ω in V can be partitioned into the disjoint union of boundaries components as it follows.

Theorem 4.31. *Notations as before.*

(i) *For any idempotent $e \in (V, \circ_\Omega)$, the intersection of $\overline{\Omega} \subset V$ with the subspace $V(e,1) \subset V$ is equal to the closure $\overline{\Omega}(e)$. In particular, $\Omega(e)$ is contained in $\overline{\Omega}$.*

(ii) *We have that*

$$\overline{\Omega} = \coprod_e \Omega(e), \tag{106}$$

where the disjoint union varies over all the idempotents $e \in (V, \circ_\Omega)$ and, for any an idempotent e, the closure of $\Omega(e)$ is a disjoint union of boundary components.

(iii) *The group $G(\Omega)^o$ of automorphisms of Ω acts on $\overline{\Omega}$ by permuting its boundary components.*

Proof. See [AMRT10, Chap. II,§3]. □

Using (ii), we can introduce an order relation on the set of idempotents of $(V, \omega_\circ)$ by saying that $e \geq e'$ if and only if $\overline{\Omega}(e) \supseteq \Omega(e')$. Indeed, it turns out that $e \geq e'$ if and only if $e = e' + e''$ for a certain idempotent e'' such that $e' \circ_\Omega e'' = 0$.

Example 4.32. (i) For $F = \mathbb{R}, \mathbb{C}, \mathbb{H}, \mathbb{O}$ and $n \geq 2$ (with the convention that $n \leq 3$ if $F = \mathbb{O}$), consider the symmetric cone $\mathcal{P}_n(F) = \text{Herm}_n^{>0}(F) \subset \text{Herm}_n(F)$ (as in Theorem 4.28) which is associated to the Euclidean Jordan algebra $(\text{Herm}_n(F), \circ)$ of Example 4.22(4.22). Every idempotent of $\text{Herm}_n(F)$ is conjugate by $G(\mathcal{P}_n(F))^o$ to the idempotent

$$e_p := \begin{pmatrix} I_p & 0 \\ 0 & 0 \end{pmatrix}$$

for some $0 \leq p \leq n$. The eigenspaces (105) of T_{e_p} are given by

$$\begin{cases} V(e_p, 1) = \left\{ \begin{pmatrix} A & 0 \\ 0 & 0 \end{pmatrix} : A \in \text{Herm}_p(F) \right\}, \\ V(e_p, 1/2) = \left\{ \begin{pmatrix} 0 & B \\ \overline{B}^t & 0 \end{pmatrix} : B \in M_{p,n-p}(F) \right\}, \\ V(e_p, 0) = \left\{ \begin{pmatrix} 0 & 0 \\ 0 & C \end{pmatrix} : C \in \text{Herm}_{n-p}(F) \right\}. \end{cases}$$

The boundary component associated to e_p is equal to

$$\mathcal{P}_n(F)(e_p) = \left\{ \begin{pmatrix} A & 0 \\ 0 & 0 \end{pmatrix} : A \in \text{Herm}_p^{>0}(F) \right\} \subset \text{Herm}_n^{\geq 0}(F) = \overline{\mathcal{P}_n(F)}.$$

Note that the above idempotents $\{e_p\}$ are such that $0 = e_0 \leq \cdots \leq e_p \leq \cdots \leq e_n = I_n$.

(ii) Consider the Lorentz cone $\mathcal{P}(1, n)$ of Theorem 4.28 which is associated to the Euclidean Jordan algebra $(\mathbb{R} \oplus \mathbb{R}^n, \circ_B)$ of Example 4.22(4.22), where $B((x_1, \ldots, x_n)) := x_1^2 + \cdots + x_n^2$ is the standard quadratic form on $\mathbb{R}^n$. By abuse of notation, we will denote also by B the symmetric bilinear form associated to the quadratic form B. Every non-trivial idempotent (i.e. different from $(0, 0)$ and $(1, 0)$) is of the form

$$e_w = \left(\frac{1}{2}, w \right) \text{ where } B(w) = \frac{1}{4}.$$

The eigenspaces (105) of T_{e_w} are given by

$$\begin{cases} V(e_w, 1) = \mathbb{R} \cdot e_w, \\ V(e_w, 1/2) = \{(0, v) : B(v, w) = 0\}, \\ V(e_w, 0) = \mathbb{R} \cdot e_{-w}. \end{cases}$$

The boundary component associated to e_w is equal to

$$\mathcal{P}_{1,n}(e_w) = \mathbb{R}_{>0} \cdot e_w \subset \overline{\mathcal{P}_{1,n}}.$$

For the symmetric cone $C(F)$ associated to a boundary component F of a bounded symmetric domain D, as in Theorem 4.10(ii), we can explicitly describe its boundary components in terms of boundary components of D that dominates F.

Theorem 4.33. *Let D be a bounded symmetric domain and let $F \leq D$ be a boundary component.*

(i) *If $F \leq F'$ then $U(F) \supseteq U(F')$ and we have the equality $\overline{C(F')} = \overline{C(F)} \cap U(F') \subset U(F)$. Moreover, $C(F')$ is a boundary component of the symmetric cone $C(F)$.*

(ii) *There is an order-reversing bijection*

$$\{F' \leq D : F \leq F' \leq D\} \xrightarrow{\cong} \{\text{Boundary components of } C(F)\} \tag{107}$$

$$F' \mapsto C(F') \subset \overline{C(F)}.$$

Proof. See [AMRT10, Chap. III, Thm. 4.8]. □

4.4 Siegel Domains

The presentation of a bounded symmetric domain D as a Siegel domain of the third kind with respect to a given boundary component $F \leq D$ (see Corollary 4.15) assumes a nicer form when the boundary component F is a point, in which case it gives rise to a presentation of D as a *Siegel domain of the second type*, or simply a *Siegel domain* (following the terminology of [Sat80]). The aim of this subsection is to introduce and study Siegel domains.

Definition 4.34. Let Ω be an open (convex and pointed) cone in a real vector space U. Let V be a complex vector space and let $H : V \times V \to U_{\mathbb{C}}$ be a Hermitian map ($\mathbb{C}$-linear in the second variable and $\mathbb{C}$-antilinear in the first variable). Assume that H is Ω-positive, i.e.

$$H(v, v) \in \Omega \setminus \{0\} \text{ for any } 0 \neq v \in V.$$

The **Siegel domain** associated to (U, V, Ω, H) is given by

$$\mathcal{S} = \mathcal{S}(U, V, \Omega, H) := \{(u, v) \in U_{\mathbb{C}} \times V : \operatorname{Im} u - H(v, v) \in \Omega\} \subset U_{\mathbb{C}} \times V. \tag{108}$$

In the special case where $V = \{0\}$, then

$$\mathcal{S} = \mathcal{S}(U, \Omega) := \{u \in U_{\mathbb{C}} : \operatorname{Im} u \in \Omega\} \subset U_{\mathbb{C}} \tag{109}$$

is called a *Siegel domain of the first kind*, or a **tube domain**.

The following result is due to Pyateskii–Shapiro [PS69] (see also [Sat80, Chap. III, Prop. 6.1]).

Theorem 4.35. *Every Siegel domain is holomorphically equivalent to a bounded domain.*

There is a nice characterization (due to Satake) of the Siegel domains that are holomorphically equivalent to a bounded symmetric domain. In order to present such a characterization, we need to introduce some notations. Consider the setting of Definition 4.34. Assume furthermore that $\Omega \subset U$ is symmetric with respect to a scalar product $\langle , \rangle$ on U (see Definition 4.16) and extend $\langle , \rangle$ to a $\mathbb{C}$-bilinear symmetric form on $U_{\mathbb{C}} \times U_{\mathbb{C}}$. Choose a base point $e \in \Omega$ such that

$$\mathbb{G}(\Omega)_e = \mathbb{G}(\Omega) \cap O(V, \langle , \rangle), \tag{110}$$

which is possible by Theorem 4.17(i). Denote by $\circ_{\Omega}$ the Jordan product on U defined by mean of (103) and extend it linearly to $U_{\mathbb{C}}$. Recall that $(U, \circ_{\Omega})$ is an Euclidean Jordan algebra with unit element e (see Theorem 4.26). Define now a positive definite Hermitian form h on V by

$$h(v, v') = \langle e, H(v, v') \rangle \text{ for any } v, v' \in V. \tag{111}$$

Using h, we can define for any $u \in U_{\mathbb{C}}$ an endomorphism $R_u \in \operatorname{End}(V)$ by mean of the formula

$$\langle u, H(v, v') \rangle = 2h(v, R_u v') \text{ for any } v, v' \in V. \tag{112}$$

It is easily checked from (112) that $\overline{R}^* = R_{\bar{u}}$, where the adjoint $*$ is with respect to the Hermitian form h. In particular, if $u \in U$ then

$$R_u \in \operatorname{Herm}(V, h) := \{f \in \operatorname{End}(V) : f^* = f\}.$$

Theorem 4.36. *Notations as above. The Siegel domain $\mathcal{S}(U, V, \Omega, H)$ is biholomorphic to a bounded symmetric domain (in which case we say that it is* symmetric*) if and only if*

(i) $\Omega \subset U$ is a symmetric cone with respect to a scalar product $\langle,\rangle$ on U;
(ii) $u \circ_\Omega H(v, v') = H(R_u v, v') + H(v, R_u v')$ for any $u \in U$ and any $v, v' \in V$;
(iii) $H(R_{H(v'',v')} v, v'') = H(v', R_{H(v,v'')} v'')$ for any $v, v', v'' \in V$.

Proof. See [Sat80, Chap. V, Thm. 3.5]. □

For a tube domain, the conditions (ii) and (iii) of Theorem 4.36 are trivially satisfies. Therefore, we get the following

Corollary 4.37. *The tube domain $\mathcal{S}(U, \Omega)$ is biholomorphic to a bounded symmetric domain if and only if $\Omega \subset U$ is a symmetric cone.*

Now we want to classify the symmetric Siegel domains, or equivalently that satisfy the three conditions of Theorem 4.36. Actually, it is possible to classify the following bigger class of Siegel domains.

Definition 4.38. A Siegel domain $\mathcal{S}(U, V, \Omega, H)$ is said to be *quasi-symmetric* if and only if it satisfies the first two conditions of Theorem 4.36.

Indeed, quasi-symmetric Siegel domains correspond to unital Jordan algebra representations of $(U, \circ_\Omega)$ into $\mathrm{Herm}(V, h)$.

Lemma 4.39. *Fix a symmetric cone $\Omega \subset U$ and keep the notations as above.*

(i) *If $H : V \times V \to U_{\mathbb{C}}$ is Ω-positive Hermitian map (as in Definition 4.34) satisfying Theorem 4.36(ii) then*

$$\rho := 2R : (U, \circ_\Omega) \longrightarrow \mathrm{Herm}(V, h) \tag{113}$$
$$u \mapsto 2R_u$$

is a unital Jordan algebra homomorphism in the sense of Remark 4.27 (we call it a complex representation of $(U, \circ_\Omega)$).

(ii) *Conversely, if we start from a unital Jordan algebra homomorphism (113) and we define a Hermitian map $H : V \times V \to U_{\mathbb{C}}$ by mean of (112), then H is Ω-positive and it satisfies Theorem 4.36(ii).*

Proof. See [Sat80, Chap. IV, Prop. 4.1]. □

Each quasi-symmetric Siegel domain is a product of irreducible quasi-symmetric Siegel domains, which we are now going to define.

Definition 4.40. (i) Let $\mathcal{S}_1 = \mathcal{S}(U_1, V_1, \Omega_1, H_1)$ and $\mathcal{S}_2 = \mathcal{S}(U_2, V_2, \Omega_2, H_2)$ two Siegel domains. The product $\mathcal{S}_1 \times \mathcal{S}_2$ is equal to the Siegel domain $\mathcal{S}(U_1 \oplus U_2, V_1 \oplus V_2, \Omega_1 + \Omega_2, H = H_1 \oplus H_2)$, where $H = H_1 \oplus H_2 : (V_1 \oplus V_2) \times (V_1 \oplus V_2) \to U_1 \oplus U_2$ is defined by $H_{|V_1 \times V_2} \equiv 0$, $H_{|V_1 \times V_1} \equiv H_1$ and $H_{|V_2 \times V_2} \equiv H_2$.
(ii) A Siegel domain is *irreducible* if and only if it cannot be written as the product of two non-trivial Siegel domains.

Theorem 4.41. (i) *A quasi-symmetric Siegel domain $\mathcal{S}(U, V, \Omega, H)$ is irreducible if and only if $\Omega \subset U$ is irreducible.*

(ii) Any quasi-symmetric (resp. symmetric) Siegel domain decomposes uniquely as the product of irreducible quasi-symmetric (resp. symmetric) Siegel domains.

According to Theorem 4.26, Lemma 4.39 and Theorem 4.41, an irreducible quasi-symmetric Siegel domain is built up from a simple Euclidean Jordan algebra $(U, \circ)$ together with a complex representation $\rho : (U, \circ) \to \mathrm{Herm}(V, h)$. Such pairs can be classified as it follows.

Theorem 4.42. *The complex representations of the simple Euclidean Jordan algebras are given as it follows:*

(i) *Type* $IV_{n;r,s}$ *(even* $n \geq 4; r \geq s \geq 0$*): the representation* $\rho_{r,s} = \mathrm{sp}_1^{\oplus r} \oplus \mathrm{sp}_2^{\oplus s}$ *of* $(\mathbb{R} \oplus \mathbb{R}^{n-1}, \circ_Q)$*, where* sp_1 *and* sp_2 *are the two spin representations (see [Sat80, Appendix, §4–6]);*

(ii) *Type* $IV_{n;r}$ *(odd* $n \geq 3$ *or* $n = 2; r \geq 0$*): the representations* $\rho_r = \mathrm{sp}^{\oplus r}$ *of* $(\mathbb{R} \oplus \mathbb{R}^{n-1}, \circ_Q)$*, where* sp *is the spin representation (see [Sat80, Appendix, §4–6]);*

(iii) *Type* $III_{n;r}$ *(* $n \geq 3; r \geq 0$*): the representations* $\rho_r = \mathrm{id}^{\oplus r}$ *of* $\mathrm{Herm}_n(\mathbb{R})$*, where* $\mathrm{id} : \mathrm{Herm}_n(\mathbb{R}) \to \mathrm{Herm}_n(\mathbb{C})$ *is the natural injection.*

(iv) *Type* $I_{n;r,s}$ *(* $n \geq 3; r \geq s \geq 0$*): the representations* $\rho_{r,s} = \mathrm{id}^{\oplus r} \oplus \overline{\mathrm{id}}^{\oplus s}$ *of* $\mathrm{Herm}_n(\mathbb{C})$*, where* $\mathrm{id} : \mathrm{Herm}_n(\mathbb{C}) \to \mathrm{Herm}_n(\mathbb{C})$ *is the identity homomorphism and* $\overline{\mathrm{id}} : \mathrm{Herm}_n(\mathbb{C}) \to \mathrm{Herm}_n(\mathbb{C})$ *is given by sending* A *into its complex conjugate* $\overline{A}$*.*

(v) *Type* $II_{n;r}$ *(* $n \geq 3; r \geq 0$*): the representations* $\rho_r = \mathrm{id}^{\oplus r}$ *of* $\mathrm{Herm}_n(\mathbb{H})$*, where*

$$\mathrm{id} : \mathrm{Herm}_n(\mathbb{H}) \longrightarrow \mathrm{Herm}_{2n}(\mathbb{C})$$

$$A + jB \mapsto \begin{pmatrix} A & B \\ -\overline{B} & \overline{A} \end{pmatrix}$$

(vi) *Type* IV_0*: the trivial representation* ρ_0 *of* $\mathrm{Herm}_3(\mathbb{O})$*.*

Proof. See [Sat80, Chap. V, §5] and the references therein. □

In Table 10, we have listed all the irreducible quasi-symmetric Siegel domains by specifying the irreducible symmetric cone and the complex representation of their associated simple Euclidean Jordan algebra. Moreover, in the last column, we have specified the quasi-symmetric Siegel domains that are also symmetric (see [Sat80, Chap. V, §5] and the references therein) together with the corresponding bounded symmetric domains (using the notations of Table 4) to which they are biholomorphic.

By looking at the last column of Table 10 and using the isomorphisms between bounded symmetric domains of small dimension belonging to different types (see Table 7), it is easy to see that every bounded symmetric domain is biholomorphic to a unique Siegel domain. As a consequence, there exists a bijection between symmetric Siegel domains and Hermitian positive JTSs, which we are now going to make explicit.

Table 10 Irreducible quasi-symmetric Siegel domains

Type	Symmetric cone	Complex representation	Symmetric cases
$IV_{n;r,s}$ $(r \geq s \geq 0)$ even $n \geq 4$	$\mathcal{P}(1, n-1)$	$\rho_{r,s} = \mathrm{sp}_1^{\oplus r} \oplus \mathrm{sp}_2^{\oplus s}$	$IV_{n;0,0} = IV_n$ $IV_{4;r,0} = I_{r+2,2}$ $IV_{6;1,0} = II_5$ $IV_{8;1,0} = V$
$IV_{n;r}$ $(r \geq 0)$ odd $n \geq 3$ or $n = 2$	$\mathcal{P}(1, n-1)$	$\rho_r = \mathrm{sp}^{\oplus r}$	$IV_{n;0} = IV_n$ $(n \geq 3)$ $IV_{2;r} = I_{r+1,1}$
$III_{n;r}$ $(n \geq 3, r \geq 0)$	$\mathcal{P}_n(\mathbb{R}) = \mathrm{Herm}_n^{>0}(\mathbb{R})$	$\rho_r = \mathrm{id}^{\oplus r}$	$III_{n;0} = III_n$
$I_{n;r,s}$ $(n \geq 3, r \geq s \geq 0)$	$\mathcal{P}_n(\mathbb{C}) = \mathrm{Herm}_n^{>0}(\mathbb{C})$	$\rho_{r,s} = \mathrm{id}^{\oplus r} \oplus \overline{\mathrm{id}}^{\oplus s}$	$I_{n;r,0} = I_{n+r,n}$
$II_{n;r}$ $(n \geq 3, r \geq 0)$	$\mathcal{P}_n(\mathbb{H}) = \mathrm{Herm}_n^{>0}(\mathbb{H})$	$\rho_r = \mathrm{id}^{\oplus r}$	$II_{n;r} = II_{2n+r}$ $(r = 0, 1)$
VI_0	$\mathcal{P}_3(\mathbb{O}) = \mathrm{Herm}_3^{>0}(\mathbb{O})$	$\rho_0 = 0$	$VI_0 = VI$

Start with a symmetric Siegel domain $\mathcal{S} = \mathcal{S}(U, V, \Omega, H)$. By Theorem 4.36(i), the cone $\Omega \subset U$ is symmetric with respect to a scalar product $\langle, \rangle$. Keeping the notations introduced before Theorem 4.36, we get a Jordan product $\circ_\Omega$ on U which we extend linearly to $U_\mathbb{C}$. Using the fact that $(U, \circ_\Omega)$ is an Euclidean Jordan algebra, it can be checked (see [Sat80, Chap. I, §6]) that $U_\mathbb{C}$ becomes a Hermitian positive JTSs with respect to the triple product

$$\{u_1, u_2, u_3\}_\Omega := (u_1 \circ_\Omega \overline{u_2}) \circ_\Omega u_3 + (u_3 \circ_\Omega \overline{u_2}) \circ_\Omega u_1 - (u_1 \circ_\Omega u_3) \circ_\Omega \overline{u_2}. \quad (114)$$

Define a triple product on $U_\mathbb{C} \oplus V$ as it follows

$$\left\{\begin{pmatrix} u_1 \\ v_1 \end{pmatrix}, \begin{pmatrix} u_2 \\ v_2 \end{pmatrix}, \begin{pmatrix} u_3 \\ v_3 \end{pmatrix}\right\}_\mathcal{S} := \begin{pmatrix} \{u_1, u_2, u_3\}_\Omega + 2H(R_{\overline{u_3}}v_2, v_1) + 2H(R_{\overline{u_1}}v_2, v_3) \\ 2R_{u_3}R_{\overline{u_2}}v_1 + 2R_{u_1}R_{\overline{u_2}}v_3 + 2R_{H(v_2,v_1)}v_3 + 2R_{H(v_2,v_3)}v_1 \end{pmatrix}. \quad (115)$$

Using that $\mathcal{S}(U, V, \Omega, H)$ is symmetric, it can be shown that $(U_\mathbb{C} \oplus V, \{., ., .\}_\mathcal{S})$ is a Hermitian positive JTS (see [Sat80, Chap. V, Thm. 6.9] and the discussion following it). Observe that the element $\tilde{e} := \begin{pmatrix} e \\ 0 \end{pmatrix} \in U_\mathbb{C} \oplus V$ is an tripotent of the Hermitian positive JTS $(U_\mathbb{C} \oplus V, \{., ., .\}_\mathcal{S})$, i.e. $\{\tilde{e}, \tilde{e}, \tilde{e}\}_\mathcal{S} = \tilde{e}$. Moreover, using the fact that $\overline{e} = e$ is the identity element of $(U_\mathbb{C}, \{., ., .\}_\Omega)$ and that $2R_e = \mathrm{id}_V$ by Lemma 4.39, we can easily compute

$$[\tilde{e} \square \tilde{e}]\begin{pmatrix} u \\ v \end{pmatrix} := \left\{\begin{pmatrix} e \\ 0 \end{pmatrix}, \begin{pmatrix} e \\ 0 \end{pmatrix}, \begin{pmatrix} u \\ v \end{pmatrix}\right\}_\mathcal{S} = \begin{pmatrix} u \\ \frac{v}{2} \end{pmatrix}. \quad (116)$$

In other words, 1 and $\frac{1}{2}$ are the only eigenvalues of $\tilde{e} \square \tilde{e}$ with associated eigenspaces

$$V(\tilde{e} \square \tilde{e}; 1) = U_\mathbb{C} \text{ and } V(\tilde{e} \square \tilde{e}; 1/2) = V. \quad (117)$$

Conversely, start with a Hermitian positive JTS $(W, \{.,.,.\})$ and choose a tripotent $e \in W$, i.e. an element of W such that $\{e, e, e\} = e$. The endomorphism $e \Box e \in \text{End}(W)$ is semisimple and it satisfies the equation $(e \Box e - 1)(2e \Box e - 1)(e \Box e) = 0$ (see [Sat80, p. 242]). Therefore, the possible eigenvalues of $e \Box e$ are 1, $1/2$ and 0. We can furthermore choose e in such a way that 0 is not an eigenvalue, in which case e is called *principal* (see [Sat80, Chap. V, §6, Ex. 5]). With this assumption, we get a decomposition

$$W = W_1 \oplus W_{1/2} = W(e \Box e; 1) \oplus W(e \Box e; 1/2) \tag{118}$$

into eigenspaces for $e \Box e$ relative to the eigenvalues 1 and $1/2$, respectively. The complex vector space W_1 becomes a Jordan algebra with unit element $e \in W_1$ with respect to the Jordan product (see [Sat80, Chap. V, Prop. 6.1])

$$a \circ b = \{a, e, b\} \text{ for any } a, b \in W_1. \tag{119}$$

Moreover, the map $a \mapsto a^* := \{e, a, e\}$ is a $\mathbb{C}$-antilinear involution on W_1 (see [Sat80, Chap. V, Prop. 6.1]). Therefore

$$\left(W_1^+ := \{a \in W_1 : a^* = a\}, \circ\right) \tag{120}$$

is a real Jordan algebra which turns out to be Euclidean (see [Sat80, p. 254]). We will denote by $\Omega_W \subset W_1^+$ its associated symmetric cone (see Theorem 4.26). The Jordan algebra $(W_1, \circ)$ comes with a unital Jordan algebra homomorphism (see [Sat80, Chap. V, Prop. 6.2])

$$\begin{aligned} 2R : W_1 &\longrightarrow \text{End}(W_{1/2}), \\ a &\mapsto 2R_a \text{ s.t. } R_a(x) := \{a, e, x\}. \end{aligned} \tag{121}$$

It can be checked (see [Sat80, p. 247, Eq. (6.21)]) that $R_{a^*} = R_a^*$, where R_a^* is the adjoint of R_a with respect to the positive definite hermitian form $h = \tau/2$ on $W_{1/2}$ with τ equal to the trace form of $(W, \{.,.,.\})$. Therefore, by restriction, we get a unital Jordan algebra homomorphism

$$2R : W_1^+ \longrightarrow \text{Herm}(W_{1/2}, h). \tag{122}$$

Consider now the Hermitian map

$$\begin{aligned} H_W : W_{1/2} \times W_{1/2} &\longrightarrow W_1, \\ (x, y) &\mapsto H(x, y) := \{e, x, y\}. \end{aligned} \tag{123}$$

It can be checked (see [Sat80, p. 247, Eq. (6.25)]) that the map H_W and the unital Jordan algebra homomorphism $2R$ satisfy formula (112), i.e.

$$\langle a, H_W(x, y) \rangle = 2h(x, R_a y) \text{ for any } x, y \in W_{1/2} \text{ and any } a \in W_1,$$

where $\langle,\rangle$ is the trace form of the Jordan algebra $(W_1, \circ)$. Therefore, from Lemma 4.39 it follows that H_W is Ω-positive and it satisfies Theorem 4.36(ii). Moreover, it can be checked (see [Sat80, p. 245, Eq. (6:15")]) that H_W satisfies Theorem 4.36(iii). Therefore, using Theorem 4.36, we infer that $\mathcal{S}(W_1^+, W_{1/2}, \Omega_W, H_W)$ is a symmetric Siegel domain.

Theorem 4.43. *Notations as above. There is a bijection*

$$\{\textit{Symmetric Siegel spaces}\} \xrightarrow{\cong} \{\textit{Hermitian positive JTSs}\}$$

$$\mathcal{S} = \mathcal{S}(U, V, \Omega, H) \longrightarrow (U_{\mathbb{C}} \oplus V, \{.,.,.\}_{\mathcal{S}}) \qquad (124)$$

$$\mathcal{S}(W_1^+, W_{1/2}, \Omega_W, H_W) \longleftarrow (W, \{.,.,.\})$$

sending irreducible symmetric Siegel domains into simple Hermitian positive JTSs.

Proof. See [Sat80, Chap. V, §6]. □

For symmetric Siegel spaces of the first kind (or symmetric tube domains), the bijection of Theorem 4.43 assumes a particular simple form. Indeed, let $\Omega \subset U$ be a symmetric cone and choose a base point $e \in \Omega$ as in (110). Consider the associated Jordan product $\circ_\Omega$ on U (see Theorem 4.26). Then the Hermitian positive JTS $\{U_\mathbb{C}, \{.,.,.\}_\mathcal{S}\}$ associated to the Siegel domain of the first kind $\mathcal{S} = \mathcal{S}(\Omega, U) \subset U_\mathbb{C}$ (as in (108)) is the one associated to the Euclidean Jordan algebra $(U, \circ_\Omega)$ as in Example 4.22(4.22).

Consider now the bounded symmetric domain $D_\Omega \subset U_\mathbb{C}$ (in its Harish–Chandra embedding) corresponding to the Hermitian positive JTS $\{U_\mathbb{C}, \{.,.,.\}_\mathcal{S}\}$ (see Theorems 2.30 and 2.42). It is possible to describe explicitly the biholomorphism between $\mathcal{S}(\Omega, U)$ and D_Ω, generalizing the Cayley transform in dimension one (see Example 2.29).

Theorem 4.44. *Notations as above. Then we have the following biholomorphism (called the generalized Cayley transform)*

$$c : D_\Omega \xrightarrow{\cong} \mathcal{S}(\Omega, U)$$
$$w \mapsto i(e + w) \circ_\Omega (e - w)^{-1}, \qquad (125)$$

The inverse is given by the map sending $z \in \mathcal{S}(\Omega, U)$ *into* $(z - ie) \circ_\Omega (z + ie)^{-1} \in D_\Omega$.

Indeed, generalized Cayley transforms have been defined for all symmetric Siegel spaces by Korányi–Wolf in [KW65, Chap. VI].

Looking at the last column of Table 10, it is easy to see which irreducible bounded symmetric domains are biholomorphic to symmetric tube domains (we call them bounded symmetric domains *of tube type*).

Corollary 4.45. *The irreducible bounded symmetric domains of tube type are the following:*

Hermitian Symmetric Manifolds

Table 11 Irreducible bounded symmetric domains of tube type

Symmetric Cone	Bounded symmetric domain of tube type	
$\mathcal{P}(1, n-1)$ $(n \geq 2)$	IV_n	if $n \geq 3$
	IV_1	if $n = 2$
$\mathcal{P}_n(\mathbb{R}) = \text{Herm}_n^{>0}(\mathbb{R})$ $(n \geq 3)$	III_n	
$\mathcal{P}_n(\mathbb{C}) = \text{Herm}_n^{>0}(\mathbb{C})$ $(n \geq 3)$	$I_{n,n}$	
$\mathcal{P}_n(\mathbb{H}) = \text{Herm}_n^{>0}(\mathbb{H})$ $(n \geq 3)$	II_{2n}	
$\mathcal{P}_3(\mathbb{O}) = \text{Herm}_3^{>0}(\mathbb{O})$	VI	

(1) $I_{n,n}$ for any $n \geq 1$;
(2) II_{2n} for any $n \geq 1$;
(3) III_n for any $n \geq 1$;
(4) IV_n for any $2 \neq n \geq 1$;
(5) VI.

We have collected the irreducible bounded symmetric domains of tube type (avoiding repetitions in small dimension, see Table 7) in the following Table 11, together with their associated symmetric cones.

4.5 Boundary Components of Irreducible Bounded Symmetric Domains

In this subsection, we describe explicitly the boundary components of each of the irreducible bounded symmetry domains (see Sect. 3). We begin with the following result.

Theorem 4.46. *Let D be an irreducible bounded symmetric domain and let $G = \text{Aut}(D)^\circ$. Then all the boundary components of rank k are conjugated by the group G.*

Proof. See [Wol72, p. 292]. □

In virtue of the above Theorem, it will be enough to describe for each irreducible symmetric domain D of rank r and each $0 \leq k < r$ a boundary component $F \leq D$ of rank k.

4.5.1 Type $I_{p,q}$ ($p \geq q \geq 1$)

Every boundary component of $\mathcal{D}_{I_{p,q}}$ of rank k (with $0 \leq k < q$) is conjugate to the following boundary component

$$\mathcal{D}_{I_{p,q}}^k := \left\{ \begin{pmatrix} Z' & 0 \\ 0 & I_{q-k} \end{pmatrix} : Z' \in \mathcal{D}_{I_{p-q+k,k}} \right\} \cong \mathcal{D}_{I_{p-q+k,k}}, \tag{126}$$

which we call the *standard boundary component of rank k*. The pair $(f_{\mathcal{D}_{I_{p,q}}^k}, \phi_{\mathcal{D}_{I_{p,q}}^k})$ of Lemma 4.6 associated to $\mathcal{D}_{I_{p,q}}^k$ is equal to

$$\begin{aligned} f_{\mathcal{D}_{I_{p,q}}^k} : \Delta &\longrightarrow \mathcal{D}_{I_{p,q}}^k \\ z &\mapsto \begin{pmatrix} 0 & 0 \\ 0 & zI_{q-k} \end{pmatrix}, \end{aligned} \tag{127}$$

$$\phi_{\mathcal{D}_{I_{p,q}}^k} : \mathbb{S}^1 \times \mathrm{SL}_2(\mathbb{R}) \longrightarrow \mathrm{SU}(p,q) = \mathrm{Hol}(\mathcal{D}_{I_{p,q}})^o$$

$$\left(e^{i\theta}, \begin{pmatrix} a & b \\ c & d \end{pmatrix} \right) \mapsto \begin{pmatrix} e^{i\theta} I_{p-q+k} & 0 & 0 & 0 \\ 0 & \frac{a+d+i(b-c)}{2} I_{q-k} & 0 & \frac{b+c+i(a-d)}{2} I_{q-k} \\ 0 & 0 & e^{-i\theta} I_k & 0 \\ 0 & \frac{b+c-i(a-d)}{2} I_{q-k} & 0 & \frac{a+d-i(b-c)}{2} I_{q-k} \end{pmatrix}. \tag{128}$$

In particular, the one-parameter subgroup of $\mathrm{SU}(p,q)$ associated to $\mathcal{D}_{I_{p,q}}^k$ as in (86) is given by

$$\begin{aligned} w_{\mathcal{D}_{I_{p,q}}^k} : \mathbb{G}_m &\longrightarrow \mathrm{SU}(p,q) \\ t &\mapsto \begin{pmatrix} I_{p-q+k} & 0 & 0 & 0 \\ 0 & \frac{t+t^{-1}}{2} I_{q-k} & 0 & \frac{i(t-t^{-1})}{2} I_{q-k} \\ 0 & 0 & I_k & 0 \\ 0 & \frac{-i(t-t^{-1})}{2} I_{q-k} & 0 & \frac{t+t^{-1}}{2} I_{q-k} \end{pmatrix}. \end{aligned} \tag{129}$$

Using the above explicit expression of $w_{\mathcal{D}_{I_{p,q}}^k}$ and Theorem 4.8, we can compute the Levi subgroup of the normalizer $N(\mathcal{D}_{I_{p,q}}^k)$ subgroup of $\mathcal{D}_{I_{p,q}}^k$ together with its decomposition as in Theorem 4.8(iv)

$$Z(w_{\mathcal{D}_{I_{p,q}}^k})^o$$

$$= \left\{ \begin{pmatrix} A & 0 & B & 0 \\ 0 & \frac{E+(E^*)^{-1}}{2} & 0 & i\frac{E-(E^*)^{-1}}{2} \\ C & 0 & D & 0 \\ 0 & -i\frac{E-(E^*)^{-1}}{2} & 0 & \frac{E+(E^*)^{-1}}{2} \end{pmatrix} \in \mathrm{SL}(p+q,\mathbb{C}) : \begin{matrix} \begin{pmatrix} A & B \\ C & D \end{pmatrix} \in \mathrm{U}(p-q+k,k) \\ \\ E \in \mathrm{GL}_{q-k}(\mathbb{C}) \end{matrix} \right\}$$

$$= \mathrm{SU}(p-q+k,k) \cdot \mathrm{GL}_{q-k}^o(\mathbb{C}) \times \mathbb{S}^1 = G_h(\mathcal{D}_{I_{p,q}}^k) \cdot G_l(\mathcal{D}_{I_{p,q}}^k) \cdot M(\mathcal{D}_{I_{p,q}}^k), \tag{130}$$

where $\mathrm{GL}_{q-k}^o(\mathbb{C}) := \{E \in \mathrm{GL}_{q-k}(\mathbb{C}) : \det E \in \mathbb{R}^*\} \subset \mathrm{GL}_{q-k}(\mathbb{C})$.

Similarly, using again the above explicit expression of $w_{\mathcal{D}_{I_{p,q}}^k}$ and Theorem 4.8, we can compute the unipotent radical of $N(\mathcal{D}_{I_{p,q}}^k)$

$$W(\mathcal{D}_{I_{p,q}}^k) = \left\{ \begin{pmatrix} I_{p-q+k} & F_1 & 0 & -iF_1 \\ -\overline{F_1}^t & I_{q-k}+iM & -i\overline{F_2}^t & M \\ 0 & iF_2 & I_k & F_2 \\ i\overline{F_1}^t & M & -\overline{F_2}^t & I_{q-k}-iM \end{pmatrix} : \begin{array}{l} F_1 \in M_{p-q+k,q-k}(\mathbb{C}), F_2 \in M_{k,q-k}(\mathbb{C}) \\ M \in M_{q-k,q-k}(\mathbb{C}) \\ \overline{F_1}^t F_1 - \overline{F_2}^t F_2 = i(\overline{M}^t - M) \end{array} \right\}. \tag{131}$$

Moreover, the center $U(\mathcal{D}_{I_{p,q}}^k)$ of $W(\mathcal{D}_{I_{p,q}}^k)$ is equal to the set of all matrices of $W(\mathcal{D}_{I_{p,q}}^k)$ such that $F_1 = F_2 = 0$, and is therefore isomorphic to the abelian unipotent Lie group underlying the vector space $\mathrm{Herm}_{q-k}(\mathbb{C})$.

The orbit of the point $o_{\mathcal{D}_{I_{p,q}}^k} = f_{\mathcal{D}_{I_{p,q}}^k}(1) = \begin{pmatrix} 0 & 0 \\ 0 & I_{q-k} \end{pmatrix}$ under the natural action of the group $G_h(\mathcal{D}_{I_{p,q}}^k) = \mathrm{SU}(p-q+k,k)$ is equal to $\mathcal{D}_{I_{p,q}}^k$ and its stabilizer subgroup is isomorphic to $K_h(\mathcal{D}_{I_{p,q}}^k) = \mathrm{SU}(p-q+k,k) \cap \mathrm{S}(\mathrm{U}_p \times \mathrm{U}_q) = \mathrm{S}(\mathrm{U}_{p-q+k} \times \mathrm{U}_k)$. Therefore, $\mathcal{D}_{I_{p,q}}^k$ is a bounded symmetric domain of type $I_{p-q+k,k}$ and it is diffeomorphic to

$$\mathcal{D}_{I_{p,q}}^k \cong \frac{G_h(\mathcal{D}_{I_{p,q}}^k)}{K_h(\mathcal{D}_{I_{p,q}}^k)} = \frac{\mathrm{SU}(p-q+k,k)}{\mathrm{S}(\mathrm{U}_{p-q+k} \times \mathrm{U}_k)}. \tag{132}$$

The action of $G_l(\mathcal{D}_{I_{p,q}}^k) = \mathrm{GL}_{q-k}^o(\mathbb{C})$ on $U(\mathcal{D}_{I_{p,q}}^k) \cong \mathrm{Herm}_{q-k}(\mathbb{C})$ is given by $(E,M) \mapsto EM\overline{E}^t$. Under this action, the orbit of the point $\Omega_{\mathcal{D}_{I_{p,q}}^k} = \frac{1}{2} I_{q-k} \in \mathrm{Herm}_{q-k}(\mathbb{C})$ is equal to the cone of positive definite Hermitian complex quadratic forms $\mathcal{P}_{q-k}(\mathbb{C})$ of size $q-k$ and its stabilizer subgroup is equal to $K_l(\mathcal{D}_{I_{p,q}}^k) = \mathrm{GL}_{q-k}^o(\mathbb{C}) \cap \mathrm{S}(\mathrm{U}_p \times \mathrm{U}_q) = \mathrm{U}^o(q-k)$, where $\mathrm{U}^o(q-k) := \{E \in \mathrm{U}(q-k) : \det E = \pm 1\}$. Therefore, $C(\mathcal{D}_{I_{p,q}}^k)$ is equal to the symmetric cone $\mathcal{P}_{n-k}(\mathbb{C})$ (see Theorem 4.28) and it is diffeomorphic to

$$C(\mathcal{D}_{I_{p,q}}^k) \cong \frac{G_l(\mathcal{D}_{I_{p,q}}^k)}{K_l(\mathcal{D}_{I_{p,q}}^k)} = \frac{\mathrm{GL}_{q-k}^o(\mathbb{C})}{\mathrm{U}^o(q-k)} = \frac{\mathrm{GL}_{q-k}(\mathbb{C})}{\mathrm{U}(q-k)} = \mathcal{P}_{q-k}(\mathbb{C}). \tag{133}$$

4.5.2 Type II_n

Every boundary component of $\mathcal{D}_{II_n}$ of rank k (with $0 \leq k < \lfloor \frac{n}{2} \rfloor$) is conjugate to the following boundary component (which we call the *standard boundary component of rank k*)

$$\mathcal{D}_{II_n}^k := \left\{ \begin{pmatrix} Z' & 0 \\ 0 & E_{n-2k-\epsilon} \end{pmatrix} : Z' \in \mathcal{D}_{II_{2k+\epsilon}} \right\} \cong \mathcal{D}_{II_{2k+\epsilon}}, \tag{134}$$

where $\epsilon = 0$ (resp. 1) if n is even (resp. odd) and for any $m \in \mathbb{N}$ we denote by E_{2m} the $2m \times 2m$-matrix formed by m diagonal blocks of the form $\begin{pmatrix} 0 & 1 \\ -1 & 0 \end{pmatrix}$.

The decomposition into factors (as in Theorem 4.8(iv)) of the Levi subgroup $L(\mathcal{D}_{II_n}^k)$ of the normalizer subgroup $N(\mathcal{D}_{II_n}^k)$ of $\mathcal{D}_{II_n}^k$ is given by (see [Sat80, p. 116])

$$L(\mathcal{D}_{II_n}^k) = G_h(\mathcal{D}_{II_n}^k) \cdot G_l(\mathcal{D}_{II_n}^k) \cdot M(\mathcal{D}_{II_n}^k)$$

$$= \begin{cases} \{1\} \cdot \mathrm{GL}_{\frac{n}{2}}(\mathbb{H}) \cdot \{1\} & \text{if } k = 0 \text{ and } n \text{ is even,} \\ \{1\} \cdot \mathrm{GL}_{\frac{n-1}{2}}(\mathbb{H}) \cdot \mathbb{S}^1 & \text{if } k = 0 \text{ and } n \text{ is odd,} \\ \mathrm{SU}(1,1) \cdot \mathrm{GL}_{\frac{n-2}{2}}(\mathbb{H}) \cdot \mathrm{SL}_1(\mathbb{H}) & \text{if } k = 1 \text{ and } n \text{ is even,} \\ \mathrm{SO}^{nc}(2n-4) \cdot \mathbb{R}^* \cdot \mathrm{SL}_1(\mathbb{H}) & \text{if } k = \frac{n-2-\epsilon}{2}, \\ \mathrm{SO}^{nc}(4k+2\epsilon) \cdot \mathrm{GL}_{\frac{n-2k-\epsilon}{2}}(\mathbb{H}) \times \{1\} & \text{otherwise.} \end{cases} \tag{135}$$

Therefore, $\mathcal{D}_{II_n}^k$ is a bounded symmetric domain of type $II_{2k+\epsilon}$ if $k \geq 1$ and it is a point if $k = 0$. Moreover, the symmetric cone associated to $\mathcal{D}_{II_n}^k$ is equal to (see Theorem 4.28)

$$C(\mathcal{D}_{II_n}^k) = \begin{cases} \mathcal{P}(1,0) & \text{if } k = \frac{n-2-\epsilon}{2}, \\ \mathcal{P}(1,5) & \text{if } k = \frac{n-4-\epsilon}{2}, \\ \mathcal{P}_{\frac{n-2k-\epsilon}{2}}(\mathbb{H}) & \text{otherwise.} \end{cases} \tag{136}$$

4.5.3 Type III_n ($n \geq 1$)

Every boundary component of $\mathcal{D}_{III_n}$ of rank k (with $0 \leq k < n$) is conjugate to the following boundary component

$$\mathcal{D}_{III_n}^k := \left\{ \begin{pmatrix} Z' & 0 \\ 0 & I_{n-k} \end{pmatrix} : Z' \in \mathcal{D}_{III_k} \right\} \cong \mathcal{D}_{III_k}, \tag{137}$$

which we call the *standard boundary component of rank k*. The pair $(f_{\mathcal{D}_{III_n}^k}, \phi_{\mathcal{D}_{III_n}^k})$ of Lemma 4.6 associated to $\mathcal{D}_{III_n}^k$ is equal to

$$f_{\mathcal{D}_{III_n}^k} : \Delta \longrightarrow \mathcal{D}_{III_n}^k$$

$$z \mapsto \begin{pmatrix} 0 & 0 \\ 0 & zI_{n-k} \end{pmatrix}, \qquad (138)$$

$$\phi_{\mathcal{D}_{III_n}^k} : \mathbb{S}^1 \times \mathrm{SL}_2(\mathbb{R}) \longrightarrow \mathrm{Sp}^{nc}(n) = \mathrm{Hol}(\mathcal{D}_{III_n})^o$$

$$\left(e^{i\theta}, \begin{pmatrix} a & b \\ c & d \end{pmatrix}\right) \mapsto \begin{pmatrix} e^{i\theta} I_k & 0 & 0 & 0 \\ 0 & \frac{a+d+i(b-c)}{2} I_{n-k} & 0 & \frac{b+c+i(a-d)}{2} I_{n-k} \\ 0 & 0 & e^{-i\theta} I_k & 0 \\ 0 & \frac{b+c-i(a-d)}{2} I_{n-k} & 0 & \frac{a+d-i(b-c)}{2} I_{n-k} \end{pmatrix}. \qquad (139)$$

In particular, the one-parameter subgroup of $\mathrm{Sp}^{nc}(n)$ associated to $\mathcal{D}_{III_n}^k$ as in (86) is given by

$$w_{\mathcal{D}_{III_n}^k} : \mathbb{G}_m \longrightarrow \mathrm{Sp}^{nc}(n)$$

$$t \mapsto \begin{pmatrix} I_k & 0 & 0 & 0 \\ 0 & \frac{t+t^{-1}}{2} I_{n-k} & 0 & \frac{i(t-t^{-1})}{2} I_{n-k} \\ 0 & 0 & I_k & 0 \\ 0 & \frac{-i(t-t^{-1})}{2} I_{n-k} & 0 & \frac{t+t^{-1}}{2} I_{n-k} \end{pmatrix}. \qquad (140)$$

Using the above explicit expression of $w_{\mathcal{D}_{III_n}^k}$ and Theorem 4.8, we can compute the Levi subgroup (together with its decomposition as in Theorem 4.8(iv)) and the unipotent radical of the normalizer $N(\mathcal{D}_{III_n}^k)$ subgroup of $\mathcal{D}_{III_n}^k$:

$$Z(w_{\mathcal{D}_{III_n}^k})^o = \left\{ \begin{pmatrix} A & 0 & B & 0 \\ 0 & \frac{E+(E^t)^{-1}}{2} & 0 & i\frac{E-(E^t)^{-1}}{2} \\ C & 0 & D & 0 \\ 0 & -i\frac{E-(E^t)^{-1}}{2} & 0 & \frac{E+(E^t)^{-1}}{2} \end{pmatrix} : \begin{pmatrix} A & B \\ C & D \end{pmatrix} \in \mathrm{Sp}^{nc}(k), E \in \mathrm{GL}_{n-k}(\mathbb{R}) \right\}$$

$$= \mathrm{Sp}^{nc}(k) \times \mathrm{GL}_{n-k}(\mathbb{R}) \times \{1\} = G_h(\mathcal{D}_{III_n}^k) \times G_l(\mathcal{D}_{III_n}^k) \times M(\mathcal{D}_{III_n}^k), \qquad (141)$$

$$W(\mathcal{D}_{III_n}^k) = \left\{ \begin{pmatrix} I_r & F & 0 & -iF \\ -\overline{F}^t & I_{n-k}+iM & -iF^t & M \\ 0 & i\overline{F} & I_k & \overline{F} \\ i\overline{F}^t & M & -F^t & I_{n-k}-iM \end{pmatrix} : \begin{array}{l} F \in M_{k,n-k}(\mathbb{C}), M \in M_{n-k,n-k}(\mathbb{R}) \\ \overline{F}^t F - F^t \overline{F} = i(M^t - M) \end{array} \right\}. \qquad (142)$$

Moreover, the center $U(\mathcal{D}_{III_n}^k)$ of $W(\mathcal{D}_{III_n}^k)$ is equal to the set of all matrices of $W(\mathcal{D}_{III_n}^k)$ such that $F = 0$, and is therefore isomorphic to the abelian unipotent Lie group underlying the vector space $\mathrm{Herm}_{n-k}(\mathbb{R})$.

The orbit of the point $o_{\mathcal{D}^k_{III_n}} = f_{\mathcal{D}^k_{III_n}}(1) = \begin{pmatrix} 0 & 0 \\ 0 & I_{n-k} \end{pmatrix}$ under the natural action of the group $G_h(\mathcal{D}^k_{III_n}) = \mathrm{Sp}^{nc}(k)$ is equal to $\mathcal{D}^k_{III_n}$ and its stabilizer subgroup is isomorphic to $K_h(\mathcal{D}^k_{III_n}) = \mathrm{Sp}^{nc}(k) \cap U(n) = U(k)$. Therefore, $\mathcal{D}^k_{III_n}$ is a bounded symmetric domain of type III_k and it is diffeomorphic to

$$\mathcal{D}^k_{III_n} \cong \frac{G_h(\mathcal{D}^k_{III_n})}{K_h(\mathcal{D}^k_{III_n})} = \frac{\mathrm{Sp}^{nc}(k)}{U(k)}. \tag{143}$$

The action of $G_l(\mathcal{D}^k_{III_n}) = \mathrm{GL}_{n-k}(\mathbb{R})$ on $U(\mathcal{D}^k_{III_n}) \cong \mathrm{Herm}_{n-k}(\mathbb{R})$ is given by $(E, M) \mapsto EME^t$. Under this action, the orbit of the point $\Omega_{\mathcal{D}^k_{III_n}} = \frac{1}{2} I_{n-k} \in \mathrm{Herm}_{n-k}(\mathbb{R})$ is equal to the cone of positive definite real quadratic forms $\mathcal{P}_{n-k}(\mathbb{R})$ of size $n-k$ and its stabilizer subgroup is isomorphic $K_l(\mathcal{D}^k_{III_n}) = \mathrm{GL}_{n-k}(\mathbb{R}) \cap U(n) = O(n-k)$. Therefore, $C(\mathcal{D}^k_{III_n})$ is equal to the symmetric cone $\mathcal{P}_{n-k}(\mathbb{R})$ (see Theorem 4.28) and it is diffeomorphic to

$$C(\mathcal{D}^k_{III_n}) \cong \frac{G_l(\mathcal{D}^k_{III_n})}{K_l(\mathcal{D}^k_{III_n})} = \frac{\mathrm{GL}_{n-k}(\mathbb{R})}{O(n-k)} = \mathcal{P}_{n-k}(\mathbb{R}). \tag{144}$$

4.5.4 Type IV_n $(n \geq 3)$

The *standard boundary component* of $\mathcal{D}_{IV}$ of rank k (for $k = 0, 1$) are given by

$$\begin{aligned} \mathcal{D}^0_{IV_n} &:= \{(-i, 0, \ldots, 0)^t \in \mathbb{C}^n\}, \\ \mathcal{D}^1_{IV_n} &:= \left\{ \left(-i\frac{1+z}{1-z}, \frac{1+z}{1-z}, 0, \ldots, 0\right)^t \in \mathbb{C}^n : |z| < 1 \right\}, \end{aligned} \tag{145}$$

see [Wol72, p. 355]. The decomposition into factors (as in Theorem 4.8(iv)) of the Levi subgroup $L(\mathcal{D}^k_{IV_n})$ of the normalizer subgroup $N(\mathcal{D}^k_{IV_n})$ of $\mathcal{D}^k_{IV_n}$ is given by (see [Sat80, p. 117])

$$\begin{aligned} L(\mathcal{D}^k_{IV_n}) &= G_h(\mathcal{D}^k_{IV_n}) \cdot G_l(\mathcal{D}^k_{IV_n}) \cdot M(\mathcal{D}^k_{IV_n}) \\ &= \begin{cases} \mathrm{SO}(1,1) \cdot \mathrm{GL}_1(\mathbb{R}) \cdot \mathrm{SO}(n-2) & \text{if } k = 1, \\ \{1\} \cdot (\mathrm{SO}(n-1,1) \times \mathbb{R}^*) \cdot \{1\} & \text{if } k = 0. \end{cases} \end{aligned} \tag{146}$$

Therefore, $\mathcal{D}^k_{IV_n}$ is a bounded symmetric domain of type $IV_1 = I_{1,1}$ if $k = 1$ and it is a point if $k = 0$. Moreover, the symmetric cone associated to $\mathcal{D}^k_{IV_n}$ is equal to (see Theorem 4.28)

$$C(\mathcal{D}_{IV_n}^k) = \begin{cases} \mathcal{P}(1,0) & \text{if } k = 1, \\ \mathcal{P}(1, n-1) & \text{if } k = 0. \end{cases} \quad (147)$$

4.5.5 Type V

Using Remark 4.11, denote by $\mathcal{D}_V^k$ (for $k = 0, 1$) the boundary component of $\mathcal{D}_V$ of rank k such that

$$o_{\mathcal{D}_V^k} = \begin{cases} \begin{pmatrix} 0 \\ 1 \end{pmatrix} & \text{if } k = 1, \\ \begin{pmatrix} 1 \\ 1 \end{pmatrix} & \text{if } k = 0. \end{cases}$$

We call $\mathcal{D}_V^k$ the *standard boundary component* of $\mathcal{D}_V$ of rank k.

The decomposition into factors (as in Theorem 4.8(iv)) of the Levi subgroup $L(\mathcal{D}_V^k)$ of the normalizer subgroup $N(\mathcal{D}_V^k)$ of $\mathcal{D}_V^k$ is given by (see [Sat80, p. 117])

$$L(\mathcal{D}_V^k) = G_h(\mathcal{D}_V^k) \cdot G_l(\mathcal{D}_V^k) \cdot M(\mathcal{D}_V^k) = \begin{cases} \mathrm{SU}(5,1) \cdot \mathrm{GL}_1(\mathbb{R}) \cdot \{1\} & \text{if } k = 1, \\ \{1\} \cdot (\mathrm{SO}(7,1) \times \mathbb{R}^*) \cdot \mathbb{S}^1 & \text{if } k = 0. \end{cases} \quad (148)$$

Therefore, $\mathcal{D}_V^k$ is a bounded symmetric domain of type $I_{5,1}$ if $k = 1$ and it is a point if $k = 0$. Moreover, the symmetric cone associated to $\mathcal{D}_V^k$ is equal to (see Theorem 4.28)

$$C(\mathcal{D}_V^k) = \begin{cases} \mathcal{P}(1,0) & \text{if } k = 1, \\ \mathcal{P}(1,7) & \text{if } k = 0. \end{cases} \quad (149)$$

4.5.6 Type VI

Using Remark 4.11, denote by $\mathcal{D}_{VI}^k$ (for $k = 0, 1, 2$) the boundary component of $\mathcal{D}_{VI}$ of rank k such that

$$o_{\mathcal{D}_{VI}^k} = \begin{pmatrix} 0 & 0 \\ 0 & I_{3-k} \end{pmatrix}.$$

We call $\mathcal{D}_{VI}^k$ the *standard boundary component* of $\mathcal{D}_{VI}$ of rank k.

The decomposition into factors (as in Theorem 4.8(iv)) of the Levi subgroup $L(\mathcal{D}_{VI}^k)$ of the normalizer subgroup $N(\mathcal{D}_{VI}^k)$ of $\mathcal{D}_{VI}^k$ is given by (see [Sat80, p. 117])

$$L(\mathcal{D}_{VI}^k) = G_h(\mathcal{D}_{VI}^k) \cdot G_l(\mathcal{D}_{VI}^k) \cdot M(\mathcal{D}_{VI}^k) = \begin{cases} SO(2,10) \cdot GL_1(\mathbb{R}) \cdot \{1\} & \text{if } k = 2, \\ SU(1,1) \cdot (SO(9,1) \times \mathbb{R}^*) \cdot \{1\} & \text{if } k = 1, \\ \{1\} \cdot (E_6(-26) \times \mathbb{R}^*) \cdot \{1\} & \text{if } k = 0. \end{cases} \quad (150)$$

Therefore, $\mathcal{D}_{VI}^k$ is a bounded symmetric domain of type IV_{10} if $k = 2$, of type $I_{1,1}$ if $k = 1$ and it is a point if $k = 0$. Moreover, the symmetric cone associated to $\mathcal{D}_{VI}^k$ is equal to (see Theorem 4.28)

$$C(\mathcal{D}_V^k) = \begin{cases} \mathcal{P}(1,0) & \text{if } k = 2, \\ \mathcal{P}(1,9) & \text{if } k = 1, \\ \mathcal{P}_3(\mathbb{O}) & \text{if } k = 0. \end{cases} \quad (151)$$

Acknowledgements These notes grew up from a Ph.D. course held by the author at the University of Roma Tre in spring 2013 and from a course held by the author at the School "Combinatorial Algebraic Geometry" held in Levico Terme (Trento, Italy) in June 2013. The author would like to thank the students and the colleagues that attended the above mentioned Ph.D. course (Fabio Felici, Roberto Fringuelli, Alessandro Maria Masullo, Margarida Melo, Riane Melo, Paola Supino, Valerio Talamanca) for their patience, encouragement and interest in the material presented. Moreover, the author would like to thank the organizers of the above mentioned School (Giorgio Ottaviani, Sandra Di Rocco, Bernd Sturmfels) for the invitation to give a course as well as all the participants to the School for their interest and feedbacks. Many thanks are due to Jan Draisma for reading a preliminary version of this manuscript and to Radu Laza for suggesting some bibliographical references.

The author is a member of the research center CMUC (University of Coimbra) and he was supported by the FCT project *Espaços de Moduli em Geometria Algébrica* (PTDC/MAT/111332/2009), by the FCT Project PTDC/MAT-GEO/0605/2012 and by the MIUR project *Spazi di moduli e applicazioni* (FIRB 2012).

References

[AMRT10] A. Ash, D. Mumford, M. Rapoport, Y. Tai, Smooth compactification of locally symmetric varieties. *Cambridge Mathematical Library* (Cambridge University Press, Cambridge, 2010)

[Bae02] J.C. Baez, The octonions. Bull. Am. Math. Soc. (N.S.) **39**(2), 145–205 (2002)

[Bor52] A. Borel, in *Les espaces hermitiens symétriques*. Séminaire Bourbaki, vol. 2, Exp. No. 62 (1952) (Soc. Math. France, Paris, 1995), pp. 121–132

[BJ06] A. Borel, L. Ji, Compactifications of symmetric and locally symmetric spaces, in *Mathematics: Theory and Applications* (Birkhäuser, Boston, 2006)

[Car26-27] É. Cartan, Sur une classe remarquable d'espaces de Riemann. Bull. Soc. Math. Fr. **54**, 214–264 (1926); ibid. **55**, 114–134 (1927)

[Car35] É. Cartan, Sur le domaines bornes homogènes de l'espace de n variables complexes. Abh. Math. Sem. Univ. Hamburg **11**, 116–162 (1935)

[FK94] J. Faraut, A. Korányi, in *Analysis on Symmetric Cones*. Oxford Mathematical Monographs. Oxford Science Publications (The Clarendon Press/Oxford University Press, New York, 1994)

[FKKLR00] J. Faraut, S. Kaneyuki, A. Korńyi, Q.K. Lu, G. Roos, in *Analysis and Geometry on Complex Homogeneous Domains*. Progress in Mathematics, vol. 185 (Birkhäuser Boston, Boston, 2000)
[FL13] R. Friedman, R. Laza, Semi-algebraic horizontal subvarieties of Calabi-Yau type. Duke Math. J. **162**(12), 2077–2148 (2013).
[Hel78] S. Helgason, in *Differential Geometry, Lie Groups, and Symmetric Spaces*. Pure and Applied Mathematics, vol. 80 (Academic, New York, 1978)
[Hua46] L.K. Hua, On the theory of Fuchsian functions of several variables. Ann. Math. **47**, 167–191 (1946)
[Kna96] A.W. Knapp, in *Lie Groups Beyond an Introduction*. Progress in Mathematics, vol. 140 (Birkhäuser Boston, Boston, 1996)
[Koe69] M. Koecher, *An Elementary Approach to Bounded Symmetric Domains* (Rice University, Houston, 1969)
[KW65] A. Korányi, J.A. Wolf, Realization of Hermitian spaces as generalized half-planes. Ann. Math. **81**, 265–288 (1965)
[LM02] J.M. Landsberg, L. Manivel, Construction and classification of complex simple Lie algebras via projective geometry. Selecta Math. (N.S.) **8**(1), 137–159 (2002)
[LM03] J.M. Landsberg, L. Manivel, On the projective geometry of rational homogeneous varieties. Comment. Math. Helv. **78**(1), 65–100 (2003)
[Loo69a] O. Loos, *Symmetric Spaces. I: General Theory* (W.A. Benjamin, New York, 1969)
[Loo69b] O. Loos, *Symmetric Spaces. II: Compact Spaces and Classification* (W.A. Benjamin, New York, 1969)
[Loo75] O. Loos, in *Jordan Pairs*. Lecture Notes in Mathematics, vol. 460 (Springer, Berlin, 1975)
[Loo77] O. Loos, in *Bounded Symmetric Domains and Jordan Pairs*. Mathematical Lectures (University of California, Irvine, 1977)
[Mil05] J.S. Milne, Introduction to Shimura varieties, in *Harmonic Analysis, the Trace Formula, and Shimura Varieties*. Clay Math. Proc., vol. 4 (American Mathematical Society, Providence, 2005), pp. 265–378
[Mil12] J.S. Milne, Shimura varieties and moduli, in *Handbook of Moduli*, vol. II, ed. by G. Farkas, I. Morrison. Advanced Lectures in Mathematics, vol. XXV (2012), pp. 467–548 (available at http://arxiv.org/abs/1105.0887)
[Mok89] N. Mok, in *Metric Rigidity Theorems on Hermitian Locally Symmetric Manifolds*. Series in Pure Mathematics, vol. 6 (World Scientific, Teaneck, 1989)
[Mor] D.W. Morris, *Introduction to Arithmetic Groups*. Preliminary version (February 27, 2013) of a book. Available at http://www.math.okstate.edu/~dwitte
[Nam80] Y. Namikawa, in *Toroidal Compactification of Siegel Spaces*. Lecture Notes in Mathematics, vol. 812 (Springer, Berlin, 1980)
[PS69] I.I. Pyateskii-Shapiro, *Automorphic Functions and the Geometry of Classical Domains*. Mathematics and Its Applications, vol. 8 (Gordon and Breach Science Publishers, New York, 1969)
[RRS92] R. Richardson, G. Röhrle, R. Steinberg, Parabolic subgroups with abelian unipotent radical. Invent. Math. **110**, 649–671 (1992)
[Roo08] G. Roos, Exceptional symmetric domains, in *Symmetries in Complex Analysis*. Contemporary Mathematics, vol. 468 (American Mathematical Society, Providence, 2008), pp. 157–189
[Sat80] I. Satake, in *Algebraic Structures of Symmetric Domains*. Kanô Memorial Lectures, vol. 4, Iwanami Shoten, Tokyo (Princeton University Press, Princeton, 1980)
[Wol64] J.A. Wolf, On the classification of Hermitian symmetric spaces. J. Math. Mec. **13**(1964), 489–495
[Wol67] J.A. Wolf, *Spaces of Constant Curvature* (McGraw-Hill, New York, 1967)
[Wol72] J.A. Wolf, Fine structure of Hermitian symmetric spaces, in *Symmetric Spaces (Short Courses, Washington Univ., St. Louis, Mo., 1969–1970)*. Pure and Applied Mathematics, vol. 8 (Dekker, New York, 1972), pp. 271–357

LECTURE NOTES IN MATHEMATICS

Edited by J.-M. Morel, B. Teissier; P.K. Maini

Editorial Policy (for Multi-Author Publications: Summer Schools / Intensive Courses)

1. Lecture Notes aim to report new developments in all areas of mathematics and their applications - quickly, informally and at a high level. Mathematical texts analysing new developments in modelling and numerical simulation are welcome. Manuscripts should be reasonably selfcontained and rounded off. Thus they may, and often will, present not only results of the author but also related work by other people. They should provide sufficient motivation, examples and applications. There should also be an introduction making the text comprehensible to a wider audience. This clearly distinguishes Lecture Notes from journal articles or technical reports which normally are very concise. Articles intended for a journal but too long to be accepted by most journals, usually do not have this "lecture notes" character.

2. In general SUMMER SCHOOLS and other similar INTENSIVE COURSES are held to present mathematical topics that are close to the frontiers of recent research to an audience at the beginning or intermediate graduate level, who may want to continue with this area of work, for a thesis or later. This makes demands on the didactic aspects of the presentation. Because the subjects of such schools are advanced, there often exists no textbook, and so ideally, the publication resulting from such a school could be a first approximation to such a textbook. Usually several authors are involved in the writing, so it is not always simple to obtain a unified approach to the presentation.

    For prospective publication in LNM, the resulting manuscript should not be just a collection of course notes, each of which has been developed by an individual author with little or no coordination with the others, and with little or no common concept. The subject matter should dictate the structure of the book, and the authorship of each part or chapter should take secondary importance. Of course the choice of authors is crucial to the quality of the material at the school and in the book, and the intention here is not to belittle their impact, but simply to say that the book should be planned to be written by these authors jointly, and not just assembled as a result of what these authors happen to submit.

    This represents considerable preparatory work (as it is imperative to ensure that the authors know these criteria before they invest work on a manuscript), and also considerable editing work afterwards, to get the book into final shape. Still it is the form that holds the most promise of a successful book that will be used by its intended audience, rather than yet another volume of proceedings for the library shelf.

3. Manuscripts should be submitted either online at www.editorialmanager.com/lnm/ to Springer's mathematics editorial, or to one of the series editors. Volume editors are expected to arrange for the refereeing, to the usual scientific standards, of the individual contributions. If the resulting reports can be forwarded to us (series editors or Springer) this is very helpful. If no reports are forwarded or if other questions remain unclear in respect of homogeneity etc, the series editors may wish to consult external referees for an overall evaluation of the volume. A final decision to publish can be made only on the basis of the complete manuscript; however a preliminary decision can be based on a pre-final or incomplete manuscript. The strict minimum amount of material that will be considered should include a detailed outline describing the planned contents of each chapter.

    Volume editors and authors should be aware that incomplete or insufficiently close to final manuscripts almost always result in longer evaluation times. They should also be aware that parallel submission of their manuscript to another publisher while under consideration for LNM will in general lead to immediate rejection.

4. Manuscripts should in general be submitted in English. Final manuscripts should contain at least 100 pages of mathematical text and should always include

   – a general table of contents;
   – an informative introduction, with adequate motivation and perhaps some historical remarks: it should be accessible to a reader not intimately familiar with the topic treated;
   – a global subject index: as a rule this is genuinely helpful for the reader.

   Lecture Notes volumes are, as a rule, printed digitally from the authors' files. We strongly recommend that all contributions in a volume be written in the same LaTeX version, preferably LaTeX2e. To ensure best results, authors are asked to use the LaTeX2e style files available from Springer's web-server at
   ftp://ftp.springer.de/pub/tex/latex/svmonot1/ (for monographs) and
   ftp://ftp.springer.de/pub/tex/latex/svmultt1/ (for summer schools/tutorials).
   Additional technical instructions, if necessary, are available on request from:
   lnm@springer.com.

5. Careful preparation of the manuscripts will help keep production time short besides ensuring satisfactory appearance of the finished book in print and online. After acceptance of the manuscript authors will be asked to prepare the final LaTeX source files and also the corresponding dvi-, pdf- or zipped ps-file. The LaTeX source files are essential for producing the full-text online version of the book. For the existing online volumes of LNM see:
   http://www.springerlink.com/openurl.asp?genre=journal&issn=0075-8434.
   The actual production of a Lecture Notes volume takes approximately 12 weeks.

6. Volume editors receive a total of 50 free copies of their volume to be shared with the authors, but no royalties. They and the authors are entitled to a discount of 33.3 % on the price of Springer books purchased for their personal use, if ordering directly from Springer.

7. Commitment to publish is made by letter of intent rather than by signing a formal contract. Springer-Verlag secures the copyright for each volume. Authors are free to reuse material contained in their LNM volumes in later publications: a brief written (or e-mail) request for formal permission is sufficient.

Addresses:
Professor J.-M. Morel, CMLA,
École Normale Supérieure de Cachan,
61 Avenue du Président Wilson, 94235 Cachan Cedex, France
E-mail: morel@cmla.ens-cachan.fr

Professor B. Teissier, Institut Mathématique de Jussieu,
UMR 7586 du CNRS, Équipe "Géométrie et Dynamique",
175 rue du Chevaleret,
75013 Paris, France
E-mail: teissier@math.jussieu.fr

For the "Mathematical Biosciences Subseries" of LNM:

Professor P. K. Maini, Center for Mathematical Biology,
Mathematical Institute, 24-29 St Giles,
Oxford OX1 3LP, UK
E-mail: maini@maths.ox.ac.uk

Springer, Mathematics Editorial I,
Tiergartenstr. 17,
69121 Heidelberg, Germany,
Tel.: +49 (6221) 4876-8259
Fax: +49 (6221) 4876-8259
E-mail: lnm@springer.com

If you have any concerns about our products,
you can contact us on
ProductSafety@springernature.com

In case Publisher is established outside the EU,
the EU authorized representative is:
**Springer Nature Customer Service Center GmbH
Europaplatz 3, 69115 Heidelberg, Germany**

Printed by Libri Plureos GmbH
in Hamburg, Germany